ALLE · ZEIT · WACH · 1842

N. H. Seemayer · W. Hadnagy (Eds.)

Environmental Hygiene III

With 125 Figures

Springer-Verlag
Berlin Heidelberg NewYork
London Paris Tokyo
Hong Kong Barcelona Budapest

Prof. Dr. NORBERT H. SEEMAYER
Medical Institute
of Environmental Hygiene
University of Düsseldorf
Auf'm Hennekamp 50
4000 Düsseldorf 1, FRG

Dipl.-Biologe WOLFGANG HADNAGY
Medical Institute
of Environmental Hygiene
University of Düsseldorf
Auf'm Hennekamp 50
4000 Düsseldorf 1, FRG

ISBN-13:978-3-642-77114-9 eISBN-13:978-3-642-77112-5
DOI:10.1007/978-3-642-77112-5

Softcover reprint of the hardcover 1st edition 1992

Typesetting: Camera ready by authors

31/3020-5 4 3 2 1 0 - Printed on acid-free paper

Preface

Serious alterations of the physico-chemical and biological environment are caused by diverse man-made activities. Limiting the pollution of soil, water and atmosphere and controlling the output of chemicals with toxic, mutagenic, carcinogenic and teratogenic activities represent goals of high-priority all over the world.
Monitoring the environment and developing new techniques and approaches to assess human health hazards poses a great task for scientists engaged in environmental research

The aim of the "European Meeting of Environmental Hygiene" is to bring together scientists from various fields of environmental research and to support and emphasize exchanges between the scientifique community, public institutions and industry.
The "First European Meeting of Environmental Hygiene" was held in Düsseldorf in May 1987 and was followed by the "Second Meeting" in Düsseldorf in May 1989. In response to the growing interest and aware of the great importance of environmental hazards the "Third European Meeting of Environmental Hygiene" was organized in Düsseldorf, June 25 - 27, 1991.

This volume contains contributions presented at the Third Meeting by leading scientists on important issues of environmental hygiene and medicine.
It covers a wide range of subject matter, from topics of molecular and cellular biology, human exposure, risks assessment up to control, legislation and prevention.

We are grateful to the "Medical Institute of Environmental Hygiene at the Heinrich-Heine-University" in Düsseldorf, particularly to the Director Prof.Dr. H.-W.Schlipköter for support and patronage of the Meeting. We would like to express our gratitude to the speakers, participants and all those who took part in organizing the Meeting.

Norbert H. Seemayer
Wolfgang Hadnagy

Contents

INTRODUCTION

Environmental medicine and risk assessment...................... 3
H.-W.Schlipköter (With 1 Figure)

EXPERIMENTAL CELL BIOLOGY

Epigenetic effects of cadmium in cellular signal transduction.. 7
D.Beyersmann, S.Hechtenberg, and C.Block (With 2 Figures)

Molecular alteration in human genital epithelial cells transformed by recombinant human papillomavirus DNA............. 11
J.A. DiPaolo, N.C.Popescu, and C.D.Woodworth

Spatial distribution as additional parameter in histochemical quantification of cell changes explored by an image analysing system........................... 15
M.Kohler, D.Ramm, J.Steinmann, M.Aufderheide, and U.Mohr (With 5 Figures)

Intraphagolysosomal pH and intracellular particle dissolution in canine and rat alveolar macrophages.............. 19
W.G.Kreyling, W.Beiskers, U.Miaskowski, M.Neuner, and P.Heilmann

UV-C and electromagnetic radiation induce premature terminal differentiation of cultured human fibroblasts.................. 23
H.P.Rodemann (With 1 Figure)

Effect of airborne particulates on phagocytosis of human and rat macrophages 27
N.H.Seemayer, A.Kiell, A.Happel, W.Hadnagy, and H.Behrendt (With 3 Figures)

CYTOTOXICITY TESTING

Testing lead neurotoxicity with an in vitro model............... 33
N.Binding, U.Altrup, M.N.Said, E.-J.Speckmann, and U.Witting (With 6 Figures)

A novel fluorometric method for monitoring virus replication in response to heavy metals......................... 37
D.Obradovic, M.Petersheim, and E.V.Orsi (With 2 Figures)

Cytoskeleton alteration in mouse tissues by postnatal exposure to heavy metal air pollutants.......................... 41
E.V.Orsi, G.Ahmad, D.Brois, B.Billings, and C.C.Orsi (With 1 Figure)

Comparison of chromate and vanadate toxicity and its relationship to oxygen radical formation........................ 45
H.H.Popper, A.Woldrich, and E.Grygar
(With 4 Figures)

Different effects of mechanical stress (high energy shock wave treatment) in normal and transformed lung epithelial cell lines of Syrian hamster.................... 49
D.Ramm, J.Knebel, S.Gärtner-Hübsch, W.Sass, and M.Aufderheide (With 4 Figures)

Effects of adrenaline and nicotine on the early chick embryo.. 53
M.Rosenbruch, J.Kniepen, and C.Weishaupt
(With 3 Figures)

MUTAGENICITY/CARCINOGENICITY STUDIES

Quantitation of 7-(2-hydroxyethyl)guanine produced by endogeneous and environmental ethylene oxide in DNA of lymphocytes.. 59
H.M.Bolt, H.Peter, M.Leutbecher, H.Ottenwälder, H.W.Goergens, and U.Föst (With 2 Figures)

New cell culture system for pulmonary carcinogenesis and inhalation toxicology.................................. 63
M.Emura, M.Riebe-Imre, A.Ochiai, G.Singh, S.L.Katyal, M.Aufderheide, U.Mohr J.Jacob, and G.Grimmer (With 2 Figures)

Stable expression of heterologous microsomal epoxide hydrolase in BHK21 cells: Influence on the mutagenicity of benzo(a)pyrene 4,5-oxide.............................. 67
H.R.Glatt, R.Becker, A.Piee, F.Oesch, and T.Friedberg (With 1 Figure)

Detection of suspicious aneugenic compounds present in environmental pollutants and their relevance in carcinogenicity.. 71
W.Hadnagy, C.Hader, and N.H.Seemayer
(With 5 Figures)

Enzyme polymorphism in human erythrocytes influences sister chromatid exchanges (SCE) induced by the fumigant methyl bromide.. 75
E.Hallier, T.Langhof, B.Gansewendt, K.Schröder, H.W.Goergens, and H.Peter (With 3 Figures)

Genotoxicity of Fe(III)-NTA and its interaction with other carcinogenic metal compounds............................ 79
A.Hartwig, R.Schlepegrell J.Kuhlmann, J.Benters, and D.Beyersmann (With 2 Figures)

Inhibition of gap-junctional intercellular communication in V79 cells by airborne pariculate matter..................... 83
G.A.H.Heussen and G.M.Alink (With 3 Figures)

The metabolism of chrysene and benzo(a)pyrene in hamster, rat and human lung cells in culture.................................. 87
J.Jacob, G.Grimmer, U.Mohr, M.Emura, M.Riebe-Imre, and G.Raab (With 5 Figures)

Analysis of transforming sequences in neoplastic Syrian hamster fetal cells induced by diverse chemical carcinogens.. 91
V.Notario, N.Popescu, and J.A.DiPaolo

Carcinogenicity testing of bladder carcinogens and other organic compounds by the intraperitoneal and intravesical route... 95
M.Roller, K.Kamino, and M.Rosenbruch (With 1 Figure)

Cytotoxicity, genotoxicity and transforming activity of 4-(methylnitrosamino)-1-(3-pyridyl)-1-butanone (NNK) in rat tracheal epithelial cells................................ 99
S.Y.Zhu, M.L.Cunningham, T.E.Gray, and P.Nettesheim (With 2 Figures)

INHALATION STUDIES

Toxicological criteria for comparison of smoke screens by an inhalative experimental study. An example: Toxic zinc oxide/hexachloroethane smoke................................ 105
U.Glaser and F.J.Mönig (With 2 Figures)

Responses of the canine respiratory tract during long-term exposure to a sulphur(IV) aerosol at low concentration.. 109
J.Heyder, I.Beck-Speier, G.Ferron, P.Heilmann, E.Karg, W.G.Kreyling, A.G.Lenz, K.Maier, H.Schulz, S.Takenaka, and T.Tuch

Acute ozone-induced change in lung permeability of genetically susceptible and resistant mice is differentially mediated by cytochrome P-450...................... 113
S.R.Kleeberger, M.A.Trush, and L.Y.Zhang (With 1 Figure)

Photocorrosion of cadmium sulfide suspensions applied in animal inhalation studies.................................... 117
H.P.König, U.Heinrich, H.Kock, and L.Peters (With 1 Figure)

Risk estimation of human lung cancer from environmental exposure to diesel exhaust: Extrapolation from particle induced experimental lung tumors?............................... 121
G.Oberdörster, W.P.Pepelko, C.P.Yu, and C.Chen (With 3 Figures)

Short term chromate inhalation in a computer-assisted inhalation chamber: immediate toxicity and late cancer development.. 127
H.H.Popper, G.Wiespainer, E.Leingartner, W.Weybora, and M.Ratschek (With 5 Figures)

Studies on the exhalation of volatile N-nitrosamines following systemic application to Sprague-Dawley rats........... 133
P.Schmezer, B.M.Schmezer, B.L.Pool-Zobel, and R.G.Klein (With 2 Figures)

The behaviour of inspired aerosol pulses in men................ 137
Ch.F.Schiller-Scotland, J.Gebhart, and R.Siekmeier (With 5 Figures)

The behavior of inhaled spherical iron oxide particles in human lungs: Magnetometric studies........................ 141
W.Stahlhofen and W.Möller (With 2 Figures)

HUMAN EXPOSURE MONITORING

Cross sectional epidemiological study on cobalt excretion in urine of children in areas with different air pollution.. 147
Th.Eikmann and A.Krönung (With 4 Figures)

Human exposure to mercury - Effect of amalgam fillings.......... 151
U.Ewers

Cytogenetic studies of chemically exposed workers in Sachsen-Anhalt... 155
E.Hüttner

Heavy metals in the blood, urine and hair of a representative population sample in the Federal Republic of Germany 1985/86.................................. 159
C.Krause, M.Chutsch, M.Henke, M.Leiske, C.Schulz, E.Schwarz, and B.Seifert

Urinary alkylpurines as markers of alkylating agent exposure in humans...................................... 163
V.Prevost, D.E.G.Shuker, and H.Bartsch

Total exposure to lead and cadmium of a group of people from a lead contaminated area................................ 167
M.Saric, M.Fugas, and M.Blanusa

The significance of asbestos bodies and occurrence of macrophages in the sputum of asbestos-exposed workers........... 171
I.Trosic, M.Saric, and L.Stilinovic (With 2 Figures)

Environmental and biological monitoring of mercury following chronic occupational exposure........................ 175
B.Widon, F.Schweinsberg, G.Holzapfel, and Ch.Brückner (With 3 Figures)

Aluminium exposure and health risks............................ 179
M.Wilhelm and F.K.Ohnesorge (With 1 Figure)

EPIDEMIOLOGY

Air pollution and bronchial reactivity in schoolchildren with a medical history of respiratory diseases.................. 185
N.Englert, W.Babisch, B.Niggemann, and U.Wahn
(With 6 Figures)

Effects of air pollution on the health of children in East- and West-Germany: Aspects of study-design and descriptive results for some questionnaire items................ 189
U.Krämer, R.Dolgner, H.J.Willer, and H.-W.Schlipköter

Air pollution and allergic reactions............................ 193
E.Marth, C.Hülße, and J.Pelech
(With 3 Figures)

Effects of the extreme environmental pollution on the general health state of the population in Bitterfeld (former G.D.R)................................... 197
U.Thielebeule, I.Winkler, R.Ebert, and Chr.Wegner
(With 3 Figures)

HAZARDOUS ENVIRONMENTAL AGENTS

1. Radon

Effects of mobility upon the distribution of lifetime radon exposures.. 203
K.-S.Liu, Y.-L. Chang, S.B.Hayward, A.J.Gadgil, and A.V.Nero (With 1 Figure)

Radon in the human environment.................................. 207
W.-U.Müller and C.Streffer

2. Dioxin

Dioxin exposure and health effects on people at Seveso, Italy.. 211
P.Mocarelli, L.L.Needham, A.Marocchi, D.G.Jr.Patterson, P.Brambilla, P.M.Gerthoux, L.Meazza, and V.Carreri

3. Mineral Dust and Fibres

Comparison of cytotoxic effects of asbestos and glass fibres on the cytoskeleton of epithelial lung cells in vitro........... 215
M.Aufderheide, M.Riebe-Imre, J.Hertes, and U.Mohr (With 3 Figures)

HL-60-cell derived macrophages are resistant to cytotoxic effects of quartz.. 219
U.Brückner-Nieder, A.Schmidt, J.Schimmelpfeng, and A.Seidel (With 5 Figures)

In vitro correlates of the pathogenicity of a long fibre sample of amosite asbestos compared to a short fibre sample... 223
K.Donaldson (With 3 Figures)

Quartz or coal mine dust treated human monocyte-derived macrophages release a proliferation factor for human lung fibroblasts (WI-38) 227
K.Hübner and N.H.Seemayer (With 3 Figures)

Dust and LPS induced stimulation of nitrite and nitrate release in bovine alveolar macrophages 231
T.Schlüter, I.Berg, and G.Gercken
(With 3 Figures)

Influence of coal mine dust and surfactant on chemiluminescence-activity of alveolar macrophages in vitro 235
J.Wackes, F.Tilkes, and E.G.Beck
(With 4 Figures)

ENVIRONMENTAL CONTROL, PREVENTION AND LEGISLATION

Industrial risk assessment and corresponding management scenarios-case study at the Aluminum Casting Plant 241
M.Boutoussov, M.H.A.Hamdy, S.Luken, W.Gaubinger, H.Friza, A.Jernelov, H.Meisterhofer, and I.Stankovich

Environmental tobacco smoke in public places 245
M.Brynnel and G.Löfroth (With 1 Figure)

Carcinogenic air pollutants: Essential legislative regulations 249
S.Dobbertin

Biological remediation processes - A challenge for the environmental hygiene 253
W.Dott and P.Kämpfer (With 2 Figures)

Polychlorinated dibenzodioxins and dibenzofurans - Measures for the protection of man and the environment 257
J.Lebsanft and A.Basler

Measurements of the environmental pollution during cleaning up of solid waste depositions of chemical weapons 261
H.-J.Moriske, M.Wiese, I.Block, R.Kneiseler, I.Trauer, H.Neumann, and H.Rüden (With 2 Figures)

Monitoring of an rDNA B.subtilis in environmental samples 265
K.Smalla, M.Isemann, G.John, K.-H.Weege, K.Wendt, and F.Niepold

Air pollution in the former German Democratic Republic - Consequences for political economy and health control 269
B.Thriene

SUBJECT INDEX 273

Contributors

You will find the addresses at the beginning of the respective contribution

Ahmad, G. 41
Alink, G.M. 83
Altrup, U. 33
Aufderheide, M. 15, 49, 63, 215

Babisch, W. 185
Bartsch, H. 163
Basler, A. 257
Beck, E.G. 235
Beck-Speier, I. 109
Becker, R. 67
Behrendt, H. 27
Beisker, W. 19
Benters, J. 79
Berg, I. 231
Beyersmann, D. 7, 79
Billings, B. 41
Binding, N. 33
Blanusa, M. 167
Block, C. 7
Block, I. 261
Bolt, H.M. 59
Boutoussov, M. 241
Brambilla, P. 211
Brois, D. 41
Brückner, Ch. 175
Brückner-Nieder, U. 219
Brynnel, M. 245

Carreri, V. 211
Chang, Y.-L. 203
Chen, C. 121
Chutsch, M. 159
Cunningham, M.L. 99

DiPaolo, J.A. 11, 91
Dobbertin, S. 249
Dolgner, R. 189
Donaldson, K. 223
Dott, W. 253

Ebert, R. 197
Eikmann, Th. 147
Emura, M. 63, 87
Englert, N. 185
Ewers, U. 151

Ferron, G.A. 109
Föst, U. 59
Friedberg, T. 67
Friza, H. 241
Fugas, M. 167

Gadgil, A.J. 203
Gärtner-Hübsch, S. 49
Gansewendt, B. 75
Gaubinger, W. 241
Gebhart, J. 137
Gercken, G. 231
Gerthoux, P.M. 211
Glaser, U. 105
Glatt, H.R. 67
Goergens, H.W. 59, 75
Gray, T.E. 99
Grimmer, G. 63, 87
Grygar, E. 45

Hader, C. 71
Hadnagy, W. 27, 71
Hallier, E. 75
Hamdy, M.H.A. 241
Happel, A. 27
Hartwig, A. 79
Hayward, S.B. 203
Hechtenberg, S. 7
Heilmann, P. 19, 109
Heinrich, U. 117
Henke, M. 159
Hertes, J. 215
Heussen, G.A.H. 83
Heyder, J. 109
Holzapfel, G. 175
Hübner, K. 227
Hülße, C. 193
Hüttner, E. 155

Isemann, M. 265

Jacob, J. 63, 87
Jernelov, A. 241
John, G. 265

Kämpfer, P. 253
Kamino, K. 95
Karg, E. 109
Katyal, S.L. 63
Kiell, A. 27
Kleeberger, S.R. 113
Klein, R.G. 133
Knebel, J. 49
Kneiseler, R. 261
Kniepen, J. 53
Kock, H. 117
König, H.P. 117
Kohler, M. 15
Krämer, U. 189
Krause, C. 159
Kreyling, W.G. 19, 109
Krönung, A. 147
Kuhlmann, J. 79

Langhof, T. 75

Lebsanft, J. 257
Leingartner, E. 127
Leiske, M. 159
Lenz, A.G. 109
Leutbecher, M. 59
Liu, K.-S. 203
Löfroth, G. 245
Luken, S. 241

Maier, K. 109
Marocchi, A. 211
Marth, E. 193
Meazza, L. 211
Meisterhofer, H. 241
Miaskowski, U. 19
Mocarelli, P. 211
Möller, W. 141
Mönig, F.J. 105
Mohr, U. 15, 63, 87, 215
Moriske, H.-J. 261
Müller, W.-U. 207

Needham, L.L. 211
Nero, A.V. 203
Nettesheim, P. 99
Neumann, H. 261
Neuner, M. 19
Niepold, F. 265
Niggemann, B. 185
Notario, V. 91

Oberdörster, G. 121
Obradovic, D. 37
Ochiai, A. 63
Oesch, F. 67
Ohnesorge, F.K. 179
Orsi, C.C. 41
Orsi, E.V. 37, 41
Ottenwälder, H. 59

Patterson, D.G.Jr. 211
Pelech, J. 193
Pepelko, W.P. 121
Peter, H. 59, 75
Peters, L. 117
Petersheim, M. 37
Piee, A. 67
Pool-Zobel, B.L. 133
Popescu, N.C. 11, 91
Popper, H.H. 45, 127
Prevost, V. 163

Raab, G. 87
Ramm, D. 15, 49
Ratschek, M. 127
Riebe-Imre, M. 63, 87, 215
Rodemann, H.P. 23
Roller, M. 95
Rosenbruch, M. 53, 95
Rüden, H. 261

Said, M.N. 33
Saric, M. 167, 171
Sass, W. 49
Schiller-Scotland, Ch.F. 137
Schimmelpfeng, J. 219
Schlepegrell, R. 79
Schlipköter, H.-W. 3, 189
Schlüter, T. 231
Schmezer, B.M. 133
Schmezer, P. 133
Schmidt, A. 219
Schröder, K. 75
Schulz, C. 159
Schulz, H. 109
Schwarz, E. 159
Schweinsberg, F. 175
Seemayer, N.H. 27, 71, 227
Seidel, A. 219
Seifert, B. 159
Shuker, D.E.G. 163
Siekmeier, R. 137
Singh, G. 63
Smalla, K. 265
Speckmann, E.-J. 33
Stahlhofen, W. 141
Stankovich, I. 241
Steinmann, J. 15
Stilinovic, L. 171
Streffer, C. 207

Takenaka, S. 109
Thielebeule, U. 197
Thriene, B. 269
Tilkes, F. 235
Trauer, I. 261
Trosic, I. 171
Trush, M.A. 113
Tuch, T. 109

Wackes, J. 235
Wahn, U. 185
Weege, K.-H. 265
Wegner, Chr. 197
Weishaupt, C. 53
Wendt, K. 265
Weybora, W. 127
Widon, B. 175
Wiese, M. 261
Wiespainer, G. 127
Wilhelm, M. 179
Willer, H.J. 189
Winkler, I. 197
Witting, U. 33
Woldrich, A. 45
Woodworth, C.D. 11

Yu, C.P. 121

Zhang, L.Y. 113
Zhu, S.Y. 99

INTRODUCTION

Environmental Medicine and Risk Assessment

Schlipköter, H.-W.
Medical Institute of Environmental Hygiene at the Heinrich-Heine-University of Düsseldorf, Auf'm Hennemkamp 50, D-4000 Düsseldorf, FRG

Public awareness of environmental problems has steadily increased during the past two decades. According to interviews in the Federal Republic of Germany and the EEC two thirds of the population consider it a political task of high priority that effective measures are taken in order to protect against environmental hazards. In this respect, air pollution by chemicals evoked the highest concern in 60 % of the subjects interviewed. As a consequence of this pronounced awareness of environmental health hazards, a more critical behaviour not only of consumers, but also of patients is developing. With increasing frequency people do not only inquire into possible adverse effects of therapeutic measures, but also express the suspicion that some of their health problems might be caused by exposure to chemicals from the environment. Hence, physicians find themselves more and more confronted with environmental issues.

The great emotive attention received by environmental issues in the public consciousness can only be understood if one considers the progress made in the analysis of noxious chemical agents. Only after lowering the detection limits of these agents it was possible to measure the low concentrations characteristic of environmental pollution. However, toxicologic assessment of the numerous chemical compounds causes considerable problems. One of these is the well-known difficulty in extrapolating data obtained in animal experiments to humans; this task becomes even more complicated if long lasting low-dose exposure of humans is to be assessed on the basis of short-term data obtained in animals. Possibilities in performing systematic studies in humans are, of course, very limited and confined to self-experimentation and exposure of volunteers under the most stringent precautionary measures. As far as possible toxic effects in man are concerned, toxicology is dependent on observations of chemical accidents or involuntary exposure. Moreover, a considerable portion of the concepts of environmental medicine are based on knowledge obtained from occupational medicine.

Research provides the scientific basis for prevention and control of environmental hazards; however, many questions regarding health effects of noxious substances, alone or as complex mixtures, remain open. Very often it takes many years for scientifically-based risk estimates. Biological reactions to environmental noxae on the cellular and subcellular level, interaction of environmental pollutants with the respiratory and immune system, adverse effects on the neurological or reproductive system as well as the contribution of environmental contaminants to chronic degenerative diseases and aging are of growing importance. Improved epidemiological studies are also essential to assess the scope of environmental health problems and to make prevention and control measures possible.

The solution of complex environmental problems exceeds the capacity of a single country. The benefits from joint efforts are manyfold. Of great benefit are the accessibility and the exchange of experimental data and information. The adoption of common methodologies allows comparability of results and makes joint evaluation possible, further strengthening the validity of data. The sharing of knowledge and experience has a considerable impact on the development of health criteria based on risk estimation and setting of standards. The widening exchange of information will promote the identification of additional topics of mutual interest in the field of environmental hygiene and preventive medicine.

Joint efforts also promote the transfer of knowledge into preventive and legislative measures.

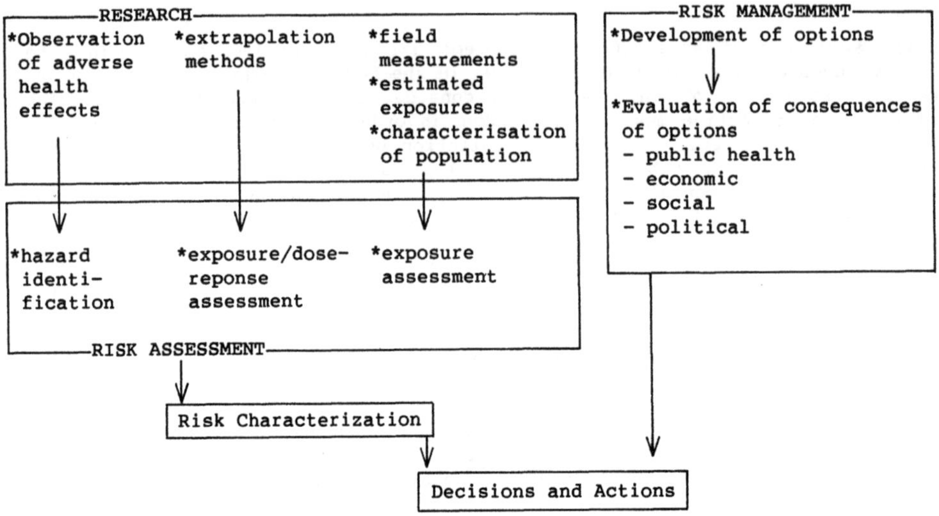

Fig. 1: Approaches to risk assessment/management (McClellan 1989, modified)

This complex system of correlations is demonstrated in figure 1. This flow-chart which was developed by a commitee of the United States National Academy of Sciences in 1983 shows the relationship between basic and applied research, risk assessment and risk management: Basic and applied research provide essential data for conducting a comprehensive and valid risk assessment. Important parts of this process are the identification of hazards and the assessment of exposure-response-relationships. Informations on population`s exposure and on dose-response-relationships allow a risk characterization, which together with other details in performing risk management leads to decisions and actions.

It is well established that multiple approaches must be used to obtain valid data and sufficient information for decision making. This variety of possible approaches is also presented in this volume showing the multifaceted aspects of environmental hygiene and medicine. Scientists from various fields present their actual results. Research using macromolecular, cellular and tissue systems offer multiple possibilities to evaluate hazards by environmental pollutants. Furthermore, data are shown on experiments using whole animals as well as from field measurements, human monitoring and epidemiological studies. Beside hazard identification and risk assessment also aspects of legislation and preventive measures are included in this volume.

References

McClellan, R.O. (1989) An introduction to inhalation toxicology. In: Concepts in Inhalation Toxicology. McClellan, R.O., Henderson, F. (eds). Hemisphere Publishing Corporation New York.

EXPERIMENTAL CELL BIOLOGY

Epigenetic Effects of Cadmium in Cellular Signal Transduction

Beyersmann, D., S. Hechtenberg and C. Block
Department of Biology and Chemistry, University of Bremen,
D-2800 Bremen 33, Germany

Introduction

The carcinogen cadmium causes direct genotoxic effects as well as epigenetic changes. Cadmium (II) evokes DNA strand breaks in mammalian cells (Snyder, 1988) and, if combined with thiol compounds, in isolated DNA, too (Müller et al., 1991; Oetken and Beyersmann, 1991). Cd^{2+} is a weak mutagen only but it enhances the mutagenicity of UV irradiation in mammalian cells by interfering with the repair of damaged DNA (Hartwig and Beyersmann, 1989). An important epigenetic mechanism in carcinogenesis is the interference with the control of cell proliferation. Here, we discuss evidence for the interaction of cadmium (II) with the inositol trisphosphate/calcium signalling pathway which controls growth and which also is susceptible to tumor promoters.

Interference with signal reception and calcium mobilization

Fig. 1 depicts the inositoltrisphosphate/calcium signalling path which is known to be triggered by extracellular signals like hormones, neurotransmitters and growth factors. Smith et al. (1989) observed that cadmium (II) trigges this pathway evoking the intracellular formation of the second messenger inositoltrisphosphate and the mobilization of calcium in human skin fibroblasts.

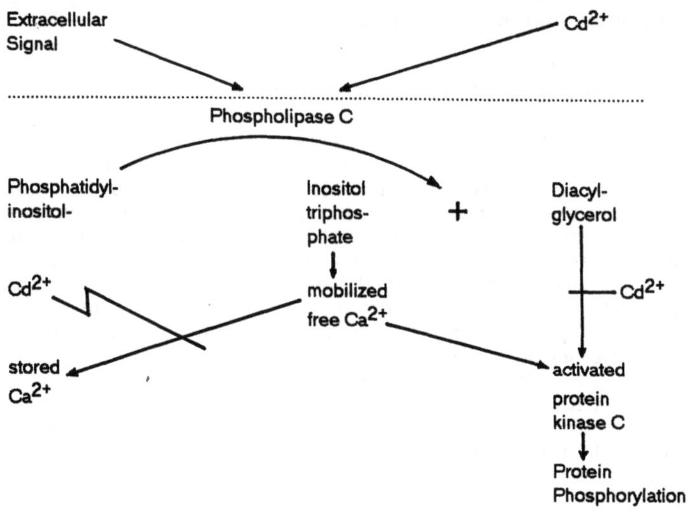

Fig. 1. Interferences of Cd^{2+} with the Inositol trisphosphate/calcium signalling path.

Another mechanism by which cadmium could cause a rise in the intracellular calcium concentration is the interference with calcium homeostasis. The cytoplasmic calcium concentration is controlled by factors that induce the release of calcium ions from calcium storing vesicles as well as by a Ca^{2+} ATPase that reloads calcium into these vesicles. As a model system, we have studied the rabbit muscle (sarcoplasmic) Ca^{2+} ATPase.
Fig. 2 shows that this enzyme is inhibited by cadmium ions at micromolar concentrations. Half inhibition is observed at 0.95 µM free Cd^{2+} concentration (Hechtenberg and Beyersmann, 1991). With a similar concentration of cadmium, the Ca^{2+} transport by the same protein into vesibles is inhibited (Hechtenberg and Beyersmann, 1991). Further studies are carried out to investigate effects of cadmium on the nuclear transport of calcium which is relevant for controlling rates of gene expression.

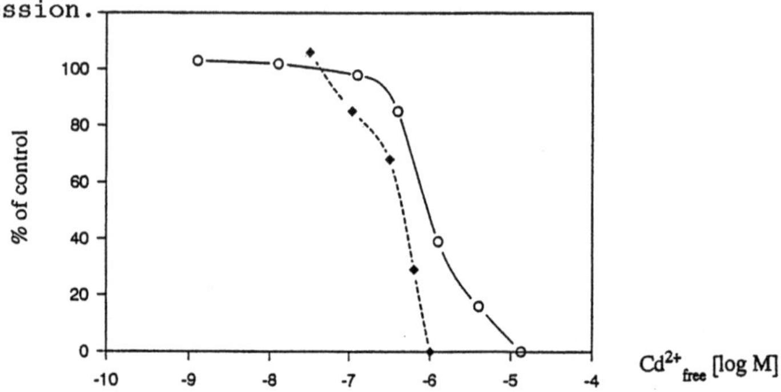

Fig. 2. Inhibition of sarcoplasmic Ca^{2+} ATPase (O) and Ca^{2+} transport (◆) by Cd^{2+}. For experimental details see Hechtenberg and Beyersmann, 1991.

Activity and translocation of Protein Kinase C

Protein Kinase C is a key enzyme in signal transduction leading to cell differentiation and proliferation. The enzyme also is a target of tumor promoting phorbol esters (Nishizuka, 1988). When the inositol trisphosphate pathway is triggered by an extracellular signal, Ca^{2+} and diacylglycerol are set free which jointly activate protein kinase C (see fig. 1). This enzyme is activated by its translocation to a membrane. Phorbol esters are able to substitute for diacylglycerol in translocation and activation ultimately resulting in tumor promotion (Nishizuka, 1988). Various reports indicate multiple metal binding centers in protein kinase C.

- At 5 µM Ca^{2+}, 300 µM Zn^{2+} enhanced the enzyme activity induced by fatty acids (Murakami et al., 1987);
- 500 µm Zn^{2+} increased the activity of protein kinase C for receptors in T-lymphocytes (Csermely et al., 1988);
- 50 µM Cd^{2+} reduced phorbol ester binding to lympohocytes (Cifone et al., 1989);
- 10 µM Zn^{2+} or Cd^{2+} inhibited the activity and phorbol ester binding of brain protein kinase (Speizer et al., 1989);
- 10 µM Zn^{2+} or Cd^{2+} augmented the binding of phorbol ester to proteinkinase C in homogenates of B-lymphocytes (Forbes et al., 1990)

The seemingly contradictory findings on metal effects on protein kinase C may be due to the complexity of the system. Protein kinase C is unusually complex in its modulation by binding to plasma

membrane, cyctoskeleton protein and possibly further cellular constituents (Forbes et al., 1990). Since we wished to elucidate possible effects of cadmium in cellular growth regulation, we have studied the protein kinase C in the nuclear fraction of mouse 3T3 cells. When these cells are stimulated by the tumor promotor TPA, cadmium but not zinc ions cause a further increase in the nuclear protein kinase C activity (Table 1). We interpret this effect as an indication of a possible increase in phosphorylation of nuclear proteins and a possible stimulation of cellular proliferation.
Finally we analyzed the effect of 50 µM $CdCl_2$ on the nuclear protein phosphorylation in Chinese Hamster Ovary cells (Kirchherr, 1991). After 3 hours incubation there was no significant change in the pattern of cytoplasmic protein phosphorylation as analyzed by polyacrylamide gel electrophoresis. However, the phosphorylation of three nuclear proteins (35, 55 and 72 KD) was reduced with respect to cells without cadmium treatment.

Table 1. Influence of Cd^{2+} on activation and translocation of nuclear protein kinase C in mouse 3T3 cells. Cells were incubated for 20 min with the agents at the concentrations given in the table. Total nuclear protein kinase C and the fraction extractable from membranes by 1 % Triton were determined.
The enzyme was assayed as described by Newton and Koshland (1989).

Agents added	Total nuclear enzyme (U/ml)	% Translocation to Triton extractable fraction
none	104	48
0.1 µM TPA	90	95
0.1 µM TPA + 50 µM Cd^{2+}	151	96
0.1 µM TPA + 100 µM Zn^{2+}	86	97

Summary

Besides immediate clastogenic effects, cadmium causes an inhibition of the Ca^{2+} ATPase loading Ca^{2+} into storage compartments of cells and causes an increase in the protein kinase C activity of the cell nucleus. Both these effects are discussed to stimulate cellular proliferation and thus proffer a possible epigenetic way of action of cadmium in cancerogenesis.

References

Cifone M.G., E. Alesse, A. Procopio, R. Paolini, S. Morrone, R. DiEugenio, G. Santoni and A. Santoni, (1988) Effects of cadmium on lymphocyte activation, Biochim. Biophys. Acta 1011, 25-32.

Csermely, P., M. Szamel, K. Resch and J. Somogy (1988) Zinc can increase the activity of protein kinase C and contributes to its binding to plasma membranes in T lymphocytes. J. Biol. Chem. 263, 6487-6490.

Forbes, I. J., P.D. Zalewski, C. Giannakis, H.S. Petkoff and P.A. Cowled (1990). Interaction between protein kinase C and regulatory ligand is enhanced by a chelatable pool of cellular zinc. Biochim. Biophys. Acta 1053, 113-117.

Hartwig A., and D. Beyersmann (1989) Comutagenicity and inhibition of DNA repair by metal ions in mammalian cells. Biol. Trace Elem. Res. 21, 359-365.

Hechtenberg, S., and D. Beyersmann (1991) Inhibition of sarcoplasmic reticulum Ca^{2+}-ATPase activity by cadmium, lead and mercury. Enzyme, in press.

Kirchherr, H, (1991). Die Streßantwort von CHO-Zellen auf Cadmium. Dissertation, Universität Bremen.

Müller, T., R. Schuckelt and L. Jaenicke (1991) Cadmium/zinc-metallothionein induces DNA strand breaks in vitro. Arch. Toxicol. 65, 20-26.

Murakami, K., M.K. Whiteley and A. Routtenberg (1987) Regulation of protein kinase C activity by cooperative interaction of Zn^{2+} and Ca^{2+}. J. Biol. Chem. 262, 13902-13906.

Newton, A.C., and D.E. Koshland (1989) High cooperativity, specifity, and multiplicity in the protein kinase C - lipid interaction. J. Biol. Chem. 264, 14909-14915.

Nishizuka, Y. (1989) Studies and prospectives of the protein kinase C family for cellular regulation. Cancer 63, 1892-1903.

Oetken, G., and Beyersmann, D. (1991) Metallothionein or cadmium and glutathione induced single strand breaks in isolated DNA. Environ. Geochem., in press.

Smith, J.B., S.D. Dwyer and L. Smith (1989). Cadmium evokes inositol polyphosphate formation and calcium mobilization. J. Biol. Chem. 256, 7115-7118.

Snyder, R.D. (1988) Role of active oxygen species in metal-induced DNA strand breakage in human diploid fibroblasts. Mutation Res. 193, 237-246

Speizer, L.A., M.J. Watson, J.R. Kanter and L.L. Brunton (1989). Inhibition of phorbol ester binding and protein kinase C activity by heavy metals. J. Biol. Chem. 264, 5581-5585.

/IOLECULAR ALTERATION IN HUMAN GENITAL EPITHELIAL CELLS TRANSFORMED BY ₹ECOMBINANT HUMAN PAPILLOMAVIRUS DNA

.A. DiPaolo, N.C. Popescu, and C.D. Woodworth. Laboratory of Biology, Natl. Cancer ıstitute, Bethesda, MD 20892.

NTRODUCTION.

;quamous cell carcinoma of the cervix is a major public health problem worldwide, as a ;ause of cancer death in women it is number two after breast cancer. The first documented lata concerning marital or sexual events was published by Rigoni-Stern in 1842 (1) who ;oncluded from a statistical analysis of death records that cancer of the uterus occurred rarely n virgins and nuns. Since then, evidence has been accumulating consistent wit the premise hat cervical cancer behaves as a venereally transmitted disease. Two important factors in he development of cancer appear to be the age of onset of sexual activity and the number ɔf partners. Today, it is accepted that a subset of human papillomaviruses (HPV) are linked vith neoplastic events in the epithelia of the female genitalia. In fact, over 95% of invasive ;ervical cancers are associated with a specific subset of HPVs (2). However, the evidence ılso indicates that while HPV is necessary it is not sufficient to induce cancer. Cervical cancer s an ideal model for studying human carcinogenesis because the normal, benign and nalignant stages can be distinguished in vivo and in vitro.

Vlost papillomas have a single host and multiply in mucosal or differentiated cutaneous ɜpithelium at specific anatomical sites. The papilloma viruses have not been grown in vitro. They appear to require differentiated keratinocytes of epithelial for vegetative replication and ·emain undetectable in the basal layer. It is presumed but not proven that the replicative ;ycle of the papilloma virus is synchronized to epithelial cell differentiation. As the cell ɔroceeds to differentiate from the basal to the spinous and granular stages, episomal DNA ·eplication occurs in the cell nucleus. Only differentiated cells contain complete virus with ɔrotein coat and thus the complete virus is found in the stratum corneum. Therefore, it is ɑppropriate and relevant to transfect HPV DNA into human keratinocytes to determine their role in carcinogenesis.

RESULTS AND DISCUSSION.

Normal (HPV negative) genital epithelial cells undergo senescence and survive 40-to-50 population doubling in serum free medium (MCDB153-LB). However, transfection of similar cells with recombinant HPVs (16, 18, 31 and 33) associated with cervical resulted in outgrowths of immortal cells, usually without crisis. Cells transfected with HPV (1a, 5, 6b, or 11 not associated with malignancy) formed small colonies that grew transiently but terminally differentiated (3). Whereas immortalized cells had integrated and rearranged HPV DNAs, those transfected with non-immortalizing HPVs had only intact virus sequences. The average copy number per cell ranged from less than one to two. Furthermore, HPV sequences migrated as episomal or linear molecules when the same DNA samples were digested with restriction enzymes that did not cleave within the transfected plasmid. Therefore, integration was not detected with non-immortalizing HPV DNAs. Thus pHPV1a, 5, and 11 failed to immortalize epithelial cells even though transfection was successful.

All cervical cancers with integrated HPV E6 and E7 ORFs which are also required for immortalization of cultured human epithelial cells. Integration is associated with such well known carcinomas as HeLa and SW756. The first demonstration of HPV-18 DNA integration near cellular proto-oncogenes and chromosomal fragile sites in HeLa cells and SW756 cells provided new insights on the role of viral integration. Evidence is accumulating that in cervical cancer the critical interaction site of human papilloma virus on human chromosomes is a non-random event because it is associated with fragile sites that may be either constitutive or heritable and also may be near the loci of known oncogenes. A single HPV-18 integration site in SW756 cells (4) was identified on chromosome 12 at band q13, integrated near Int-1 proto-oncogene in contrast to HeLa cells (5) that have three integration sites in chromosomes 8, 9 and 22 corresponding to the location of c-myc, c-abl and c-sis proto-oncogenes, respectively. The HPV sequence of the recombinant HPV-18 was interrupted in ORF E1 by insertion of pSV2neo vector which contains a strong polyadenylation signal. Thus, the possibility exists that E6 and E7 are the only intact genes expressed because the other ORFs are separated from the upstream regulatory region. To test this possibility RNA from HPV-18-immortalized HKc was hybridized to two subgenomic fragments of HPV-18 DNA (3). The first included the intact E6 and E7 ORFs and a partial E1, the second E2, E4, E5, L2 and most of L1. The E6/E7 probe hybridized to the same transcripts in three HPV-18-immortalized cell lines as did the intact HPV genome. In contrast, the second failed to hybridize to any messages in one line and hybridized very weakly in the two other lines. Therefore, at least one HPV-18-immortalized HKc line transcribes RNAs only from the intact E6 and E7 ORFs. This suggests that the immortalization function is limited to one or both of these genes. Recombinant E6 and E7 genes can efficiently immortalize human genital keratinocytes (6); however, in rare instances E7 alone at high level is sufficient. The products of these two genes are often referred to as oncoproteins. Because the tumor- suppression proteins pRb and p53 can be complexed by the E7 and E6 proteins, respectively (7), it has been postulated that they are relevant targets of HPVs. Mutation of Rb as well as p53 have been reported for diverse cancers. In a series of HPV-positive cervical cell lines, all the pRb and p53 proteins appeared normal as detected by sequence analysis. Furthermore, immunobot analysis showed that p53 levels were low to undetectable relative to normal cells (7).

Chromosome analysis verified the human origin of these cells and provided evidence that the transfected keratinocytes undergo drastic chromosomal alterations. The non-treated cells are diploid and normal by G banding (8). Epithelial cell lines, immortalized by HPV-16 were cytogenetically abnormal, exhibiting numerical deviations and altered chromosomes due to translocations, deletions, achromatic lesions, or partial duplications. Furthermore, all lines had cells with either homogeneously staining regions (HSR) or double minute chromosomes, alterations which reflect DNA amplification associated with malignancy or drug resistance. By in situ chromosome hybridization, viral sequences were identified on abnormal chromosomes at the site of chromosome translocations, at achromatic lesions and within HSR and duplicated chromosome segments. Thus, for the first time HPV-16 integration into the cellular genome was shown to be associated with the induction of chromosome alterations.

The immortality property of HPV containing cells can be uncoupled from differentiation (8). The addition of calcium > 0.3 mM or calcium-free serum to the medium will induce differentiation. However, a subpopulation of transfected cells will continue to grow in the presence of calcium or serum. Evidence for the origin and maintenance of epithelial properties of these cells is demonstrated by keratin analysis of controls (9). The non-treated controls exhibit a number of keratins that are found in non-keratinizing squamous stratified epithelia. With culturing and transfection, there is an increase of several keratins associated with culturing. Furthermore, after transfection as well as growth in serum, there is an abundant amount of keratin 19 which is a major component of simple epithelia which is found in cultured keratinocytes and squamous cell carcinomas. Thus, evidence indicates that the recipient cells as well as the transfected cells which become immortalized are keratinocytes.

Immortality is an important step in the multistage process leading to malignancy. One mechanism by which HPV might contribute to the carcinogenic process is by altering normal differentiation referred to as cervical intraepithelial neoplasia. Therefore, an objective was to develop and animal model for studying normal or abberant cervical differentiation. Barrandon et al. (10), demonstrated that normal human epidermal cells form a well-differentiated stratified squamous epithelium when monolayer cultures were transplanted intact beneath a skin-muscle flap in nude mice. This model is useful for studying differentiation of normal human cervical cells derived from the transformation zone, and cells immortalized by HPV DNAs associated with cervical cancer (11). Two to three weeks after transplantation, cervical cell culture produced non-keratinizing thicker squamous epithelia morphologically reflecting epithelia from which they were derived. Both early (<65 population doublings) and late (>200 population doublings) immortal cultures were examined. Most cell lines immortalized by HPV-18 DNA and grafted at early passages formed dysplastic epithelia. These grafts contained altered mitoses, increased nuclear to cytoplasmic ratio, and frequently a total lack of squamous differentiation. However, no invasion of the basement membrane was observed. Clinical evidence indicates that HPV-18 is associated with a more aggressive and rapidly progressive form of cervical neoplasia. In contrast, cell lines of similar passage immortalized by HPV-16 DNA formed thin squamous epithelia without noticeable dysplastic changes. The basal layer was often flattened and grafts were covered by extensive layers of keratinized cells. A precursor of the cornified envelope, involucrin, is a marker for squamous differentiation. Expression of involucrin was confined to the suprabasal layers in normal cervical or foreskin grafts. HPV-immortalized cells lines at early passage formed a thin squamous basal cell layer also expressed involucrin. the more severely dysplastic grafts stained variably, often with a random pattern. This random pattern of involucrin expression resembled that of an invasive cervical carcinoma cell line, C4-1. Similar alterations in pattern of involucrin expression occur during the development of dysplastic cervical differentiation in situ.

Repeated tumorigenicity tests with cells grown in serum-free medium or serum-resistant cells that mimic carcinomas in morphology in culture have been negative in athymic nude mice. Invasive cervical carcinomas often contain an activated Harvey-ras gene (12). The activated ras gene was found in conjunction with both HPV-16 and 18 DNA suggesting that ras activation is involved in the conversion of the papillomavirus-containing cervical cells to a malignant state. In some cervical tumors ras activation is also accompanied by amplification and/or overexpression of the c-myc gene. HPV-immortalized cells transfected with v-Ha-ras failed to terminally differentiate and rapidly formed tumors subcutaneously in nude mice (13). The tumors were cystic squamous cell carcinomas which is consistent with the derivation of cells from the squamous-columnar junction that represents the target for malignant transformation. Thus, this is the first in vitro model with normal cervical cells for studying the sequential development of cervical carcinoma cells for studying the sequential development of cervical carcinoma. Southern blot analyses of high molecular weight DNA confirmed that the specific HPV-16 BamHI fragments including at least one HPV-16 intact copy were identical before and after ras addition. Furthermore, tumors returned to culture also had the identical pattern HPV-16 specific BamHI fragments. Thus, addition of the ras gene was not responsible for any rearrangements of HPV-16 sequences. Cells contained and expressed the cotransfected viral Ha-ras gene. Intact and non-intact viral Ha-ras BamHI fragments were detected in the transformed line as well as the tumor derived lines. As expected, the ras transfected cervical cells differed in the expression of ras protein (p21) compared to the parental line indicating that the transfected v-Ha-ras was expressed. Expression of the activated v-Ha-ras paralleled the tumorigenicity observed.

CONCLUSION.

Experimental data supports the hypothesis that HPV is an etiologic agent of cervical cancer. However, HPVs are not sufficient by themselves to produce cancer. Transfection with HPVs induces dysplastic lesions similar to preinvasive cervical neoplasia. Integrated HPV DNA sequences are stably maintained in immortalized epithelial cell lines, and integration site of HPV may vary, but generally occurs within E1 or E2 ORFs thereby deregulating the expression of the E6 and E7 ORFs. The ability to study aberrant differentiation after HPV transfection and to obtain invasive squamous cell carcinoma in a two-stage carcinogenesis in vitro model indicates that some HPVs may have a critical role in the development of human cancer.

REFERENCES.

1. zur Hausen H. (1989) Papillomaviruses in Anogenital Cancer as a Model to Understand the Role of Viruses in Human Cancers. Cancer Res. 49:4677-4681.
2. Rotkin ID. (1973) A comparison review of key epidemiological studies in cervical cancer related to current searches for transmission agents. Cancer Res., 33:1353-1367.
3. Woodworth CD, Doniger J, DiPaolo JA. (1989) Immortalization of Human Foreskin Keratinocytes by Various Human Papillomavirus DNAs Correspond to Their Association with Cervical Carcinoma. J. Virology 63:159-164.
4. Popescu NC, Amsbaugh SC, and DiPaolo JA. (1987) Human papillomavirus 18 DNA is integrated at a single chromosome site in cervical carcinoma cell line SW756, J. Virol. 51:1682.
5. Popescu NC, DiPaolo JA, and Amsbaugh SC. (1987) Integration sites of human papillomavirus 18 DNA sequences on HeLa cell chromosomes. Cytogenet. Cell Genet. 44:58.
6. Munger K, Phelps WC, Bubb V, Howley PM, and Schlegel R. (1989) The E6 and E7 genes of the human papillomavirus type 16 together are necessary and sufficient for transformation of primary human keratinocytes. J. Virol. 63:4417-4421.
7. Scheffner M, Werness B, Huibregtse J, Levin A, and Howley P. (1990) The E6 oncoprotein encoded by human papillomavirus types 16 and 18 promotes the degradation of p52. Cell 63: 1129-1136.
8. Pirisi L, Creek KE, Doniger J, DiPaolo JA. (1988) Continuous cell lines with altered growth and differentiation properties originate after transfection of human keratinocytes with human papillomavirus type 16 DNA. Carcinogenesis 9:1573-1579.
9. Woodworth CD, Bowden PE, Doniger J, Pirisi L, Barnes W. Lancaster WD, and DiPaolo JA. (1988) Characterization of normal human exocervical epithelial cells immortalized in vitro by papillomavirus types 16 and 18 DNA. Cancer Res. 48:4620-4628.
10. Barrandon Y, Li V, and Green H. (1988) New techniques for grafting of cultured human epidermal cells onto athymic animals. J. Invest. Dermatol. 91:315.
11. Woodworth CD, Waggoner S, Barnes W, Stoler MH, and DiPaolo JA. (1990) Human cervical and Foreskin Epithelial Cells Immortalized by Human Papillomavirus DNAs exhibit Dysplastic Differentiation in Vivo. Cancer Res. 50:3709-3715.
12. Riou GF, Barrois M, Sheng Z, Duvillard P, and Lhomme C. (1988) Somatic deletions and mutation of c-Ha-ras gene in human cervical cancers. Oncogene 3:329.
13. DiPaolo JA, Woodworth CD, Popescu N, Notario V, and Doniger J. (1989) Induction of human cervical squamous cell carcinoma by sequential transfection with human papillomavirus 16 DNA and viral Harvey ras. Oncogene 4:395.

Spatial Distribution as Additional Parameter in Histochemical Quantification of Cell Changes Explored by an Image Analysing System *

Kohler M., Doris Ramm, J. Steinmann, Michaela Aufderheide, U. Mohr
Institute of Experimental Pathology, Hannover Medical School, Konstanty-Gutschow-Str. 8, Hannover, FRG.

Introduction

Both Golgi apparatus and lysosomes are associated with turnover of cell compartments and extracellular processes. The role of the microtubules maintaining the location of Golgi vesicles as well as the intracellular transport of lysosomes is demonstrated in many experiments dealing with microtubule poisons. Due to the sensitivity of the microtubule system to chemicals, a direct correlation between the spatial distribution of the lysosomal apparatus and the biochemical condition of the cell can be assumed.
In this study, a fetal Syrian hamster respiratory epithelial cell line, M3E3/C3, was treated with demecolcine in concentrations ranging from $1*10^{-5}$ to $1*10^{-7}$M. The histochemical appearance of the lysosomal apparatus was examined by staining the marker enzyme 'acid phosphatase'.

Material and Methods

Fetal Syrian hamster respiratory epithelial cells, M3E3/C3 (EMURA, 1982), were seeded on cover slips in 8 cm^2 dishes in a concentration of $5*10^3$ cells per dish. Before treatment they were cultivated for 4 days under normal conditions (37^{0}C, 10% CO_2, 98% r.h.) in RPMI 1640 supplemented with 20% fetal bovine serum. The cells were treated with demecolcine in concentrations of $1*10^{-5}$, $1*10^{-6}$ and $1*10^{-7}$ M, soluted in medium for one hour. After fixation in 2% neutral buffered p-formalin with 0,2% glutaraldehyde for 30 min, the enzyme 'acid phosphatase' was stained according to BARKA and ANDERSON (1962). For image analysing no staining was done, neither of the cytoplasm nor of the nucleus.
The colorimetric determination of acid phosphatase was done by SIGMA-Kit (No. 104-AS).
The image analysis was carried out by the following steps. Images of the cells were scanned by a color TV camera and stored in the memory of the image analysing system IBAS. The cells were discriminated by several automatic steps but only those cells being in contact with each other could be separated interactively. Another image memory contains the discriminated B/W-image of the acid phosphatase areas. To estimate the amount of the enzyme 'acid phosphatase' per cell, the computer has to accumulate the integrated optical density (I.O.D.) of all objects identified as histochemically active regions of the enzyme lying in the area of each isolated cell. This was achieved by masking the image of the discriminated enzyme-active areas by the areas of the accompanying cells one by one.

The computer identified all objects lying in this area as separated objects and measured the following parameters:

- area and
- perimeter of the cell
- sum of the areas of all 'acid phosphatase' regions
- I.O.D. of these regions
- number of separate areas of 'acid phosphatase'.

The effects of all these parameters were analysed by an analysis of variance (ANOVA).

Results

Only slight differences could be seen in the spatial distribution of the acid phosphatase in the microscopic image. Control cells showed a compact juxtanuclear polarised agglomeration of the enzyme with few granules

* This projekt was granted by the Minister of Health: 417-6080-1/44

in the peripheral cytoplasm. Treated cells showed a dose-dependent increase of enzyme-active granules in the cytoplasm with a dissolving of the unfocused perinuclear area of acid phosphatase.
There is no significant difference (shown by the significance limits (ELSNER, 1991) at $p \leq 0.05$) for the quantitative colorimetric determination of acid phosphatases in the cell suspension between control and dose groups (figure 1), but there is a difference between the lowest concentration and the higher dose groups.

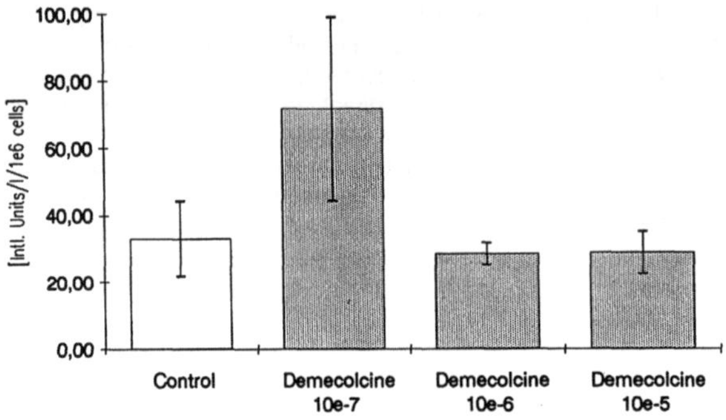

Figure 1: Quantitative colorimetric Determination of Acid Phosphatase; mean ± significance limits

The data evaluated by the image analysing system showed slightly different results:
No significant difference could be shown between the control and the dose groups neither for the area nor the perimeter of the cells (figure 2 and 3). These findings demonstrate that shape and size of the cells do not change throughout the treatment.

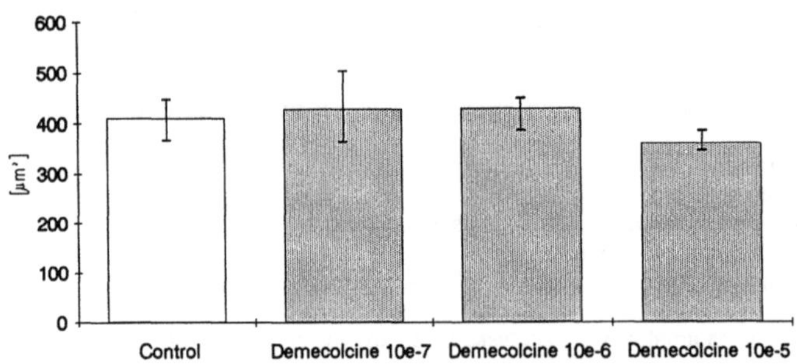

Figure 2: Area of cells; mean ± extremes (n=4)

The area of acid phosphatase shows significant differences between control and treated groups, but no differences between dose groups. It seems that the substance only enhances the spreading of the enzyme around the cytoplasm.
For the I.O.D. of the acid phosphatase we also found a significant difference between the groups of 10^{-5} and 10^{-6}M, while there was no difference between 10^{-6} and 10^{-7} M. These findings correlate to the microscopic images of the indirect immunofluorescence of the microtubule-system (MT). While we observed normal structure of MT's in both of the lower dose groups no MT's could be seen in the 10^{-5} M group.

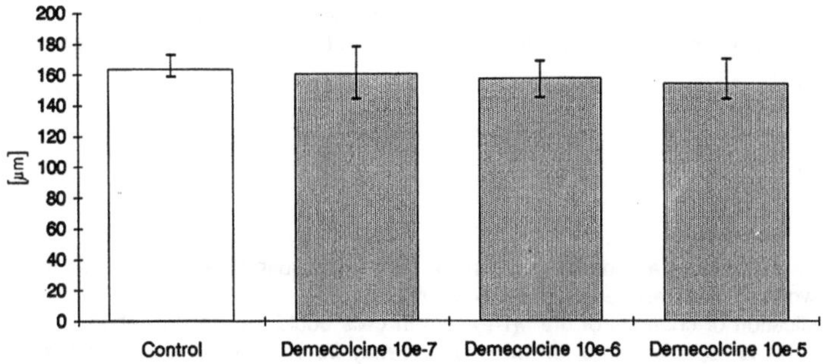

Figure 3: Perimeter of cells; mean ± extremes (n=4)

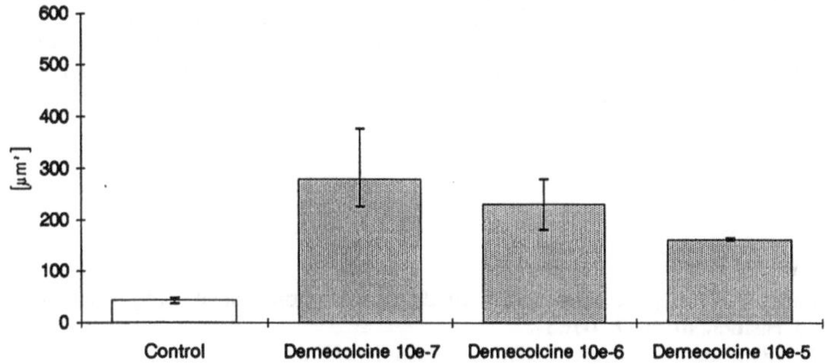

Figure 4: Area of acid phosphatase; mean ± extremes (n=4)

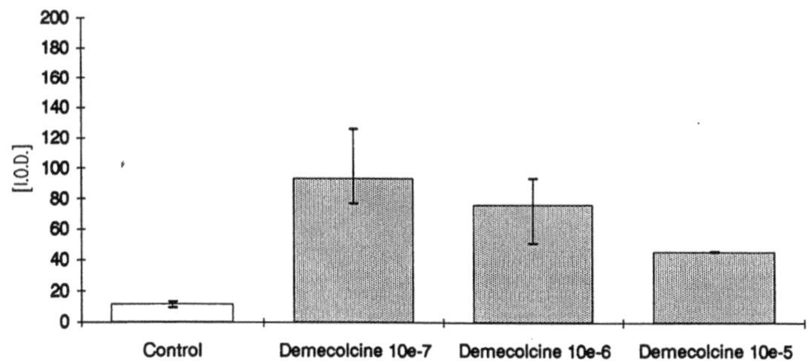

Figure 5: Integrated optical density of acid phosphatase; mean ± extremes (n=4)

In addition to these data we found a significant difference to the control group in the number of acid phosphatase-active areas per cell only in the group of 10^{-5} M (5.6 areas/cell in contrast to 2.2-2.5 areas/cell in the other groups).

Discussion

In this study we evaluated a sensitive and simple method to quantify small changes in the MT-system by chemicals. Two main facts resulted from our experiment:

1. The quantification of changes of the MT-system in cells could not be described exactly by biochemical methods. This is especially true for low doses. It may be an effect of the error through the relation of colorimetric measurements to an estimated number of cells. A better understanding of the effect of poisons affecting the MT-system could be depicted by image analysing only a few (n < 200) single cells.
2. Using image analysis we found higher enzyme concentrations and a larger area covered by acid phosphatase after treatment of cells with low doses of demecolcine. Increasing the doses up to 10^{-5} M causes the MT-system to collapse, which in fact accompanies fragmentation and dispersal of the Golgi apparatus from the nuclear region. This was observed in the increasing number of acid phosphatase areas per cell, while the amount of the enzyme decreases. These data correspond to the hypothesis discussed about the role of MT-supported transport of vesicles (KELLY, 1990).

The role of the MT-system in vesicle transportation for endo- and exocytosis is known to be important only for the movement between the Golgi network and the cell surface (HOLTZMAN, 1989; KELLY, 1990). If the MT-system is destabilized by low concentrations of demecolcine, the transport of lysosomes to the cell surface is inhibited, while synthesis of protein remains still active. Thus the loaded vesicles accumulate near the Golgi apparatus which is localized perinuclear.

References

Barka T, Anderson PJ (1962) Histochemical methods for acid phosphatase using hexazonium pararosanilin as coupler. J. Histochem. 10: 741-753

Dustin P (1984) Microtubles. Springer Berlin Heidelberg New York Tokyo: 112 ff.

Emura M, Richter-Reichhelm HB, Böning W, Eichinger R, Schoch C, Althoff J, Mohr U (1982) A fetal respiratory epithelial cell line for studying some problems of transplacental carcinogenesis in syrian golden hamsters. J. Cancer Res. Clin. Oncol. 104: 133-14

Holtzman E (1989) Lysosomes. Plenum Press New York: 84-87

Kelly RB (1991) Microtubles, Membrane Traffic, and Cell Organization. Cell 61: 5-7

Lowry OH, Rosenbrough NJ, Farr AL, Randall R.J (1951) Protein measurement with the Folin-phenol reagent. J. Biol. Chem. 193: 265

Rohen JW, Lütjen-Drecoll E (1990) Funktionelle Histologie: kurzgefaßtes Lehrbuch der Zytologie, Histologie und mikroskopischen Anatomie des Meschen nach funktionellen Gesichtspunkten. Schattauer Verlag Stuttgart New York:13-48

Intraphagolysosomal pH and Intracellular Particle Dissolution in Canine and Rat Alveolar Macrophages

W.G. Kreyling[1], W. Beisker[2], U. Miaskowski[1], M. Neuner[1] and P. Heilmann[1]

GSF - Forschungszentrum für Umwelt und Gesundheit, [1]Projekt Inhalation, [2]Institut für Biophysikalische Strahlenforschung, D-8042 Neuherberg, FR Germany

Introduction

Intracellular dissolution of a large variety of inhaled inorganic environmental particles - which are not readily soluble in the epithelial lining fluid of the lungs - is the first step of an important long-term clearance mechanism of the lungs by which the dissolved material is eliminated to blood (Kreyling, 1992). Intracellular particle dissolution (IPD) occurs in phagolysosomal vacuoles of phagocytes (Kreyling et al., 1990). IPD was shown to be more effective for a number of materials than dissolution in extracellular body fluids as recently reviewed (Kreyling, 1992). Since the intraphagolysosomal pH (PLpH) at about 5 is lower than that of the surrounding cell plasma at about 7.2 (Nyberg et al., 1989a, 1989b), PLpH is believed to be an important co-factor of IPD and thereby of an important long-term clearance mechanism of the lungs.

Materials and Methods

For PLpH measurements, AM were obtained from partial bronchoalveolar lavages (BAL) of 15 Beagle dogs utilizing fiberoptic bronchoscopy and in situ total lung lavages of 18 anesthetized and exsanguinated Wistar rats via the cannulated trachea (Heilmann et al., 1992). Two dogs received two sequential BALs 50 days apart from each other. For IPD measurements, AM were obtained from 16 rats which were not entirely the same as those used for PLpH measurements. Similarly for IPD measurements, AM were obtained from 7 sequential BALs of each of 8 dogs carried out at 7 weeks interval as described previously (Ferron et al., 1990).

Fluorescence microscope photometry (FMP) and dual laser flow cytometry (FCM) of PLpH in canine and rat alveolar macrophages (AM) were compared using fluoresceinisothiocyanate labeled, amorphous silica test particles (FSP) as described previously (Nyberg et al., 1989b, Heilmann et al., 1992). FMP was carried out on monolayers of adherent AM in full medium (RPMI 1640 + fetal bovine serum + antibiotics) in 96 well cell culture plates together with FSP to allow for phagocytosis and fusion of lysosomes with the particle containing phagosomes in order to obtain maturation of fully developed phagolysosomes during 24 h and up to several days. Particle to cell ratio was in the range 1-10 : 1. The ratio of fluorescence intensities measured at wavelengths >520 nm when excited either at 436 $\pm$ 5 nm or at 488 $\pm$ 8 nm wavelengths of single phagocytized FSP were determined by inverted microscopy and photometry using a small aperture (6 μm diameter). A calibration curve for the pH value X as a function of the fluorescence ratio R(X) was separately determined on free FSP in standard buffers of pH values between 3-7 and were approximated by an exponential function using a least squares method:

$$R = a * \exp(b * X) + c \qquad (1)$$

For FCM, phagocytosis of FSP was allowed in cell suspensions in full medium contained in polypropylene tubes for 24 h and up to several days of incubation. Particle to cell ratio was in the range 1-10 : 1. In order to classify different cell populations and to discriminate free or attached beads from phagocytized FSP using dual laser flow-cytometry (FACStar-Plus, Becton & Dichinson), six parameter list mode data were recorded from forward scatter, side scatter and fluorescence intensities at 530 nm excited at wavelengths of 457 nm and 488 nm. Also, the logarithmic fluorescence intensity of the DNA marker propidium iodide (PI) added to the cell suspension immediately prior to PLpH determination was measured at wavelengths >550 nm when excited at a wavelength of 488 nm, to distinguish viable from lysing or dead cells. In this paper results obtained only from viable AM are reported. Similar to FMP, a calibration curve for the pH value was determined on free FSP in standard buffer suspensions of pH values between 3-7.

IPD of porous monodisperse and chemically homogeneous $^{57}Co_3O_4$ test particles of two particle sizes (0.5 and 0.7 μm geometric mean diameter) was carried out in canine and rat AM cultures during 2 weeks of incubation and IPD rates (dissolved particle mass per day normalized to the initially incubated particle mass) for both particle sizes were calculated as described recently (Kreyling et al., 1990).

Result and Discussion

For both methods FMP and FCM, calibration data were reproducible. For FMP the approximated parameters of the calibration curve (equation 1) were: a=2.26; b=0.34; c=2.24. Due to the variable alignment of the laser illuminations of FCM resulting in different amplifications of the two fluorescence intensities, calibration curves could not be normalized and averaged. Therefore, calibration data were always determined immediately prior to PLpH measurements without any changes of the parameters of the flowcytometer.

Determinations of the PLpH between 1 - 7 days of incubation of AM obtained from five dogs varied very little around each of the average PLpH values. Normalizing these data to each of the average PLpH values, the standard deviation of the normalized data was similar for both methods 0.0037 (FMP, n=25) and 0.0041 (FCM, n=13). The small variation of PLpH during incubation indicated that PLpH is a constant functional parameter of AM during in vitro culture. Since FSP were incubated together with AM either 24 h prior to PLpH measure-ments at various time points of incubation or up to seven days of incubation, FSP proved to be stable probes for this cell function within the period of observation. However, when com-paring PLpH of AM obtained from the same BAL as determined either by FMP or FCM the former method yielded PLpH values which were slightly higher 0.30 $\pm$ 0.25 than the latter. This was found in AM obtained from eight BALs in 18 determinations of each method. It is not clear whether this difference was instrumental or whether it was due to the different culture conditions of adhering AM to the well bottom of cell culture plates versus suspended AM in full medium.

PLpH of AM obtained from sequential BALs of two dogs varied within the range of PLpH found in the whole population of 15 dogs, similar to earlier sequential observations on AM of three dogs (Heilmann et al., 1992). The intersubject variability suggested that PLpH was not a constant AM function of a given animal but may vary with time.

The mean PLpH (± standard deviation S.D.) of on AM is given in Table 1 for all animals of both species dogs and rats as determined by FMP and FCM, respectively. The mean PLpH obtained from AM of both species using FCM were comparable to previous measurements (Heilmann et al., 1992). Interestingly, no significant difference in PLpH of AM of rats was observed using the two methods FMP and FCM as was seen in canine AM. The mean PLpH found in AM of rats but was significantly ($p < 0.05$) higher than in AM of dogs as determined by both methods FMP and FCM.

Mean values of in vitro IPD rates of both sizes of $^{57}Co_3O_4$ test particles in canine and rat AM are also given in Table 1. The mean IPD rates of both particle sizes obtained from canine AM were significantly ($p < 0.01$) by a factor of about 2.5 larger than those of rat AM.

Table 1. Mean PLpH values (± S.D.) measured either by FMP or FCM and mean IPD rates (± S.D.) for both particle sizes of rat and canine AM. Underneath each value the number of BALs is given in brackets from which AM were obtained.

	PLpH (FMP)	PLpH (FCM)	IPD rate (0.5 μm) $*10^{-2}$ d^{-1}	IPD rate (0.7 μm) $*10^{-2}$ d^{-1}
dog	4.73 ± 0.24 (17)	4.28 ± 0.26 (10)	1.49 ± 0.21 (54)	0.83 ± 0.14 (51)
rat	5.06 ± 0.29 (18)	5.13 ± 0.30 (14)	0.53 ± 0.10 (16)	0.29 ± 0.03 (14)

The mean PLpH found in AM of rats was similar to that reported on rabbit AM (Nyberg et al., 1989a and 1989b). The mean PLpH found in canine AM using FMP were the same as that of canine AM observed at a recent interspecies comparison of PLpH (Kreyling et al., 1991). In the previous study PLpH was also determined on adhering monolayers of AM. Preliminary results obtained from baboon and bovine AM which were kindly provided by Dr.

J.-L. Poncy, CEA, Bruyères-Le-Châtel France, and Dr. I. Berg, University of Hamburg, Hamburg FRG, respectively, indicated a similar correlation between PLpH and IPD as shown here for rat and canine AM. Baboon AM showed similarly low PLpH and similarly high IPD rates as those of canine AM whereas bovine AM showed similarly high PLpH and similarly low IPD rates as those of rat AM.

The higher PLpH in rat AM correlated with the smaller IPD rate in rat AM when compared to the dog data, but the slight change of about 0.5 in PLpH could not result in the drastic change by a factor of about 2.5 of the IPD rate as was shown in studies of the dissolution of these test particles in media at various pH values without AM (Kreyling et al., 1988). Hence, PLpH is an important co-factor for IPD and thereby for an important long-term clearance mechanism of the lungs. Moreover, other factors affecting IPD are obviously also different between AM of both species.

References

Ferron G.A., Kreyling W.G., Fürst G., Neuner M., Schumann G., and Heyder J. (1990) Long-term exposure of dogs to a sulfite aerosol: III Effect of lung clearance. J. Aerosol Sci. 21:S479-S482.

Heilmann P., Beisker W., Miaskowski U., Camner P. and Kreyling W.G. (1992) Intraphagolysosomal pH in canine and rat alveolar macrophages: flow cytometric measurements. Environm. Hlth. Perspect. (in press).

Kreyling, W.G., Neuner M. and Matejkova E. (1988) Analysis of intraphagolysosomal dissolution of test particles in canine alveolar macrophages. J. Aerosol Sci.19:1071-1074.

Kreyling, W.G., Godleski, J.J., Kariya, S.T., Rose, R.M. and Brain, J.D. (1990) In vitro dissolution of uniform cobalt oxide particles by human and canine alveolar macrophages. Am. J. Respir. Cell Mol. Biol. 2:413-422.

Kreyling W.G., Nyberg K., Nolibé D., Collier C.G., Camner P., Heilmann P., Lirsac P., Lundborg M. and Matejkova E. (1991) Interspecies comparison of phagolysosomal pH in alveolar macrophages. Inhalation Toxicology 3:91-100.

Kreyling W.G. (1992) Intracellular particle dissolution in alveolar macrophages. Environm. Hlth. Perspect. (in press)

Nyberg, K., Johansson, A. and Camner, P. (1989a) Intraphagosomal pH in alveolar macrophages studied with fluorescein-labeled amorphous silica particles. Exp. Lung Res. 15: 49-62.

Nyberg, K., Johansson, A., Rundquist, J. and Camner, P. (1989b) Estimation of pH in individual alveolar macrophage phagolysosomes. Exp. Lung Res. 15: 499-510.

UV-C and Electromagnetic Radiation Induce Premature Terminal Differentiation of Cultured Human Fibroblasts

H. Peter Rodemann
Section of Radiobiology, Eberhard-Karls-University, Hoppe-Seyler-Str. 3, D-7400 Tübingen, Germany

Introduction

Over the recent years the bio-hazardous effects of UV-C- as well as electromagnetic radiation have increasing impact in discussions concerning the enviromental hygiene. At the cellular level considerable data exist that UV-C is a severe hazard for eukaryotic cells in terms of cell killing and/or induction of skin cancer. However, little is known about the biological effects of UV-C- and electromagnetic radiation as inducers of terminal differentiation processes in cells, especially mammalian cells. Although, low frequency electromagnetic fields (EMF) have been administered in bone fracture healing with reasonable success for almost 20 years (Bassett et al 1982; Kraus and Lechner 1972), the cellular and/or molecular basis of the mechanisms involved are still unknown.

On the basis of this knowledge the effects of UV-C and low frequency electromagnetic radiation were investigated in the terminally differentiating cell system of human skin fibroblasts. As accessory cells, fibroblasts of all connective tissues fulfill important functions, e.g. in tissue homeostasis. The cellbiological and biochemical parameters of the differentiation processes occuring in these cells have been described in detail recently (Bayreuther et al. 1988, Rodemann et al. 1989). Fibroblasts in vivo and in vitro differentiate along a terminal cell lineage: MFI-MFII-MFIII-PMIV-PMFV-PMFVI. MFI, MFII and MFIII are the mitotically active progenitor cells of the fibroblast cell system; PMFIV and PMFV are postmitotic transition cell types to the terminally differentiated cell type PMFVI (Bayreuther et al. 1988; Bayreuther et al. 1989; Rodemann et al. 1989; Rodemann 1989). After a genetically determined lifespan the cell type PMFVI degenerates presumably by the onset of apoptosis (Bayreuther personal communication). Based on these facts the fibroblast cell system is the cell system of choice for investigations into the mechanisms of radiation induced alterations in terminal differentiation processes at the molecular level.

Material and Methods

<u>Cells:</u> Human skin fibroblasts (cell strain HH8) were routinely cultured in DME-medium (DMEM) supplemented with 10% fetal calf serum (fcs) and subcultured once per week as described in detail elsewhere(Rodemann 1989).

<u>Radiation experiments:</u> MFII-type fibroblasts were seeded at low cell density (10^3 cells/cm^2) in 10 cm culture dishes. For EMF-irradiation cells were exposed to EMF (frequency 20Hz; magnetic induction 6 mT) for 2 x 6 h per day for up to 21 days. Survival kinetic experiments applying UV-doses of 0-35 Jm^{-2} revealed that 15 Jm^{-2} resulted in a sufficient inhibition of cell growth without considerable amount of cell killing (<5 %). After UV-C-irradiation cells were incubated again in DMEM plus 10%fcs for up to 21 days.

<u>Cell type analysis, clonogenic activity and cell doubling potential:</u> At various times after exposure to EMF- or UV-C-irradiation, cell cultures were analysed for the frequency of MF- and PMF-type cells by specific staining procedures and subsequent morphological classification (Bayreuther et al. 1988). The clonogenic potential of radiation exposed and control cells was evaluated by cloning 500 cells per well in 6-well culture dishes and incubating them without further radiation exposure. The cell doubling potential was analysed by determinations of the cell numbers at various time points after exposure. Cumulative population doublings (CPD) were calculated as recently described (Bayreuther et al. 1988).

Biochemical analysis: As biochemical marker of an induced terminal differentiation the expression of protein PIVa was analysed by high resolution 2-D-PAGE (Rodemann 1989). The biosynthetic rates of collagen subtypes were studied by incorporation of 35S-methionine and subsequent SDS-PAGE as described earlier (Rodemann et al. 1984, Rodemann and Bayreuther 1986; Rodemann and Müller 1991).

Results and Discussion

As demonstrated earlier (Rodemann et al. 1989) human skin as well as lung fibroblasts differentiate along a terminal cell lineage of 3 mitotically active progenitor fibroblasts (MFI,MFII,MFIII) and three postmitotic cell types (PMFIV,PMFV,PMFVI). The lifespan of each of these cell types seems to be regulated by a genetic mechnanism as it can be concluded from subcloning experiments (Rodemann et al 1991) and studies demonstrating a cell type specific and sequentially induced expression of marker proteins with distinct half lifetimes (Rodemann 1989). For one of these proteins, protein PIVa, a regulatory function has been postulated based on molecular studies into the induction of this protein and the concommittant repression of the proto-oncogene c-fos in experimentally induced PMF-cell types (Rodemann et al. 1991).

In the present study both, UV-C-irradiation and EMF-exposure at certain intensities induced the premature terminal differentiation of MFII-type progenitor fibroblasts to postmitotic cells of the type PMFVI. A single dose of 15 Jm^{-2} UV-C was sufficient to induce PMF-type cells at an efficiency rate of app. 95-98% within 4-14 days of post-radiation incubation. When UV-C irradiated cells were tested for clonogenic activity the percentage of cells capable of forming colonies bigger than 16 cells within 14 days of incubation decreased to 0% within 4 days after irradiation (Fig. 1A).

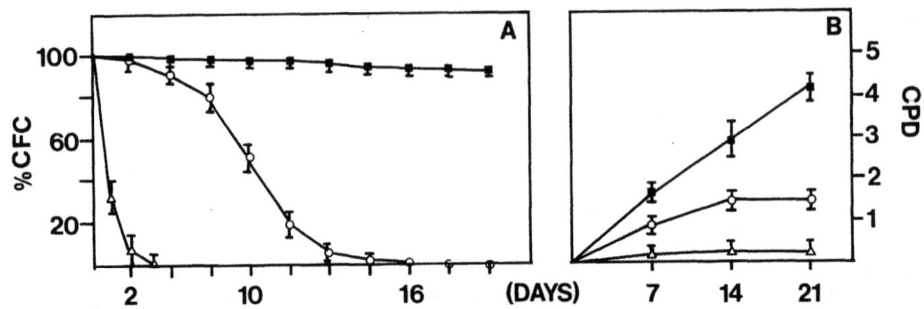

Fig. 1: Clonogenic and cell doubling potential of human skin fibroblasts. A: Clonogenic potential; **B:** Cell doubling potential. o Control; o EMF-exposed cells; o UV-C-irradiated cells; CFC, colony forming cells; CPD, cumulative population doublings.

Likewise, the number of cell doublings of the UV-C-irradiated cultures was only about 0.2-0.3 CPD within a 21 day incubation period after irradiation (Fig. 1B). Thus, the UV-C-stimulated induction of the terminal differentiation occured in the majority of the cells without further cell division.

Within 14-21 days of exposure to low frequency EMF normal skin and lung fibroblast cultures composed of MFII-type cells differentiated into PMFVI-type cultures at a rate of 85-90%. The percentage of the clonogenic cells decreased exponentially to 0% within 14 days of exposure

(Fig. 1A). During a 21 day exposure to EMF the average number of CPD was app. 1.5 (Fig. 1B). Thus, the induction of terminally differentiated PMFVI-cells from MFFII-progenitor fibroblasts occured after 1-2 cell cycles. The number of cell cycles necessary for spontaneously occuring terminal differentiation from MFII to PMFVI-type cells is in the range of 30-40 cycles. The data presented in this paper most likely indicate that the underlying mechanisms of the UV-C- and EMF-induced premature terminal differentiation are not necessarily identical since UV-C-radiation is effective without a further cell cyle while EMF needs at least 1-2 additional cell cycles. However, both treatments result in the same phenotype. Based on the biochemical analysis of the expression of protein PIVa as well as the synthesis of the interstitial collagen types I, III and V both, UV-C- and EMF-induced PMF-cell types are identical to the spontaneously arising PMF-cells during normal differentiation (Table 1).

Table 1: Collagen synthesis in spontaneously arising, UV-C- and EMF-induced PMF-type human skin fibroblasts.

	Total Collagen ($cpm/SEMx10^{-5}$ cells)	Relative Amounts of Collagen Subtypes		
		I	III	V
MFII-Control	851/94	89.0%	10.5%	0.5%
PMF-spont.	12713/1603*	71.5%*	25.0%*	3.5%*
UV-C-PMF	9465/1056*	67.3%*	28.0%*	4.7%*
EMF-PMF	11892/1187*	69.5%*	26.0%*	4.5%*

* significant differences from MFII-Control, $p < 0.01$, n=4

As analysed by 2-D-PAGE after pulse-labelling the cells with 35S-methionine for various times (15min-12h) the resulting PMF-type cells do not show any appearance of abnormal proteins which could result from UV-C- or EMF-induced mutations. Furthermore, the rate of degradation of short-lived proteins which would indicate an enhanced number of abnormal proteins is not changed under both conditions. Data from other studies indicate that MFII-progenitor cells and naturally arising as well as experimentally induced PMF-cells are identical in the rate of excission repair for UV-induced pyrimidine-photo-dimers (Niggli et al 1989).

Especially the UV-C- and EMF-induced expression of protein PIVa seems to play a key role in the induction of PMF-cells. Under both irradiation conditions, the appearance of protein PIVa preceeds the morphological differentiation parameters, i.e. PMF-type morphology with characteristic organization of the actin fibres (Bayreuther et al 1988), by about 3-10 days. Thus, under both conditions the induced expression of protein PIVa may trigger the molecular events resulting in a premature terminal differentiation. This in turn results in a premature aging process of the cell populations irradiated with UV-C or EMF in the given doses due to the differential depletion of the mitotically active progenitor cells. Thus, besides the already known effects of UV-C and EMF their additional potential bio-hazardous effect at least at the doses and intensities indicated may primarely be based on their capability to stimulate progenitor cells of regenerating tissue into terminal differentiation which leads to a depletion of the regenerative capacity of the affected tissue. However, under certain circumstances, e.g. bone fractures, this effect of especially EMF-radiation, may induce premature terminal differentiation of a certain number of tissue forming cells and may, thus, be advantageous for the woun wound healing process, e.g. during bone fracture healing. Further molecular biological studies are in progress to identify the underlying mechanisms regulating the induced terminal differentiation.

Acknowledgement

The excellent technical assistance of Petra Wecke, Developmental Biology, University of Bielefeld, and the scientific and technical support concerning the EMF-application system of W. Kraus, Institute für Medizinische Physik, München, is gratefully acknowledged. This study was supported by grants from the Deutsche Forschungsgemeinschaft (SFB 223 B05; Ro 527/2-2).

References

Bassett C.A.L., S.N. Mitchell and S.R. Gaston (1982) Treatment of ununited tibial diaphyseal fractures with pulsed electromagnetic fields. J. Bone Joint Surg. 63A:511-523

Bayreuther K., H.P. Rodemann, R. Hommel, K. Dittmann, M. Albiez, and P.I. Francz (1988) Human skin fibroblasts in vitro differentiate along a terminal cell lineage. Proc. Natl. Acad. Sci. USA 85:5112-5116

Bayreuther K., H.P. Rodemann, P.I. Francz, and K. Maier (1989) Differentiation of fibroblast stem cells. J. Cell Sci. Suppl. 10:115-131

Kraus W. and F. Lechner (1972) Heilung von Pseudathrosen und Spontanfrakturen durch strukturbildende elektrodynamische Potentiale. Münch. Med. Woschr. 42:1814-1819

Niggli H., K. Bayreuther K., H.P. Rodemann, R Röthlisberger , and P.I. Francz (1989) Mitomycin C-induced postmitotic fibroblasts retain the capacity to repair pyrimidine photodimers formed after UV-irradiation. Mut. Res. 219:231-240

Rodemann H.P. and K. Bayreuther (1984) Abnormal collagen metabolism in cultured skin fibroblasts from patients with Duchenne muscular dystrophy. Proc. Natl. Acad. Sci. USA 81: 5130-5134

Rodemann H.P. and K. Bayreuther (1986) Differential degradation of 35S-methionine polypeptides in Duchenne muscular dystrophy skin fibroblasts in vitro. Proc. Natl. Acad. Sci. USA 83:2086-2090

Rodemann H.P., K. Bayreuther, K. Dittmann, P.I. Francz, and M. Albiez (1989) Selective enrichment and biochemical characterization of seven human skin fibroblast cell types in vitro. Exp. Cell Res. 180:84-93

Rodemann H.P. (1989) Differential degradation of intracellular proteins in human skin fibroblasts of mitotic and mitomycin C (MMC)-induced postmitotic differentiation states. Differentiation 42:37-43

Rodemann H.P. and G.A. Müller (1991) In vitro growth, differentiation and collagen synthesis of fibroblasts from kidneys with interstitial fibrosis. Am. J. Kid. Dis. XVIII:684-686

Rodemann H.P., H.-P. Peterson, K. Schwenke, and K.-H. von Wangenheim (1991) Terminal differentiation of human fibroblasts is induced by radiation. Scann. Microscopy in press

Rodemann H.P., P. Wecke and M. Jäschke (1991) Low frequency electromagnetic fields induce premature terminal differentiation of in vitro cultured human skin and lung fibroblasts. J. Electroanal. Chem. in press

Effect of Airborne Particulates on Phagocytosis of Human and Rat Macrophages

Seemayer, N.H., Angelika Kiell, A. Happel, W. Hadnagy and Heidrun Behrendt
Medical Institute of Environmental Hygiene at the Heinrich-Heine-University of Düsseldorf, Gurlittstr. 53, D-4000 Düsseldorf, F.R.G.

Introduction

Industrial processes, power generation, traffic, waste incineration and coal or fuel combustion for space heating are the most important sources of atmospheric pollutants (Fishbein, 1990). Between 500 - 2,000 mostly organic compounds have been detected as atmospheric pollutants, among them potential and recognized mutagens and carcinogens (Helmes et al., 1983; Schlipköter, 1983; Fishbein, 1990). Beside gaseous pollutants airborne particulates are of special medical concern. They are composed of a very large number of chemical substances and with a diameter smaller than 5 µm they represent respirable dust, which by inhalation reaches broncho-alveolar space in human lung. Alveolar space in human lung contains an "inner surface" of 100 - 150 m^2 (Burri, 1982). Each day this "inner surface" of human lung is exposed to a volume of air of 12 - 14 m^3 and contaminants such as airborne particulates, noxious gases, infectious agents, immunogenic and allergic particles (Green et al., 1977). Alveolar macrophages, which are found on the "inner surface" serve as an important line of defense, keeping alveoli clean by ingesting inhaled particles and micro-organisms (Brain, 1986). Adverse effects on alveolar macrophages could impair the pulmonary defense system contributing to the causation of diseases. There is evidence that particulate pollutants affect the process of phagocytosis by macrophages. A strong inhibition of phagocytosis of alveolar macrophages of the Syrian golden hamster was reported by Beck et al. (1988) after intratracheal instillation of respirable particles of waste oil heater. Van Houdt and Rietjens (1988) observed a strong dose-dependent inhibition of phagocytosis of rat alveolar macrophages in vitro by polluted outdoor and indoor samples of particulates. According to Ruiz et al. (1988) peripheral blood macrophages of school children from a highly polluted city (Santiago de Chile) revealed a lower phagocytic index than those from a rural village. This paper reports about effects on macrophages induced by organic extracts of airborne particulates collected in the heavily industrialized Rhine-Ruhr district (F.R.G.). Inhalation experiments were carried out utilizing a combined in vivo/in vitro method to evaluate the pulmonary toxicity of airborne particulates. Furthermore, in comparison in vitro studies on human monocyte-derived macrophages were performed.

Material and Methods

<u>Airborne particulate sampling:</u> Airborne particulates were collected with a high volume sampler "Draeger Box Micron Filter MB 1700" (air flow 1500 m^3/h, 10 membrane filter lamellae of 1.5 m^2, collection period 215 hours, total air volume 322,000 m^3) in the city of Duisburg (F.R.G.) in winter of 1983/84. Particulates were extracted with methanol and quantitatively transferred to Lutrol E 400 (Polyethylene glycol, MW 400, Roth, Karlsruhe No. 32861). A benzo(a)pyrene quantity of 20 ng corresponded to an air volume of particulate collection from 6.5 m^3 of air. The stock solution of extract consisted of extracted substances from particulates corresponding to 648 m^3 of air per ml in Lutrol. For in vitro experiments the dosage was calculated according to the benzo(a)pyrene content of the extract.
<u>Inhalation experiments:</u> Inhalation experiments were performed with 6 - 8 weeks old female Wistar rats in an inhalation equipment for "nose-only exposure"(TSE Company, Kronberg, F.R.G.) consisting of an inhalation chamber, an aerosol generating system and a control "Multi-Function Unit". Details of inhalation conditions were previously outlined (Hadnagy et al., 1991; Kiell et al., 1991). For aerosol generation extract in Lutrol E 400 (2 ml/h) was mixed with an air stream of 600 litres per hour utilizing a nozzle. Exposure time per day lasted for 5 hours. Animals were exposed to aerosol produced from the stock solution for 1, 2, 3 and 4 days.
<u>Rat alveolar macrophages:</u> Immediately after corresponding inhalation period animals were sacrificed by Nembutal-anaesthesia and alveolar macrophages were isolated by broncho-alveolar lavage as described elsewhere (Thissen et al., 1991). After determination of cell viability by the dye exclusion test with trypan blue rat alveolar macrophages were seeded into LAB-Tek Tissue Culture Chambers No. 4808 in medium RPMI 1640 with 10% fetal calf serum at a cell density of $4 - 5 \times 10^5$ cells per cm^2.
<u>Human macrophage cultures:</u> Human macrophages were obtained by cultivation of monocytes

from the peripheral blood in medium RPMI 1640 containing 10% fetal calf serum and 10% human AB serum for 10 - 14 days as earlier described in detail (Seemayer and Braumann, 1988).

Phagocytosis assay: Test of phagocytosis was carried out on alveolar macrophage cultures of exposed animals as well as after in vitro exposure of human monocyte-derived macrophage cultures to various concentrations of extract at a final concentration of 1% Lutrol. Phagocytosis was performed by addition of opsonized Zymosan A particles (Sigma No. Z 4250), latex beads (diameter 4.72 µm) or Polychromatic Fluoresbrite Microspheres (Polyscience Inc., No. 19111, diameter 5.72 µm) to cell cultures of human or rat macrophages in serum-free RPMI 1640 medium at a cell-particle ratio of 1 : 10 - 100. Following an incubation period of one hour at 37°C cell cultures were carefully rinsed 3-times with PBS, fixed with Bouin's solution and stained with hematoxylin and eosin. The phagocytic activitiy (% of cells showing more than 2 particles) and phagocytic capacity (number of particles per cell) was determined.

Electron microscopy: For analysis by transmission electron microscopy (TEM) broncho-alveolar macrophages from the rat were processed as described in detail elsewhere (Behrendt et al., 1987).

Results and Discussion

Based on the inhalation conditions selected including the actual aerosol concentration in the chamber and the respiratory volume of the animals, rats were exposed to extractable substances of airborne particulates from 1 m³ of air per hour (Hadnagy et al., 1991; Kiell et al., 1991). After inhalation of aerosols containing extractable substances from airborne particulates rat alveolar macrophages revealed functional and morphological alterations in a dose or time-dependent manner. As criteria of adverse effects we determined the cell viability of macrophages by dye exclusion test as well as the phagocytic activity (% of cells containing more than 2 particles) and the phagocytic capacity (number of particles per cell) utilizing Micropheres or Zymosan A particles. Fig. 1a demonstrate a strong phagocytosis of Microspheres by alveolar macrophages in culture from a rat exposed to the vehicle control "Lutrol" for 20 (4 x 5) hours. In contrast, alveolar macrophages from a rat after 20 (4 x 5) hours of inhalation of an aerosol containing extractable substances from airborne particulates disclose a strong inhibition of phagocytosis (Fig. 1b). The recovery of broncho-alveolar lavage cells was similar in control and exposed animals amounting to 5 - 10 x 10^6 cells. Furthermore, as shown on Fig. 2a alveolar macrophages of control and exposed rats revealed no differences in cell viability of 80 to 90%.

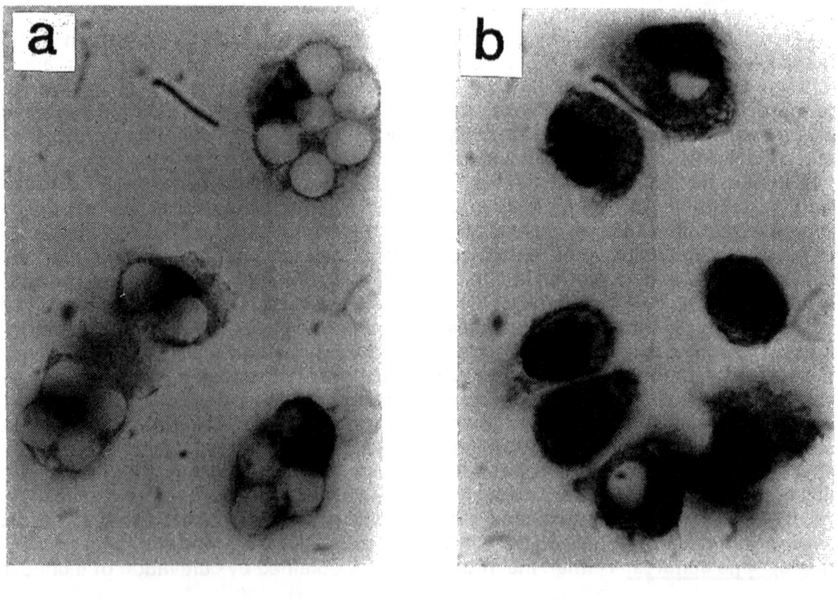

Fig.1: Phagocytosis of Microspheres by rat alveolar macrophages a) vehicle control aerosol "Lutrol" and b) aerosol containing extractable substances from airborne particulates (Magnification 400x)

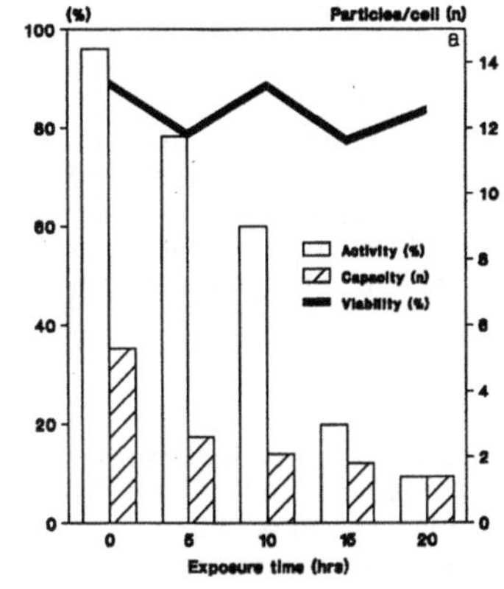

Fig.2: Cell viability, phagocytic activity and capacity of a) rat alveolar macrophages after aerosol inhalation of extract from airborne particulates b) human macrophages incubated in vitro for 2 hrs with extract of airborne particulates

On the other hand, despite the well preserved cell viability, phagocytic activity as well as phagocytic capacity decreased in a concentration dependent manner after exposure to aerosols generated from organic extracts of airborne particulates (Fig. 2a). In comparison the respective vehicle controls exerted no impairment of phagocytosis over the exposure period (data not shown). Analysis by TEM revealed morphological alterations after inhalation exposure of rats for 5 to 20 hours (Fig. 3). Electron dense granula indicating a lysosomal activation increased in a concentration and time dependent manner. The incidence of vacuoles in the cytoplasm suggest an injury of the intracytoplasmatic membrane system as had also been described for mouse macrophage cultures exposed to extract of airborne particulates in vitro (Behrendt et al., 1986).

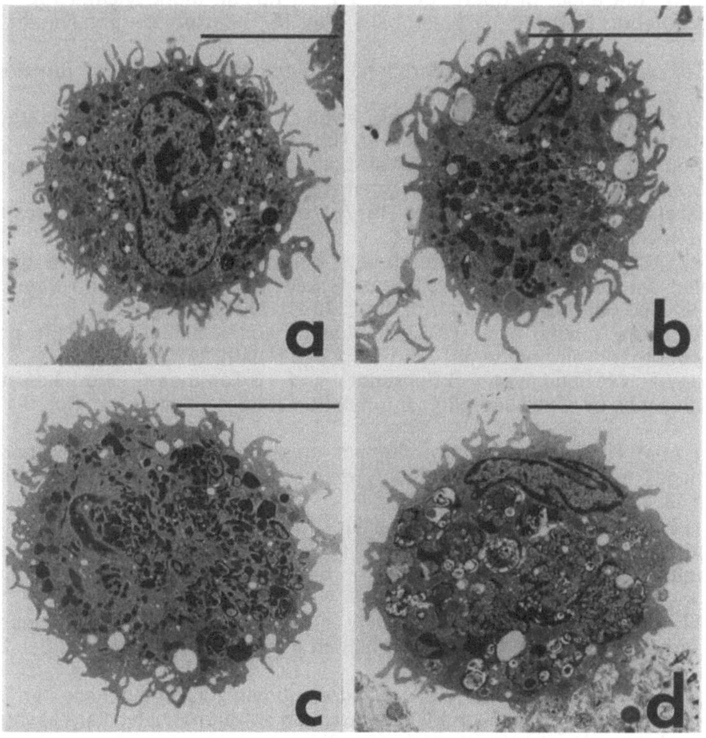

Fig.3: Ultrastructural alterations of rat alveolar macrophages exposed to aerosols from extract of airborne particulates. Inhalation period a) 5 hrs b) 10 hrs c) 15 hrs d) 20 hrs. Bars represent 5µm.

In further experiments we exposed monocyte-derived human macrophages in culture to the same extract in Lutrol, corresponding to extractable substances from airborne particulates of 0.8 - 6.5 m³ of air per ml. Quantitativ data after an incubation period of 2 hours are presented on Fig. 2b. A strong dose-dependent inhibition of phagocytic activity and capacity of human macrophages can be seen. In presence of substances of particultes from 6.5 m³ of air phagocytic activity and capacitiy was completely abolished, while the cell viability was preserved at nearly 90%. It is remarkable, that after an incubation period of 24 hours in presence of extract, followed by a recovery period in absence of extract for 24 hours, human macrophages as well as rat alveolar macrophages regained again their phagocytic activity and capacity (Seemayer et al., 1990).

Conclusion

Data presented reveal that substances contained in airborne particulates from polluted regions could strongly impair phagocytosis of macrophages after in vitro treatment as well as after inhalation. In conclusion, airborne particulates have the capability of adversely affecting alveolar marophages, a crucial element in lung defense and therefore increasing the risk of the host to pulmonary disease.

This work was supported in part by Bundesministerium für Forschung und Technologie, Contract No.07 ALL 049

References

Beck,B., J.D.Brain and S.F.Wolfthal (1988) Assessment of lung injury produced by particulate emissions of space heaters burning automotive waste oil. Ann. Occup. Hyg. 32: 257-265.

Behrendt,H., N.H.Seemayer and R.Tomingas (1986) Evidence for specific ultrastructural alterations of macrophages induced by airborne particulates. AEROSOL: Formation and Reactivity, 2nd Int. Aerosol Conf. Berlin, Pergamon Journals Ltd., pp. 212-214.

Behrendt,H., N.H.Seemayer, A.Braumann and M.Nissen (1987) Electron microscopy investigations on the effect of quartz dust DQ12 on human monocyte/macrophages in vitro. In: Silicosis Report North-Rhine Westphalia, Vol. 16, Steinkohlebergbauverein Essen, pp. 171-183.

Brain,J.D. (1986) Toxicological aspects of alterations of pulmonary macrophage function. Ann Rev. Pharmacol. Toxicol. 26: 547-65.

Burri,P.H., (1983) Cellular components of the alveolar unit. In: G. Favez, A.Junod and P.Leuenberger (eds.). The cell of the alveolar unit. H.Huber Publ. Bern, 11-22.

Fishbein,L. (1990) Sources,nature and levels of air pollutants. In: L.Tomatis (ed). Air pollution and human cancer, Springer-Verlag Berlin, Heidelberg, 9-34.

Green,G.M., G.J.Jakab, R.B.Low and G.S.Davis (1977) Defense mechanisms of the respiratory membrane. Am. Rev. Respir. Dis. 115: 479-514.

Hadnagy,W., A.Kiell, N.H.Seemayer, H.Behrendt and R.Tomingas: Cytotoxic and genotoxic effects of inhaled aerosols from a highly industrialized region in rats. VIIth International Symposium in Inhaled Particles, Edingburgh, Sept. 16-20, 1991 (in press).

Helmes,C.T., D.T.Atkinson, J.Jaffer et al. (1982) Evaluation and classification of the potential carcinogenicity of organic air pollutants. J. Environ. Sci. Helath A17: 321-389.

Kiell,A., W.Hadnagy, N.H.Seemayer, H.Behrendt and R.Tomingas: 3rd International Symposium" Advances in controlled clinical inhalation studies", Hannover, Oct. 9-11, 1991 (in press).

Ruiz,F., L.A.Videla et al. (1988) Air pollution impact on phagocytotic capacitiy of peripheral blood macrophages and antioxidant activity of plasma among school children. Arch. Envrion. Health 43: 286-291.

Schlipköter,H.-W. (1983) Lufthygienische Probleme der Großstadt. Arcus 5: 244-255.

Seemayer,N.H. and A.Braumann (1988) Effects of particle size of coal mine dust in experimental anthracosilicosis: In vitro studies on human macrophages. Ann. Occup. Hyg. 32: 1178-1180.

Seemayer,N.H., A.Happel, H.Behrendt, W.Hadnagy and R.Tomingas (1990) Comparison of cytotoxicity of airborne particulates to rat and human macrophages. J. Aerosol Sci. 21: 387-391.

Thissen,J., N.H.Seemayer and W.Hilscher (1991) Langzeitstudien zur Pathogenese der Silikose: Wirkung geringer Dosen von Quarz- und Grubenstäuben auf bronchoalveoläre Lavagezellen der Ratte. Silikosebericht Nordrhein-Westfalen 18: 253-268.

van Houdt,J.J. and I.M.C.M.Rietjens (1988) Toxicity of airborne particulate matter to rat alveolar macrophages: A comparative study of five extracts collected indoors and outdoors. Toxic. In Vitro 2: 121-123.

CYTOTOXICITY TESTING

Testing lead neurotoxicity with an in vitro model

Binding,N.[1], U.Altrup[2], M.N.Said[3], E.-J.Speckmann[3], U.Witting[1]

([1]Institut für Arbeitsmedizin der Universität Münster, FRG; [2]Institut für Experimentelle Epilepsieforschung der Universität Münster, FRG; [3]Institut für Physiologie der Universität Münster, FRG)

Numerous hazardous substances show neurotoxic effects after acute incorporation as well as after chronic exposure. Some organic solvents for example are known to cause polyneuropathy (e.g. Konietzko 1981, Gregersen et al. 1984, McCunney 1988). Exposure to heavy metals such as mercury or lead may result in an impairment of the central as well as of the peripheral nervous system (e.g. Zielhuis 1975, Triebig et al. 1981). Since knowledge about neurotoxicity is rather incomplete, the need for experimental test systems to investigate the mechanisms involved is quite obvious (Triebig et al. 1989). Furthermore, a test system allowing for the prediction of neurotoxicity would be of great importance with regard to the prevention of - in some cases irreversible - neuronal damages.

Only few attempts have been made to develop in vitro models for testing neurotoxicity of hazardous substances (e.g. Audesirk 1987). Therefore the suitability of neurons from the isolated buccal ganglia of the land snail Helix pomatia as test system for neurotoxicity has been evaluated in a pilot study. These neurons have been used as in vitro models in experimental epilepsy research for rather a long time (Speckmann and Caspers 1973, Altrup et al. 1990a), thus representing a well-established test system for basic neuronal cell functions.

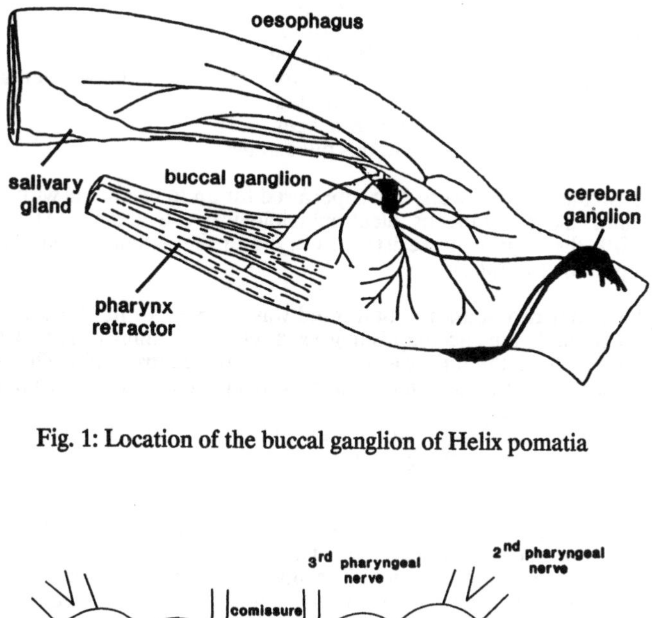

Fig. 1: Location of the buccal ganglion of Helix pomatia

The buccal ganglion of Helix pomatia is located near the oesophagus (Fig. 1). In the ganglion there are four individually identified neurons labeled B1 to B4 (Schulze et al. 1975, Peters and Altrup 1984, Altrup 1987, Altrup et al. 1990b)(Fig.2). B1 to B3 were used in this investigation. After preparation the isolated neurons are active for several hours when superfused with a suitable Tris-HCl-buffered (pH 7.35 - 7.45) bath fluid containing sodium chloride, potassium chloride,

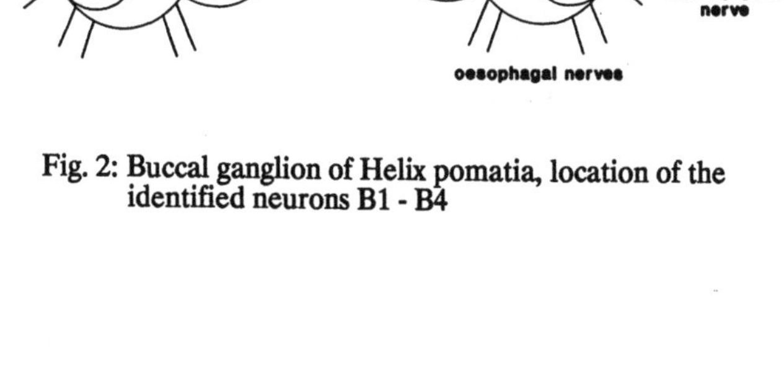

Fig. 2: Buccal ganglion of Helix pomatia, location of the identified neurons B1 - B4

and calcium chloride. Besides their local and functional identification one main advantage of these neurons is their size, allowing to impale several glass microelectrodes (Fig. 3) for monitoring elementary electrophysiological parameters. Two electrodes are used, one to monitor the bioelectrical potentials, the other to inject currents into the cell. The reference electrode is placed in the bath fluid.

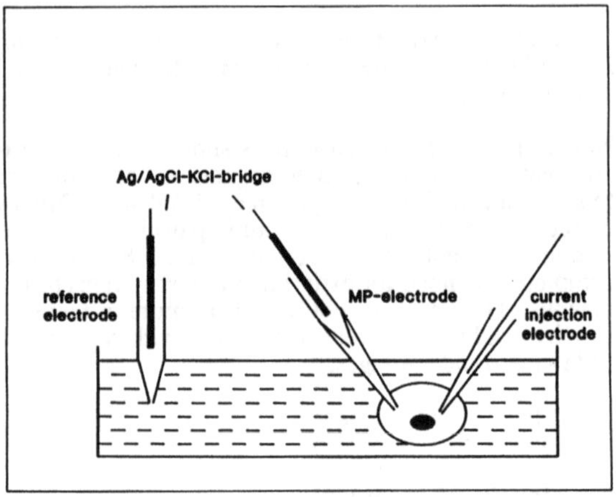

Fig. 3: Experimental set-up

Several neurological effects of lead, such as a decrease in the velocity of nerve conduction (Seppäläinen et al. 1979), loss of neuromuscular coordination (Hanninen et al. 1979, Valcukias et al. 1978, Arnvig et al. 1980), and even dementia after massive intoxication (Landrigan 1989) or neurobehavioral dysfunctions after moderate exposure (Araki et al. 1986) have been reported. Therefore this heavy metal has been chosen as model substance in this pilot study.

The isolated ganglia were first superfused for up to two hours with standard control bath fluid to monitor the spontaneous electrophysiological cell activity. The resting membrane potential (RMP), the membrane resistance (R_M) and the action potentials (AP) were recorded during this period (control 1).

After about two hours the bath fluid was changed to fluid containing certain amounts of lead (lead period), thus accomplishing extracellular administration of the heavy metal. Except the additional lead, the fluid was identical with the control fluid. The basic electrophysiological parameters of the neuronal membranes were again monitored for about two hours.

The fluid was again changed to standard control bath fluid (control 2) after the lead period to investigate the reversibility of lead effects. The bioelectrical activity of the cell was recorded for again two hours.

In a first series of experiments, bath fluid containing 0.7 mg/l lead was applied. This concentration has been chosen in correspondence to the German biological threshold limit value for the blood lead level of occupationally exposed workers. The first effect observed when monitoring a single neuron was an increase in the frequency of spontaneous action potentials as compared with control 1. It was interpreted as a lead-induced decrease in the action potential threshold of the individual neuron which was completely reversible in the control 2 phase. Amplitude, slope and duration of action potentials remained unaffected.

An additional effect of lead was observed when the membrane resistance was determined by injecting hyperpolarizing current pulses of 15 nA via the second electrode: Fig. 4 shows an increase of the membrane resistance of about 60 %. This increase also was fully reversible when the cell was washed by superfusing with control bath fluid (control 2).

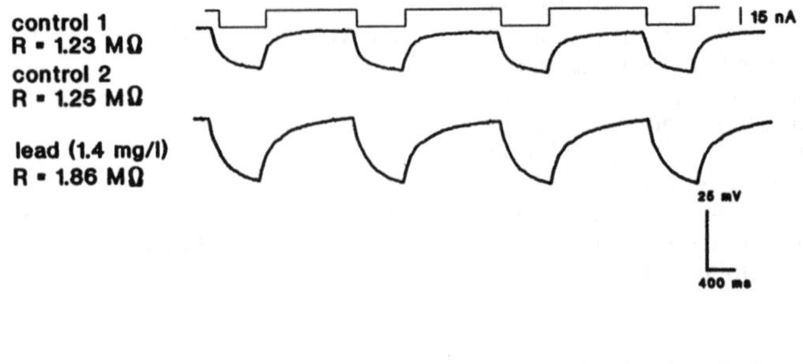

Fig. 4: Increase of membrane resistance during lead period

Doubling the lead concentration to 1.4 mg/l resulted in additional effects. Again the frequency of spontaneous AP's was increased. Furthermore amplitude and duration of action potentials were markedly influenced. Fig. 5 shows three typical AP's with an amplitude of about 80 mV during the control 1 phase. Changing to the lead-containing bath fluid resulted in a time-dependent decrease in the potential amplitude of up to 10 mV after 50 minutes. This was accompanied by a more than threefold prolongation of AP duration. In the control 2 phase, the AP amplitude increased again to its control value after about 1 hour. Compared with the lead phase, a decrease in AP duration was found, but not to its initial value. This holds even for washing periods of more than two hours with control bath fluid. Simultaneous recording of the three neurons B1, B2, and B3 revealed that the AP's became synchronized with lead application (Fig. 6). This indicates that not only voltage dependent membrane currents were altered by lead administration, but also ligand operated as well as electrotonic ones.

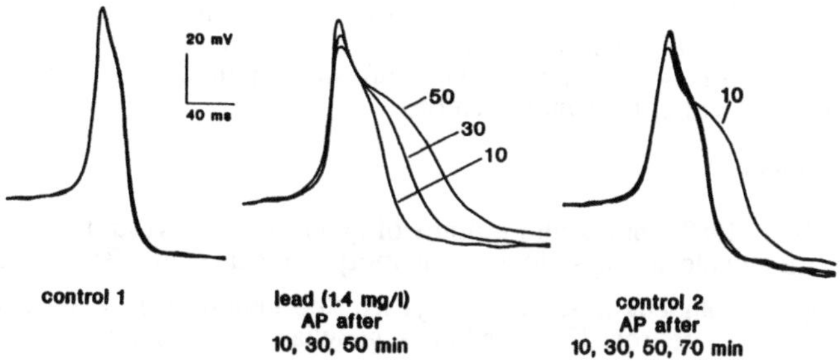

Fig. 5: Changes in AP amplitude and duration during lead administration

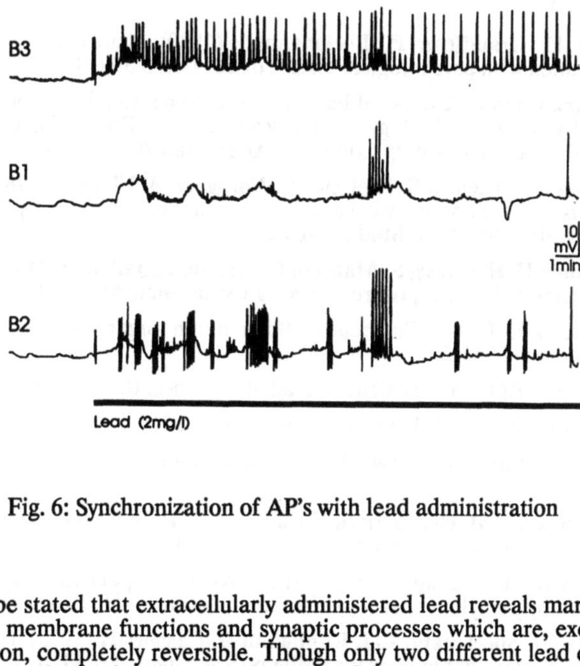

Fig. 6: Synchronization of AP's with lead administration

To summarize the results, it can be stated that extracellularly administered lead reveals marked interferences with basic neuronal membrane functions and synaptic processes which are, except for the prolongation of AP duration, completely reversible. Though only two different lead concentrations were used, there is every indication that the alterations found are dose dependent. An adequate dose-response relationship will be established in the near future.

It is too early to give a substantiated interpretation of the lead-associated alterations of neuronal functions found in this pilot study. Further investigations on this question are in progress using the voltage clamp technique to get informations about interferences of lead with single membrane channels and ion transport mechanisms through the membrane. First results point at an extracellular and not an intracellular site of lead action. In principle, the effects found in this study may be due to interactions of lead with calcium dependent channels. This interpretation would even correspond to some of the impairments of neuronal functions found in humans after lead intoxication such as the reduction of the velocity of nerve conduction.

A main result of this pilot study is that neurons of the isolated buccal ganglia of Helix pomatia are a promising in vitro monitoring system for the influence of neurotoxic substances on the basic electrophysiological parameters of neuronal membranes. It may enhance the ability to predict neurotoxic potency and provide information about the elementary mechanisms involved in the impairment of neuronal functions.

References

Altrup U (1987) Inputs and outputs of giant neurons B1 and B2 in the buccal ganglia of Helix pomatia: an electrophysiological and morphological study. Brain Res 414:271-284

Altrup U, Lehmenkühler A, Madeja M, Speckmann E-J (1990b) Morphology and function of the identified neuron B3 in the buccal ganglia of Helix pomatia. Comp Biochem Physiol 97A:65-74

Altrup U, Madeja M, Speckmann E-J (1990a) Die Buccalganglien der Weinbergschnecke (Helix pomatia) als Modellnervensystem in der experimentellen Epilepsieforschung. EEG-Labor 12:12-25

Araki S, Yokoyama K, Aono H, Murata K (1986) Psychologic performance in relation to central and peripheral nerve conduction in workers exposed to lead, zinc and copper. Am J Ind Med 9:535-542

Arnvig E, Grandjean P, Beckmann J (1980) Neurotoxic effects of heavy lead exposure determined with psychological tests. Toxicol Lett 5:399-404

Audesirk G (1987) Effects of lead exposure on neuronal physiology and morphology in Lymnaea stagnalis. In: Neurobiology - molluscan models (Boer HH, Geraerts WPM, Joosse J (eds)). North-Holland Publishing Company, Amsterdam/Oxford/New York, 143-149

Gregersen P, Angelsø B, Nielsen TE, Nørgaard B, Uldal C (1984) Neurotoxic effects of organic solvents in exposed workers: an occupational, neurophysiological, and neurological investigation. Am J Ind Med 5:201-225

Hanninen H, Hernberg S, Mantere P, Vesanto R, Jalkanen M (1979) Psychological performance of subjects with low exposure to lead. J Occup Med 20:683-689

Konietzko H (1981) Polyneuropathien durch organische Lösemittel. Arbeitsmed Sozialmed Präventivmed 247-249

Landrigan PJ (1989) Toxicity of lead at low dose. Br J Ind Med 46:593-596

McCunney RJ (1988) Diverse manifestations of trichloroethylene. Br J Ind Med 45:122-126

Peters M, Altrup U (1984) Motor organization in pharynx of Helix pomatia. J Neurophysiol 52:389-409

Seppäläinen AM, Hernberg S, Kock B (1979) Relationship between blood lead levels and nerve conduction velocities. Neurotoxicology 1:313-332

Speckmann E-J, Caspers H (1973) Paroxysmal depolarization and changes in action potentials induced by pentylenetetrazol in isolated neurons of Helix pomatia. Epilepsia 14:397-408

Schulze H, Speckmann E-J, Kuhlmann D, Caspers H (1975) Topography and bioelectrical properties of identifiable neurons in the buccal ganglion of Helix pomatia. Neurosci Lett 1:277-281

Triebig G, Bleecker M, Gilioli R, O'Flynn RR (1989) International working group on the epidemiology of the chronic neurobehavioral effects of organic solvents. A need for further research. Int Arch Occup Environ Health 61:423-424

Triebig G, Schaller K-H, Valentin H (1981) Untersuchungen zur Neurotoxizität von Arbeitsstoffen. I. Messung der motorischen und sensorischen Nervenleitgeschwindigkeit bei beruflich Quecksilber-belasteten Personen. Int Arch Occup Environ Health 48:119-129

Valciukas JA, Lilis R, Fischbein A, Selikoff IJ, Eisinger J, Blumberg W (1978) Central nervous system dysfunction due to lead exposure. Science 201:465-467

Zielhuis RL (1975) Dose-response relationships for inorganic lead. II. Subjective and functional responses - chronic sequelae - no-response levels. Int Arch occup environ Health 35:19-35

A Novel Fluorometric Method for Monitoring Virus Replication In Response to Heavy Metals

Obradovic, D., M. Petersheim* and E.V. Orsi
Biology and Chemistry* Departments, Seton Hall University , South Orange, New Jersey, 07079, U.S.A.

INTRODUCTION

Earlier, we found with Epstein Barr virus and Type 2 Herpes simplex that copper acetate added after virus attachmment reduced significantly virion production as measured by immunofluorescence microscopy on fixed monolayers (Orsi et al 1988, Obradovic & Orsi 1990). However with similar conditions lead acetate did not modify virus replication. Our microscopy methods provided insufficient emission intensity for measuring the low levels of fluorescence associated with viral DNA formation and the early stages of virion maturation. We then devised a method based on procedures used for fixed cell visualization (Narin 1964, Brigati et al 1983). In our method, cell monolayers were infected, metal added, cells fixed and permeabilized for addition of the fluorescent probe which was then extracted and the emission intensity measured by fluorometry.

MATERIALS

I.Cell Cultures: Balb/3T3 clone A31 line (ATCC) in MEM + 10% FBS
2.Cell Monolayer Dispersion:.
a) For routine cell line propagation: Trypsin-EDTA (Gibco)
b) For monolayer dispersion prior to fluorometry: The above mixture but without the usual phenol red indicator.
3.Virus: Herpes simplex type 2 (G strain, ATCC)
4.Materials for Virus Particles Quantitation
a) Phosphate buffered saline (PBS) pH 7.2 + 0.05% Tween 20
b) Fluorescein isothiocyanate (FITC) labeled rabbit anti-Herpes simplex type 2 IgG (DAKO Corp.)
5.Materials for Virus DNA Quantitation (In Situ hybridization)
a) PBS with and without EDTA
b) Deionized formamide (Fluka Chemika)
c) Dextran sulfate (Sigma Chemical Co.)
d) The hybridization mixture: the above plus the DNA probe (Enzo Diagnostic, Color Gene kit) which is obtained from a from a mixture of two clones of Herpes simplex (type 1 and type 2) DNA sequences in the bam HI site of pBR322.
e) Blocking buffer: PBS + 0.1 % Triton X-100 (Fisher)
f) Antibody diluting buffer: PBS + 0.2 % bovine serum albumin (Sigma Chemical Co.)
g) IgG fraction rabbit anti-biotin (Enzo Diagnostic)
h) FITC goat anti-rabbit IgG (Sigma Chemical Co.)
6.Heavy Metals and Sodium Acetate Controls in PBS
a) Lead Acetate: 7.818 E-9 M. Sodium acetate: 1.564 E-8 M
b) Copper Acetate: 4.673 E-7 M Sodium acetate: 9.346 E-7 M
8.Fluorescence Measurements: SPEX fluoremeter (Spex Industries) was used for all extracted FITC probe measurements. The fluorometer was connected to a computer (DATAMATE) which permitted data saving as well as graphic data display. In addition the fluorometer had a system for providing ratios of the signals from the two detectors to compensate for any light instability.

METHODS

1. Cell Line Propagation:

a) Cells were passaged by standard methods in closed 25 cm^2 flasks (Corning Glass) for both maintaining stock cultures and producing the Herpes simplex type 2 virus inoculum.

b) Cells for fluorometry and virus plaque assay were grown in the appropriate vessels with a conventional open incubating system provding 5% CO_2 and 95% air at 90% humidity.

2. Virus Inoculum Assay: Standard Herpes simplex plaque assay for infectious virus (PFU) procedures in which inhibitory agar was replaced by methylcellulose were used (Merchant et al 1960).

3. Viral Particle Fluorometry Procedures:

a) Quantitation without metal utilization used cells grown in two well Lab-Tek culture slides (Nunc Inc.). After a monolayer was obtained the media was aspirated from all the wells. Some wells received virus at an m.o.i. of 10 PFU/cell and other wells got the same 1.0 ml of MEM + 5% FBS but without virus. After two hours all wells received an additional 1.5 ml of medium. Infection was stopped by aspirating the medium and adding 100% methanol (at room temperature for 10 minutes and followed by washing once with PBS and air drying. Each infected and control well then received 0.5 ml of the FITC HSV 2 conjugate. After 1 hr at 37° C, all wells were rinsed with the various PBS. Each well then received 1.0 ml trypsin-EDTA and incubated for 12 hours in a humid shaker at 37° C with shielding from light. The fluorescein released from the dispersed cells was transferred to a microcuvette wth additional 0.5 ml of PBS. Appropriate trypsinization as well as negative controls were employed. Intensity of fluorescence was measured with the fluorometer at an excitation wavelength of 495 nm and an emission wavelength of 517 nm.

b) Quantitation in the presence of metals used the same procedure except that after the two hours for virus attachment the MEM + 5% FBS medium that was added contained 50 ul of the metal acetate or sodium acetate solution.

4. Viral DNA Fluorometry procedures.

a) Quantitation without metal utilization employed cell monolayers in 10 ml glass beakers so that after aspirating the medium, 2 ml of 100% acetone was added for fixing followed by air drying. Selected infected and non-infected cultures got 100 ul of the complete hybridization mixture which contained the HSV DNA probe.Other appropriate negative and positive staining controls were employed. All monolayers were covered with round coverslips which insured uniform spreading of the mixture and elimination of air bubbles. The beakers were then placed on a heating block for exactly 3 minutes (92° C +/- 2°) followed by one hour at room temperature. The coverslips were removed, monolayers washed with appropriate buffers,followed by 0.2 ml rabbit anti-biotin antibody (to react with the biotin incorporated in the DNA probe) for 1 hr at 37° C in a humid atmosphere. After washing, 0.4 ml of FITC goat anti rabbit antibody was added in a similar manner. The monolayers were then treated with trypsin -EDTA for 12 hours in a humidified shaker at 37° C with each beaker covered with a 35 mm petri dish. The contents of each beaker were then handled in the same way as for viral particle fluorometry assay.

b) Quantitation in the presence of metals used the same procedure except that the medium contained 50 ul of the respective metal acetate or sodium acetate control solution.

RESULTS

In figures 1 and 2 we see the time associated increase of viral DNA and virus particles pattern typical of Herpes simplex (Ginsberg 1980, Roizman and Sears 1990).

Fig. 1. HERPES SIMPLEX TYPE 2 DNA FITC EMISSION

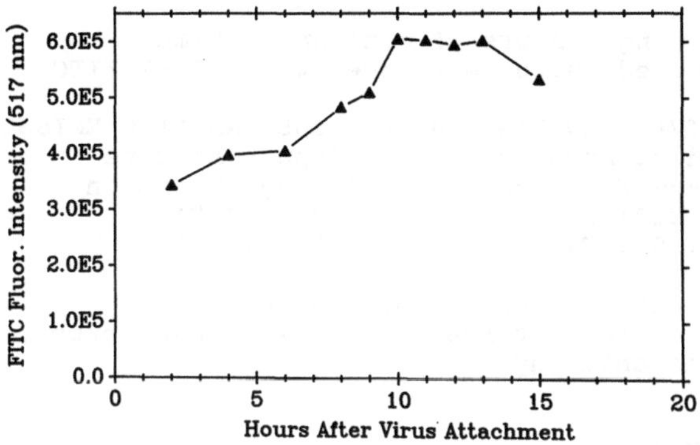

Fig. 2 HERPES SIMPLEX TYPE 2 VIRUS PARTICLE FITC EMISSION

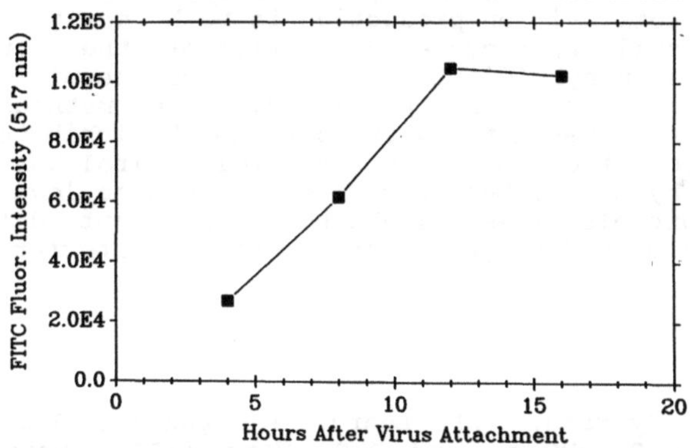

Preliminary tests with lead and copper acetate (table 1) showed no response to lead but a significant lower virus particle yield with copper acetate.

Table 1. FLUORESCENCE INTENSITY OF VIRUS PARTICLE FITC AFTER 7 HOURS INCUBATION

Test # 1	Sodium Acetate	Lead Acetate	P
Number of Cultures	10	10	
Average Relative Emission	7.36203 E3	7.47602 E3	> 0.05
Test # 2	Sodium Acetate	Copper Acetate	
Number of Cultures	4	4	
Average Relative Emission	4.67235 E4	2.34607 E4	< 0.05

Consequently further studies (Tables 2,3) were carried out with copper acetate and sodiumn acetate controls on Herpes type 2 viral DNA and viral particle formation. In these tests negative controls (uninfected cells with all probe components and infected cells without the virus specific probe component) were again utilized. Also, nonspecific emission contributed by copper and sodium acetate was subtracted from the total emission values.

Table 2. RELATIVE QUANTITATION OF VIRUS PARTICLE FITC EMISSION

Time	#	Sodium Acetate Average Emission	#	Copper Acetate Average Emission	P
9	6	8.4127 E5	6	5.4949 E5	< 0.01
12.5	6	1.6599 E6	6	1.3204 E6	< 0.001
18	6	1.7552 E6	6	1.5177 E6	< 0.01

Time = Hours following 2 hours of attachment time
# = Number of infected cultures treated with HSV-2 FITC

Table 3. RELATIVE QUANTITATION OF VIRUS DNA FITC EMISSION

Time	#	Sodium Acetate Average Emission	#	Copper Acetate Average Emission	P
9	5	3.2201 E4	5	2.3958 E4	< 0.001
12.5	5	3.5839 E4	2	3.5216 E4	> 0.05

Time = Hours following 2 hours of attachment time
# = Number of infected cultures treated with DNA probe, primary antibody and second antibody

SUMMARY AND CONCLUSIONS

Extraction of a fluorescent dye antibody marker permitted measurement of total virion production in both early and late stages of the replication cycle. Extraction of the fluorescent moiety of an in situ hybridization HSV-2 probe permitted assay of viral DNA by a non-radioisotope method. These methods showed that addition of copper after virus attachment inhibited total virus particle production for the entire cycle. Viral DNA synthesis was delayed by copper but then reached the same level as the sodium acetate controls. These slight but significant differences obtained by this method suggest that copper is interfering with virus maturation.

REFERENCES

Brigati, D.J.,D. Myerson, J.J. Leary, B. Spaholz, S.Z. Travis, C.K.Y. Fong, G.D. Hsiung and D.C. Ward (1983) Detection of viral genomes in cultured cells and paraffin-embedded tissue sections using biotin-labeled hybridization probes. Virology. 126: 32-50.

Ginsberg, H. (1980) In: Virology., Dulbecco R. and Ginsberg H. eds., Harper and Row, Hagerstown.

Merchant, D.J., R.H. Kahn and W.H. Murphy, Jr.(1960). In: Handbook of Cell and Organ Culture. Burgess Publishing Co., Minneapolis.

Narin, R.C. (1964) In: Fluorescent protein tracing. The Williams and Wilkins Co., Baltimore.

Orsi, E.V., D. Koval and M. Bavlsik (1988) Virus production and cytoskeleton alteration in Epstein-Barr virus (EBV) transformed cells by heavy metals. Amer. Soc. Microbiology. 88th Meeting Abstract Q-173.

Obradovic, D. and E.V. Orsi (1990) Unpublished Data.

Roizman, B. and A.E. Sears (1990) In: Fundamental Virology, 2nd ed., Fields, B.N. and Knipe, D.M. eds. Raven Press, New York.

Cytoskeleton Alteration in Mouse Tissues by Postnatal Exposure to Heavy Metal Air Pollutants

Orsi, E.V., G. Ahmad, D. Brois, B. Billings and C.C. Orsi*
Biology Department, Seton Hall University, So. Orange, N.J. 07079,
B-W Laboratory*, Wilmington, N.Y. 12997, U.S.A..

INTRODUCTION

In past experiments (Orsi et al 1988, Orsi et al 1990) mice born and raised in an urban area (SHU) showed less weight gain, decreased lysosome enzyme activity and higher levels of lead and copper in their hearts and lungs than those born and raised in cleaner (BW) air. Studies of the effects of lead on cytoskeleton depletion in cell cultures (Orsi et al 1987) suggested that the (SHU) progeny might show a similar depletion. Consequently, fixed heart and lung tissues saved from the third experiment (Orsi et al 1990) and those from this experiment a year later were utilized for comparing cytoskeleton levels in progeny from high and low air pollution regions.

MATERIALS AND METHODS

I. Air and Animal Monitoring

In this present study the same protocol was used as in the previous three experiments (Orsi et al 1988, Orsi et al 1990) to insure that exposure to air pollution was the only variable. Air and tissue samples were collected and processed as before (Orsi et al 1987, Orsi et al 1988) and metal levels measured by flame atomic absorption spectrophotometry (Perkin Elmer 2280).

<u>II.Third Experiment Animals:</u> Cytoskeleton Assay by Measurement of Immuno-fluorescence Decay (Fluorescence Intensity Half-life).

A total of 54 slides of heart and lung respectively were obtained from five male and five female pups and two dams from each site. The sections were deparaffinized with "Histoclear" (National Diagnostics Co.), heated for one hour at 60° C and rehydrated in 100% and 95% ethyl alcohol. Tissue permeabilization for actin or vimentin then followed (Vielkind and Swierga 1989). To minimize background staining each section was overlaid with 1% bovine serum albumin (37°C, 15 minutes) followed by the primary antibody, i.e. rabbit anti-actin (ICN Immunobiologics) or goat anti-vimentin (Miles Scientific) for one hour at 37° C. After three washes with 1% bovine serum albumin the sections were similarly incubated with biotin labeled anti-rabbit or anti-goat antibody (Biomeda Corp). Sections were washed as above, and then received phycoerythrin-avidin fluorescent conjugate (Biomeda Corp.)for 30 minutes at 37°C The sections were again washed as above and followed by rinsing with phosphate buffered saline . The sections were then mounted with "Airvol" (Air Product. Chemical Inc.). Fluorescence intensity decay values were obtained by adapting a photomicroscopy method (Orsi et al 1987) for fluorescence half life regression analysis. Continuous images, obtained by standard reflected U V illumination were recorded on videotape (Dage-Mit Corp. camera). The fluorescence intensity decay rate was obtained by playing back the tape, scanning the monitor image with a photodiode probe (Oriel Corp.) and plotting the decrease in amplitude from the monitor to a chart recorder (Figure 1). Each organ was evaluated by taping three different fields for four minutes each. A minimum of three cells from each field were scanned, thereby generating a total of nine plots for both the actin and vimentin components of each organ. Manual measurement of amplitude decrease permitted utilization of a rapid computer based half life calculation method for comparing the fluorescence intensity and hence the relative amount of cytoskeleton protein present.

Figure 1: FLUORESCENCE DECAY HALF-LIFE OF A LUNG SECTION STAINED WITH BIOTIN LABELED ANTI-ACTIN PRIMARY ANTIBODY FOLLOWED BY AVIDIN PHYCOERYTHRIN FLUORESCENT CONJUGATE.

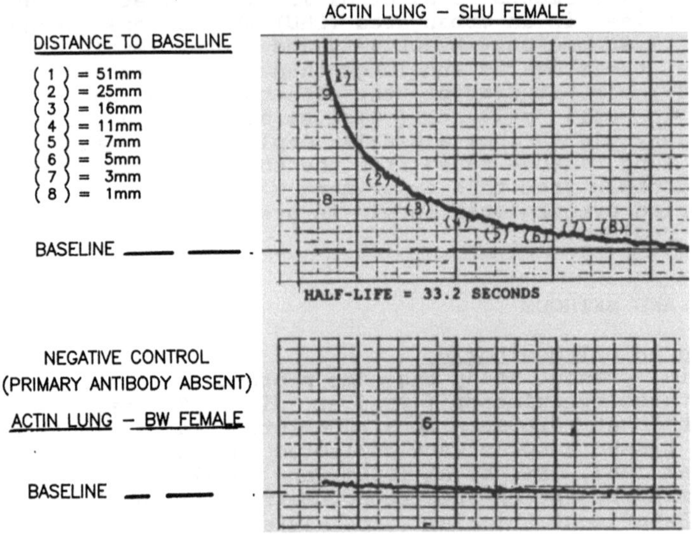

III. Fourth Experiment Animals: Cytoskeleton Assay by SDS Poly-Acrylamide Gel Electrophoresis.

Mouse heart and lung tissue extracts were assayed for actin and myosin in the fourth experiment by conventional SDS-PAGE procedures (Weber and Osborn 1969, 1975, Fehrnstrom and Moberg 1977). Proteins were separated with a LKB electrophoretic unit Model 2117. After a gel was completed it was scanned with a Bio-Rad video densitometer model 620, in order to quantitate the relative amount of actin and myosin. Confirmation of actin and myosin was by western blot procedures (Towbin et al 1979, Salinovich and Montelaro 1986). Primary antibodies and horseradish peroxidase anti-Ig conjugate from Sigma Chemical Co. were used. Each site provided 24 progeny (12 female and 12 male) from which heart and lung protein extracts were obtained.

RESULTS

I.Third Experiment Animals:

As seen in table, 1 there were no differences in the heart and lung actin levels between male and female progeny from the (BW) site . In contrast, female (SHU) progeny had significantly less actin and vimentin in their hearts and lungs (P = < 0.05) than their male litter mates.

Table 1. CYTOSKELETON DISTRIBUTION IN HEARTS AND LUNGS OF MALES AND FEMALES AT THE SAME SITE

Fluorescence Half Life SHU (Urban Site)		
HEART	Females	Males
Actin	39.5	42.2
Vimentin	33.5	45.7
LUNG	Females	Males
Actin	41.5	47.6
Vimentin	40.8	49.5

Fluorescence Half Life BW (Adirondack Mts)		
HEART	Females	Males
Actin	41.6	42.6
Vimentin	44.9	43.6
LUNG	Females	Males
Actin	46.9	46.1
Vimentin	44.9	48.5

Note:Half life (seconds) : Average of actin and vimentin values of five males and five females from each site. Actin and vimentin sections were from the same animal.

<u>II.Fourth Experiment Animals:</u> Assay with SDS-PAGE (Table 2) also showed less lung cytoskeleton in female urban progeny compared to their male litter mates. The Adirondack progeny again showed no differences in male and female lung cytoskeleton levels.

Table 2. CYTOSKELETON DISTRIBUTION IN HEARTS AND LUNGS OF MALES AND FEMALES AT THE SAME SITE

ACTIN LUNG		
* SHU male 12.89 +/- 0.45		SHU female 10.54 +/- 0.73
BW male 12.89 +/- 0.57		BW female 12.02 +/- 1.03
ACTIN HEART		
* SHU male 18.11 +/- 0.69		SHU female 13.63 +/- 1.05
* BW male 18.53 +/- 0.53		BW female 13.59 +/- 1.23
MYOSIN LUNG		
* SHU male 26.34 +/- 0.65		SHU female 19.43 +/- 1.37
BW male 24.15 +/- 0.89		BW female 22.22 +/- 1.34
MYOSIN HEART		
SHU male 36.11 +/- 2.50		SHU female 37.48 +/- 3.16
BW male 37.11 +/- 2.21		BW female 45.14 +/- 3.16

* = Significant difference (P = < 0.05)

<u>III.Air Quality at SHU and BW:</u> Table 3 summarizes the consistently higher levels of airborne particles (CN) and heavy metals in the urban air obtained during the third (III) and fourth (IV) studies.

Table 3, AIRBORNE PARTICLES AND HEAVY METAL LEVELS

	III (May-June 1988)		IV (Oct-Nov.1989)	
	SHU	BW	SHU	BW
	PPM	PPM	PPM	PPM
LEAD*	8.71	0.67	0.91	0.38
COPPER*	10.40	1.07	0.58	0.32
	CN	CN	CN	CN
INDOOR	6,325	4,675	12,027	2,620
OUTDOOR	39,437	5,213	34,923	3,167

*: Indoor Levels
CN: Condensation Nuclei/CC Air

SUMMARY AND CONCLUSIONS

Two studies one year apart using the same protocol and CF-1 mouse strain showed significantly less distribution of actin, myosin and vimentin in the lungs of female urban progeny compared to their male litter mates. Heart actin levels were lower for females at both sites. Male and female progeny at the Adirondack site did not show any site related differences. There was no significant difference in the number of births and in the very low level of cannibalism at both sites. All progeny showed normal histology of lungs and heart tissue by conventional staining.

Apparently normal progeny from dams of the same strain chosen randomly at the same time, with the same food, water and bedding will show slight but significant cytoskeleton depletion if born and raised in a heavily air polluted urban area in contrast to those raised in cleaner environment. Site associated inhibition of actin, myosin and vimentin formation was observed primarily in the lungs of females and never in the hearts or lungs of males.

REFERENCES

Fehrnstrom, H. and U. Moberg (1977) In "SDS and Conventional Polyacrylamide Electrophoresis with LKB 2117 Multiphor", LKB Produkter AB, Bromma , Sweden

Orsi, E.V., M. Bavlsik, R. Viera, M. Petersheim and O.F. Baturay (1987) Lysosome response and cytoskeleton alteration in cell structures exposed to airborne lead. Experientia 51 (suppl): 243-248.

Orsi, E.V., G. Ames and M. Bavlsik (1988) Postnatal mouse heart and lung lysosome RNAase levels in response to air pollutants. In: Seemayer NH & Hadnagy W (eds) Environmental Hygiene, Springer-Verlag, Berlin, p. 125-128.

Orsi, E.V., M. Bavlsik, C.C. Orsi, and O.F. Baturay (1990) Postnatal exposure to airborne pollution and heavy metal levels in hearts and lungs of mice. In: Seemayer NH & Hadnagy W (eds) Environmental Hygiene II. Springer-Verlag, Berlin, p. 225-228.

Salinovich, O. & R. Montelaro (1986) Reversible staining and peptide mapping of proteins transferred to nitrocellulose after separation by SDS-PAGE. Anal. Biochem. 156: 341-347.

Towbin, H.T., T. Staehlin & J. Gordon (1979) Electrophoretic transfer of proteins from polyacrylamide gels to nitrocellulose sheets. Proc. Natl. Acad. Sci. (USA) 76: 4350-4354.

Vielkind , U. & S.H. Swierenga (1989) A simple fixation procedure for immunofluorescent detection of different cytoskeletal components within the same cell. Histochemistry 91: 81-88.

Weber, K, and M. Osborn (1975) Proteins and sodium dodecyl sulfate molecular weight determination on polyarylamide gels and related procedures. In "The Proteins" vol.1, Third edition, Neurath H, & Hill,RL (eds) Academic Press, New York, p.179-223.

ACKNOWLEDGEMENTS

This study was supported in part by a Seton Hall University Research Council grant awarded to E.V. Orsi.

We are grateful to John L. Horward (University of Medicine & Dentistry of NJ) for expert sectioning of mouse tissue samples.

Air sampling assistance was ably provided by Veronica Orsi-Mitta and Catherine A. Prezioso.

We thank Roman Razdro for his most valuable computer program.

Comparison of chromate and vanadate toxicity and its relationship to oxygen radical formation

Popper HH, A Woldrich, E Grygar
Institute of Pathology and Lab. Environ.Pathol., University of Graz, Medical School, Auenbruggerplatz 25, A-8036 Graz, Austria

Introduction

In previous investigations chromates were proven to be highly toxic for different cell systems of rat and human origin (Popper et al., 1990). In addition it causes severe organ injury in the respiratory and alimentary tract and in the spleen (see Popper et al., this volume). Different chromates and especially water soluble dichromates exert their toxic effects on mammalian cell systems at least in part by the liberation of oxygen radicals (Popper, Woldrich, 1991). Our knowledge about VC's is very limited. There are some reports on the toxicity of VC's (Rhoads et al., 1985, Waters, 1977) and some dealing more specifically with the mechanisms of VC toxicity (Zychlinski et al., 1990, Owusuyaw et al., 1990, Roldan et al., 1990). There is one report which suggests a similar mechanism of vanadium-5-oxide reduction as it is described for dichromates (Shi et al., 1990). We therefore investigated different VC's in a pneumocyte cell culture system and compared the results with the effects of dichromate.

Material and methods

Pneumocyte II assay

Rat pneumocytes type II (Li et al., 1983) were cultured as monolayers in Ham's F12 medium. The cells were seeded at a density of 2×10^2 / 5 ml medium (total volume 5 ml) into cell culture flasks, and were incubated for 1 day in the incubator (37^oC, 95% relative humidity, 5% CO2) until all cells settled down. Than the cells were incubated with dichromate and VC's in the below mentioned concen-trations for 12 hours. After incubation the cells were supplemented with fresh medium and cultivated for further 16 days without these metal compounds. The culture was stopped by formaldehyde fixation, the cells were stained with Diff Quick® stain and the colonies were counted (1 colony > 35 cells). To each flask either potassium-dichromate, vanadium-3-chloride or vanadium-3-, -4-, or -5-oxide were added in one of the following concentrations: K2Cr2O7 0/ 0.0268/ 0.1352/ 0.6764/ 3.382/ 16.8932/ 84.5683 µM, VCl3 or V2O3/ V2O4/ V2O5 0/ 0.00044/ 0.002/ 0.01/ 0.05/ 0.27/ 1.37µM.

Indirect oxygen radical assay

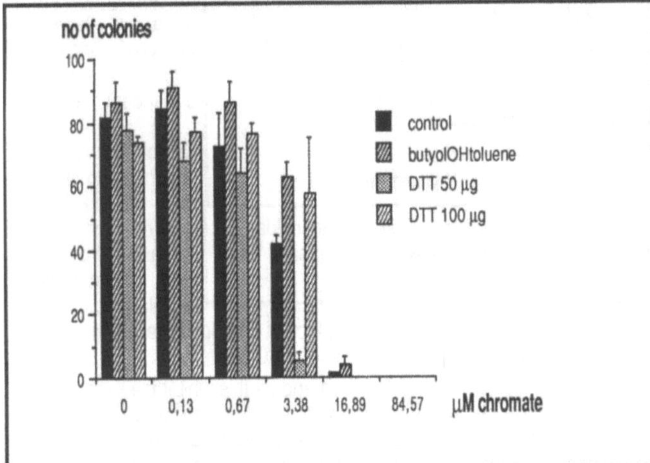

Fig. 1 Toxicity of potassium dichromate (controls) and antitoxic effect of butylated hydroxy-toluene. Similar results were obtained by the other antioxidants; N = 8, mean ± SEM

To the pneumocyte cell assay different concentrations of dichromate or VC's together with different concentrations of SOD (12.5 - 37.5 - 112.5 µg/ml DMEM), catalase (16 - 64 µg/ml), reduced glutathion (4 - 8 - 16 µg/ml), dithiothreitol (50 - 100 µg/ml), DMSO (0.1 - 1 - 10%), EDTA (50 - 100 µM), butylated hydroxy-toluene (10 - 50 - 100 µM), desferal (20 - 100 µM) or trolox® (50 - 200 µM) were added. Culture conditions were as described in the pneumocyte culture section and the colonies were counted on day 16.

Statistical analysis

For statistic analysis either the Wilcoxon Signed Rank or the paired Student t-test were used, with the help of a MacIntosh SE® computer and the programs Data Desk Professional® and CricketGraph®.

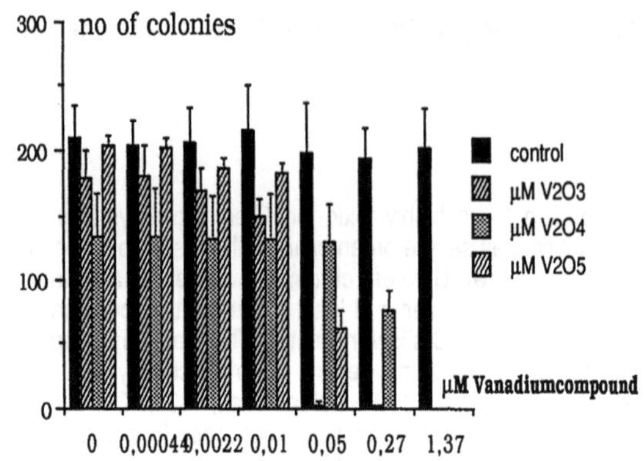

Fig. 2: Comparison of toxicity of different vanadium oxides. Cytotoxicity depends on solubility of each compound, but generally is within a narrow range; N = 8, mean ± SEM

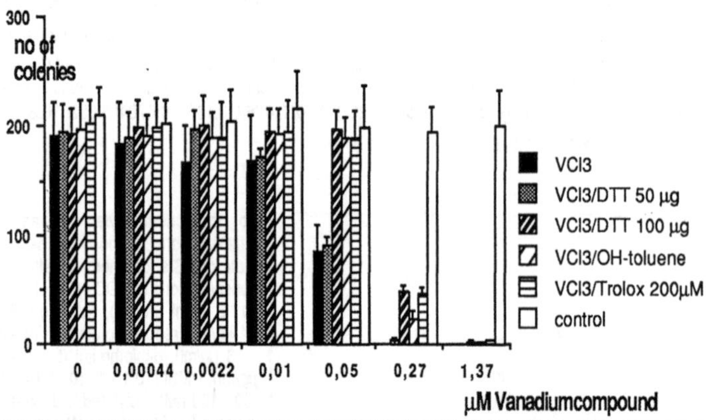

Fig. 3: Toxicity of vanadium-3-chloride and protective effects of different antioxidants on the survival of pneumocytes type II. Most effective are trolox and dithiotreitol. The results for desferal are not shown; N = 8, mean ± SEM

Results and Discussion

Vanadium compounds have a similar toxic effect on pneumocytes than dichromates (Figs: 1-3). Although generally less soluble the vanadium compounds used in this study were ingested or phagocytosed by the cells. Vanadium-5-oxide was the most toxic substance. Generally the toxic effects of VC´s were in the range of 0.01 to 0.27 µM. The least soluble compound, vanadium-4-oxide had the least toxic effect.

Superoxide dismutase, reduced glutathione, EDTA, and DMSO when added to the culture medium had no effect on survival of pneumocyte colonies, whereas dithiotreitol (Fig. 1, 3, 4), desferal (not shown), butylhydroxytoluene (Fig. 1, 3) and trolox (water soluble α-tocopherol, Fig. 3) were effective in preventing dichromate and vanadium compound cytotoxicity in a dose dependent manner.

Dithiotreitol (DTT) in contrast to glutathion can pass intact cell membranes and is intracellulary metabolized to glutathion. By this the intracellular glutathion pool can be increased. The antioxidant effect of DTT in our experiments is therefore an evidence for the participation of oxygen radical in di-chromate and VC cytotoxicity.

EDTA primarily can prevent oxygen radical cytotoxicity by complexing iron and by this inhibit the Fe^{2+}/ Fe^{3+} re-dox system, which is closely linked to oxygen radical formation. The fact that chromate or VC toxicity is not decreased by EDTA leads to the assumption that the iron-redox system is not required by chromate.

Superoxide dismutase does not pass intact cell membranes, but can inhibit accumulation of oxygen radicals released either by phagocytosing or necrobiotic cells. Again the results show that this is not the way by which either chromate nor VC´s act.

α-tocopherol (Trolox), desferal (desferoxamine mesylate) and butylhydroxy-toluene are well known antioxidants which act either by inhibition of oxygen radical formation/ accumulation or are scavengers of oxygen radicals, esp. hydroxyl radical and peroxides. α-tocopherol did not prevent chromate toxicity in pre-

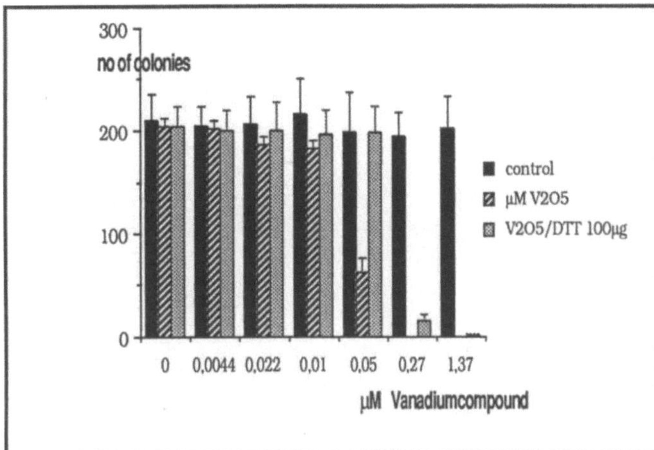

Fig. 4: Comparison of toxicity of vanadium-5-oxide and the antioxidant effect of dithiotreitol (DTT). Between 0.05 and 1.37µM V2O5 significant more colonies survived in the DTT group. The results with the other antioxidants are not shown on account of clarity of the graph; N = 8, mean ± SEM

vious experiments (Popper, Woldrich, 1991). Therefore we used Trolox, a water soluble derivate of α-tocopherol. This compound increased the number of surviving pneumocyte colonies in a dose dependent manner (Fig. 3).
The presented results are in our opinion an indirect evidence that chromates and vanadium compounds act in a similar way on pneumocytes by the formation of oxygen radicals. Chromate and probable also vanadium oxides can pass the cell membrane via the PO4 anion mechanism and might be reduced to other valence forms. This was already shown for dichromate and chromate anions. As chromates also vanadium-5-oxide are reduced to intermediate forms by the action of thiol groups (Shi et al., 1990). This thiols are than oxidized to thiyls, which might be another indirect evidence of oxygen radical formation. There is still the unanswered question by which mechanisms chromate or vanadium compounds act in the process of oxygen radical formation: Is there an intracellular inhibition of protective enzymes like glutathion reductase, superoxide dismutase or catalase, which is more unlikely from our results; are there protective substances oxidized by chromates or VC's, like reduced glutathion, which seems most likely from our results; is there a direct mechanism for chromates or VC's to generate oxygen radicals. At the moment we are unable to answer these questions, but hope to clarify some of these by other investigations.

Acknowledgements

This investigation was supported by the Jubiläums Research Fund of the Austrian National Bank Project 3400 to HHP, which is gratefully acknowledged. The pneumocyte cell line (LEC) used in this research was developed at the Lovelace Inhalation Toxicology Research Institute, Albuquerque, NM, USA, with support from the U.S. Department of Energy's Office of Health and Environmental Research.

References

Owusuyaw J, Cohen MD, Fernando SY, Wei Cl: An Assessment of the Genotoxicity of Vanadium. Toxicol.Let. 50, 327-336, 1990
Popper HH, Grygar E, Ratschek M, Weybora W, Ingolic E, Wiesspeiner G, Wawschinek O. Cytotoxicity of chromium compounds: Comparison of in vitro and in vivo tests. Environmental Hygiene, Vo. 2, 73-76, 1990, Springer Verlag, Berlin, Heidelberg, New York
Popper HH, Woldrich A: Oxygen radical formation - a probable mechanism for chromate toxicity. In: Histo- and Cytochemistry as a Tool in Environmental Toxicology, eds. W. Graumann, J. Drukker. Prog. Histochem.Cytochem. 23, 220-226, 1991
Rhoads K, Sanders CL: Lung clearance, translocation, and acute toxicity of arsenic, beryllium, cadmium, cobalt, lead, selenium, vanadium, and ytterbium oxides following deposition in rat lung. Environ.Res. 36, 359-378, 1985
Roldan RE, Altamirano LMA: Chromosomal Aberrations, Sister-Chromatid Exchanges, Cell-Cycle Kinetics and Satellite Associations in Human Lymphocyte Cultures Exposed to Vanadium Pentoxide. Mut.Res. 245, 61-65, 1990
Shi XL, Sun XY, Dalal NS: Reaction of Vanadium(V) with Thiols Generates Vanadium(IV) and Thiyl Radicals. FEBS Lett. 271, 185-188, 1990
Waters MD: Toxicology of Vanadium. A Review. Adv.Mod.Tox. 2, 147-189, 1977
Zychlinski L, Byczkowski JZ: Inhibitory Effects of Vanadium Pentoxide on Respiration of Rat Liver Mitochondria. Arch.Environ.Contam.Toxicol. 19, 138-142, 1990

Different effects of mechanical stress (high energy shock wave treatment) in normal and transformed lung epithelial cell lines of Syrian hamster

Ramm, Doris; Jan Knebel; Susanne Gärtner-Hübsch; Wolfgang Sass*; Michaela Aufderheide
Institute of Experimental Pathology, Hannover Medical School, Konstanty-Gutschow-Str. 8, D-3000 Hannover 61, FRG
*Department of Experimental Surgery, University of Kiel, Michaelisstr. 5, D-2300 Kiel 1, FRG

Introduction

Extracorporeal shock wave treatment is mainly used to destroy renal-, ureteral-, and gall-stones (Chaussy et al. 1983, 1984; Hood et al. 1988; Sackmann et al. 1988; Staritz et al. 1989). This application raises the question whether shock waves also exert effects on the surrounding tissue, especially since mechanical stress may induce cellular damage and alterations in intracellular organization. Studies with various tumor cell lines revealed different sensitivities of these cells to high energy shock waves (HESW) implicating that HESW might also be a tool for treatment of tumor tissue (Berens et al. 1989; van Dongen et al. 1989; Russso et al. 1985, 1986). Whereas these investigations were performed with different cell types, it is useful to study cell lines originating from the same clone for direct comparison of effects.
Therefore in this study three epithelial cell lines of fetal hamster lung were used as a model to evaluate the effects of HESW treatment by investigating cell viability and colony-forming efficiency. Additionally, visualization of cytoskeletal structures allows the qualification of intracellular alterations induced by mechanical stress in normal and transformed cells.

Material and methods

The nontumorigenic epithelial cell line M3E3/C3 was established from normal fetal hamster lung (Emura 1982). A further two cell lines derived from this clone: the cell line SA17 was transformed by the chemical carcinogen MNNG (N-methyl-N-nitro-N-nitrosoguanidine) (Emura 1982), whereas the cell line PrT1 was transformed by Printex particles. Cells were cultivated in RPMI 1640 medium (Gibco) supplemented with 20% fetal bovine serum (FBS, Boehringer) in a humidified incubator (Cytoperm, Heraeus) at 37°C with 10% CO_2 in air. Confluent cultures were harvested by trypsinization (0.12% trypsin, 0.02% EDTA in PBS).
For treatment with HESW, cell suspensions in a density of $4*10^5$cells/ml were transferred into cryotubes (Costar). These polypropylene tubes show the same acoustic impedance as water and tissue. Electromagnetic shock waves were generated by a lithotriptor (Lithostar, Siemens). Samples, fixed by a tube holder vertical to the shock wave generator, were submerged in a special waterbath at a temperature of 37°C. The focus of the HESW was adjusted by x-ray monitoring to the bottom of an empty tube before shock wave application. Treatment was performed changing kV-setting in the range from 13.3kV to 18.1kV and the number of shock waves from 200 to 800. Untreated control cells were placed in the waterbath outside the focus. All experiments were performed in triplicate.
For evaluation of colony-forming efficiency, 200 viable cells from each sample, determined by trypan blue exclusion (Freshney 1990), were plated in dishes (Falcon) and cultured by the above-mentioned conditions for 7 days without change of medium. Following methanol fixation and Giemsa-staining the cell colonies were counted using a stereo microscope (Wild Leitz).
After treatment cells were seeded on coverslips, cultivated for two days, and prepared for immuno-fluorescent staining of the cytoskeleton. For labeling of vimentin-filaments, the cells were fixed with methanol (10min) and counterfixed with acetone (2min) at -20°C. After incubation with FBS (10% in PBS, 15min), vimentin structures were labeled by primary mouse-anti-vimentin-IgG (Boehringer) and secondary goat-anti-mouse-IgG combined with the fluorescent dye AMCA (aminomethylcoumarin acetic acid, Dianova) each for 45min, rinsed with PBS and mounted with Moviol (Hoechst). For staining actin filaments, respectively microtubules, the cells were fixed in paraformaldehyde (4% in PBS, 37°C, 20min) and permeabilized with Nonidet P-40 (1% in PBS, Shell Chemicals) for 10min. After incubation with FBS, cells were labeled with mouse-anti-actin-IgM (Amersham) and goat-anti-mouse-IgM-AMCA (Dianova) respectively with mouse-anti-a-tubulin-IgG (Sigma) and goat-anti-mouse-IgG-AMCA (Dianova). The fluorescent-labeled structures were visualized by a fluorescence microscope (Axiophot, Zeiss).

Results and Discussion

In all cell lines, evaluation of cell viability shows only slight decreases corresponding to increasing shock wave treatment. In all dose groups the cell viability does not drop below 80% surviving cells. No obvious differences in cell viability could be detected between the different cell lines. In contrast, evaluation of the colony-forming efficiency demonstrates different effects between normal and transformed cells. This finding implicates that evaluation of cell viability alone is not a sufficient parameter to elucidate effects of mechanical stress. The normal cell line M3E3/C3 shows an increase of colony-forming ability with the increasing number of shock waves at each energy level (figure 1a). The rising plating efficiency dependent on the number of shock waves may be explained by the following fact: cells disrupted by membrane defects during treatment are recognized by trypan blue exclusion, whereas sublethal damaged cells are tracked down only by means of plating efficiency. Thus, treatment with low numbers of shock waves induces a larger amount of sublethal damaged cells, which no longer have proliferative capacity. In contrast, observation of increasing energy levels alone leads to a decrease in proliferative activity (figure 1a) indicating a higher sensitivity of normal cells to HESW at high energy levels.

The transformed cell line SA17 reveals a high rate of sublethal damaged cells at low numbers of shock waves, whereas in contrast to findings in M3E3/C3 cells, an increasing colony-forming ability is connected with rising energy level (figure 1b).

The cell line PrT1 shows the weakest effects of HESW in colony-forming efficiency compared to the other cell lines (figure 1c). In this case there is no marked influence from energy level nor from number of shock waves.

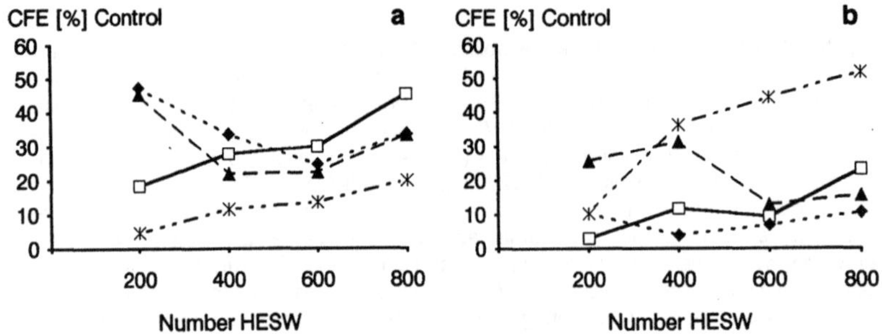

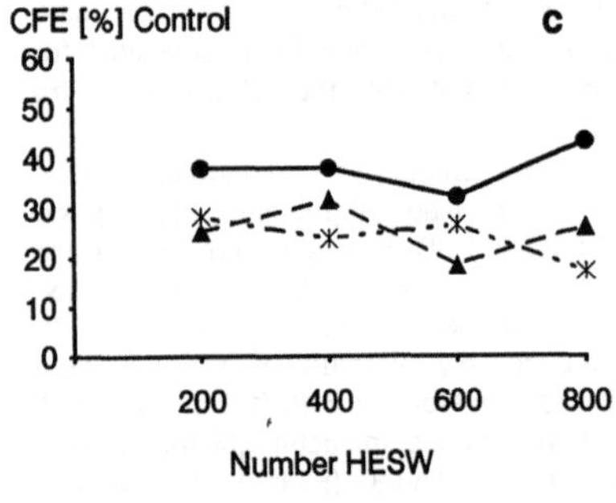

Figure 1: Colony forming efficiency (CFE) of different lung epithelial cell lines dependent on number and intensity of high energy shock waves (HESW)
a: M3E3/C3 cells; b: SA17 cells; c: PrT1 cells

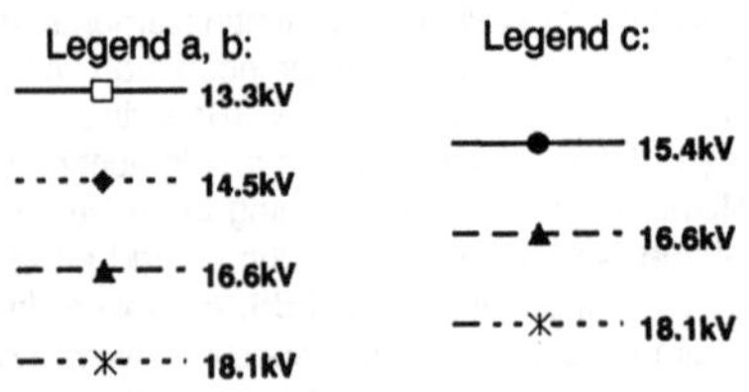

Observation of cytoskeletal proteins reveals no differences between normal and transformed cell lines regarding the structural organization of microtubules and vimentin filament system. Furthermore, these two components of cytoskeleton were not influenced by treatment with shock waves. Distinct alterations are observed only in the organization of actin filaments in the different cell types.
Normal lung cells (M3E3/C3) express formations of actin filament bundles seen as prominent stress fibers crossing the cytoplasm of the cells (figure 2a).Treatment of these cells with mechanical stress causes a reduction of stress fibers combined with an increasing number of actin granules in the cytoplasm (figure 2b). Especially at energy levels of 14.5kV and 16.6kV and 200 shock waves cells show an evident loss of stress fibers indicating distinct alterations in cellular organization and highest sensitivity of the treated cells in these dose groups. These findings may be correlated to increasing proliferative activity of these treatment groups, as seen in figure 1a. The reorganization of actin filaments to granulated actin structures may be related to transformation of cells (Gobert-Bohlen 1989, Lombardi 1990). Thus the granular appearance of actin in treated M3E3/C3 cells similar to transformed SA17 cells indicates a possible transformation process accompanied by a stimulation of proliferation caused by mechanical treatment.

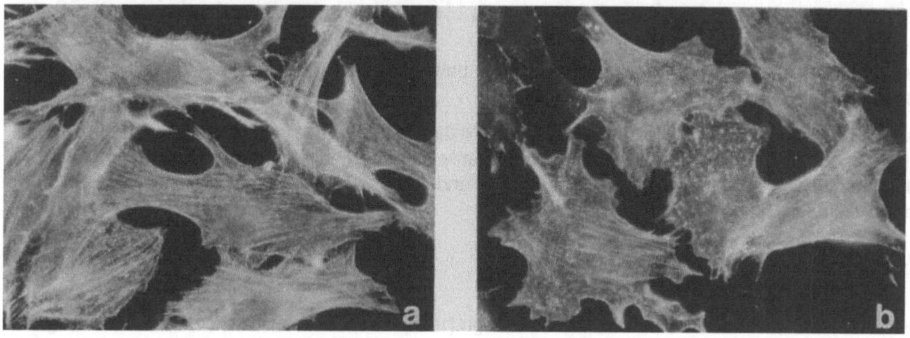

Figure 2: Actin organization in M3E3/C3 cells, scale 1:160; a: untreated M3E3/C3 cells; b: M3E3/C3 cells treated with 600 shock waves at 16.6kV

The chemically transformed cell line SA17 only shows cytoplasmic actin-containing granules (figure 3). These transformed cells do not build any stress fibers as seen in normal cells. In SA17 cells no visible changes of actin structure could be observed after shock wave treatment, in contrast to PrT1 cells.

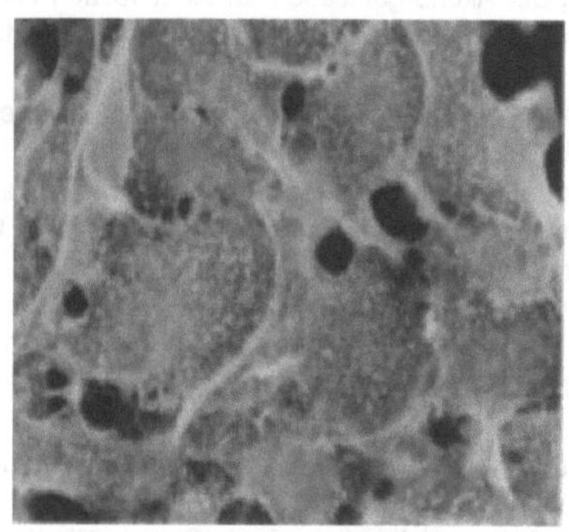

Figure 3: Actin organization in untreated SA17 cells, scale 1:160

The cell line PrT1 transformed by particles demonstrates actin formation ranging between structure of normal and chemically transformed cells. PrT1 cells contain less stress fibers than normal cells but actin granules as in the chemically transformed cells (figure 4a). In PrT1 cells, additional reduction of stress fibers can be observed after increasing shock wave treatment. In the highest application groups there is a total loss of stress fibers combined with an increase of actin granules in cytoplasm (figure 4b). These findings indicate that alterations in the actin filament system of epithelial lung cells after transformation by particles are not endpoint alterations. Further reduction of stress fibers can be induced in PrT1 cells by shock wave treatment. This additional reorganization of actin structure may lead to increasing loss of substrate adhesion of PrT1 cells.

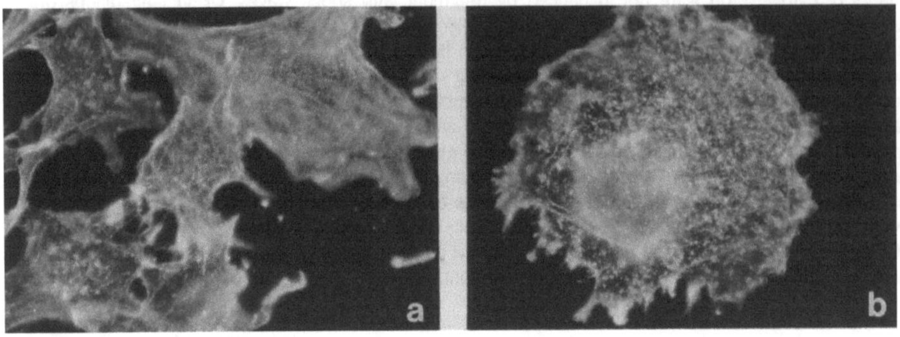

Figure 4: Actin organization in PrT1 cells, scale 1:160; a: untreated cells; b: cells treated with 800 shock waves at 18.1kV

These findings indicate that HESW treatment induces cell damage as well as intracellular alterations in normal and transformed cells in vitro, leading to altered proliferative activity. Thus further investigations are necessary to elucidate the effects of mechanical stress in vitro and in vivo to assess the usefulness of HESW for treatment of tumor tissue.

References

Berens, M.E., C.E. Welander, A:S: Griffin, D.L. McCullogh (1989): Effect of acoustic shock waves on clongenic growth and drug sensitivity of human tumor cells in vitro. J. Urol. 142, 1090-1094

Chaussy, C., E. Schmiedt (1983): Shock wave treatment for stones in the upper urinary tract. Urol.Clin.North Am. 10,743

Chaussy, C., E. Schmiedt (1984): Extracorporeal shock wave lithotripsy (ESWL) for kidney stones.An alternative to surgery? Urol.Radiol. 6,80

Dongen, van J.W., G.J. van Steenbrugge, J.C. Romijn, F.H. Schröder (1989): The cytocidal effect of high energy shock waves on human prostatic tumour cell lines. Eur.J.Cancer Clin. Oncol. 25/8,1173-1179

Gobert-Bohlen, A. (1989): Untersuchung der Aktinorganisation an einer fetalen Hamsterlungenepithelzellinie nach neoplastischer Transformation durch chemische Karzinogene, Dissertation an der Universität Hannover, FRG

Hood, K.A., A. Keightley, R.H. Dowling, J.A. Dick, C.N. Mallinson (1988): Piezo-ceramic lithotripsy of gallbladder stones: initial experience in 38 patients. Lancet 1322

Lombardi, L., D. Ballinari, I. Bongarzone, M. Migliari, P. Mondellini, C. Traversari, S. Modina (1990): Ultrastructural cytoskeleton alterations and modifications of actin expression in the NIH/3T3 cell line after transformation with Ha-ras-activated oncogene. Cell Motil. Cytoskeleton 15/4, 220-229

Russo, P., R. Stephenson , W.D.W. Heston, W.R. Fair (1985): The in vitro effect of high energy shock waves on human prostate cancer cell line PC-3. Proceedings of AACR, Cancer Res. 26, 267

Russo, P., R. Stephenson , C. Mies , R. Huryk, W.D.W. Heston, M.R. Melamed, W.R. Fair (1986): High energy shock waves supress tumor growth in vitro and in vivo. J. Urol. 135, 626-628

Sackmann, M., M. Delius, T. Sauerbruch, J. Holl, W. Weber, E. Ippisch, U. Hagelauer, O. Wess, W. Hepp, W. Brendel, G. Paumgartner (1988): Shock-wave lithotripsy of gallbladder stones. N.Engl.J.Med. 318, 393-397

Staritz, M., A. Rambow, P. Mildenberger, M. Goebel. Th. Scherfe, A. Grosse, Th. Junginger, R. Hohenfeller, M. Thelen, K.-H. Meyer zum Büschenfelde (1989): Electromagnetically generated extracorporeal shock waves for gallstonelithotripsy: in vitro experiments and clinical relevance. Eur.J.Clin.Inv.19,142-145

Effects of Adrenaline and Nicotine on the Early Chick Embryo

Martin Rosenbruch, Jürgen Kniepen and Christel Weishaupt
Div. Environm. Pathology, Medical Institute of Environmental Hygiene,
D- 4000 Düsseldorf 1 Gurlittstr. 53 (Germany)

Introduction

During the early stages of development, the chick embryo with the surrounding yolk sac blood vessels offers a good model for vascular and circulatory studies. At days 3-6 of incubation the test-system mainly consists of the chambered heart with embryonic and extraembryonic blood vessels. The system has been proofed to be a suitable model for the assessment of acute toxicity of irritant substances (Rosenbruch and Holst, 1990).
The study presented here was carried out to examine whether functional parameters of the cardio-vascular system can also be studied in this model and, if reproducible results can be obtained after intravascular injection into the extraembryonic yolk sac blood vessels. The experiments were done with nicotine as a vascular toxic substance and with adrenaline because of its well known effects on the chicken embryo heart (Kolesari et al., 1988).

Materials and Methods

White Leghorn eggs (n=325) were used for the experiments. Details of incubation and opening of the eggs have been described previously (Rosenbruch and Holst, 1990). 20µl of nicotine solution of different concentrations (0.01M, 0.1M, 1.0M, corresponding 0.032mg, 0.32mg, 3.2mg; Aldrich Chemie 18.637-6) were injected into the yolk before incubation (20 eggs per concentration). At day 5 of incubation the rate of undeveloped embryos was determined. Furthermore, groups of 20 embryos were treated daily at days 3-7, 4-8 and 5-9 of incubation with 50µl 0.1M and 0.01M nicotine and an electrolyte solution. The test substances were applied topically on the yolk sac system. Control eggs remained untreated. After the half of the incubation period, the embryos were sacreficed and the body weight and the heart weight was determined.
The intravascular injection of adrenaline (suprarenin, Hoechst, Frankfurt; 1µl≈1µg) was carried out at day 5 of incubation (stage 24, Hamburger and Hamilton, 1951) using an electronic injection equipment with a hamilton syringe and a glass drain. To evaluate functional effects, the heart rate and the percentage of arrhythmia in the chick embryos were determined. The results were compared with electrolyte injection and the physiological decrease of the heart rate in room temperature out of the incubator. Additionally, the mortality rate of the chick embryos following nicotine and adrenaline application was recorded.

Results

A single injection of nicotine in to the yolk prior to incubation results in a dose-related increase of undeveloped chick embryos at day 5 of incubation (figure 1).
The rate increases constantly from the control group up to the highest concentration of nicotine. Evaluating the mortality rate of the chick

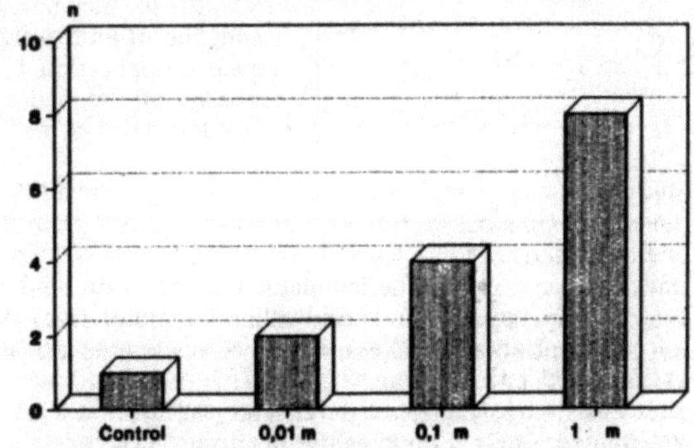

Figure 1 Undeveloped chick embryos (day 5 of incubation) after different doses of nicotine injection before incubation (20/group)

embryos in this experiment, the dose response effect is confirmed: 30 % (6/20) mortality rate due to 0.01M nicotine, 65 % (13/20) due to 0.1M and 95 % (19/20) due to 1.0M.

The topical application of 50µl of 0.1M nicotine on the test-system leads to capillary hyperaemia of the yolk sac blood vessels followed by focal haemorrhages after approximately 5 minutes. Additionally, haemorrhages in the embryo itself occur after some hours, preferentially in the limb buds, the eyes and the nervous system. In some areas of the yolk sac blood vessel system, the blood flow stops and stasis and edema are obvious (figure 2).
After the second application of 50µl 0.1M nicotine solution, all embryos die.

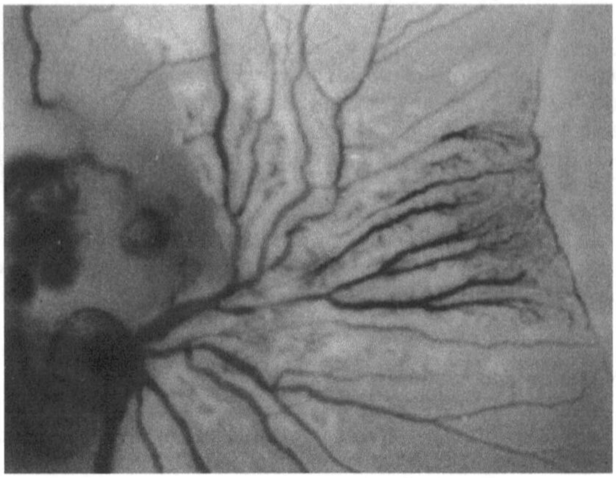

Figure 2 Stasis with edema and growth inhibition in yolk sac blood vessels after application of nicotine (50µl, 0.1M; day 6 of incub.)

The data of the body weight of surviving embryos after repeated nicotine application of the lower concentration (0.01M), determined at the half of the incubation period, are show in table 1. As compared with the electrolyte control, a clear reduction of the body weight can be detected. Applying the nicotine during earlier stages of development (days 3-7), the weight reduction reaches a level up to 20%.

Table 1

Reduction of body weight (%of electrolyte control) after repeated application of nicotine (50µl, 0.01M) at different days of incubation (I: day 3-7, n=8; II: day 4-8, n=14; III: day 5-9, n=5)

I	II	III
81,3	93,0	89,3

During the stay out of the incubator the heart rate of the chick embryo decreases constantly due to temperature differences within 3 minutes from 154/min to 127/min (fig. 3, control line). The intravascular (iv) injection of saline does not change this heart rate pattern in any way.
But, after iv-application of adrenaline, the heart rate even increases up to one minute after injection and after 3 minutes the heart rate levels are still significantly (student t-test) higher as compared with the control group (figure 3). Additionally, the percentage of chick embryos with heart arrhythmia, i.e. temporary asystoly or ventricular dysrhythmic phenomena, was determined. Again a dose-response effect is detectable: 27% (6/22) after 2µl, 38% (9/24) after 4µl and 58% (11/19) after 8µl. The mortality rate of chick embryos after iv-injection of adrenaline is about 3-fold higher as compared with the saline group.

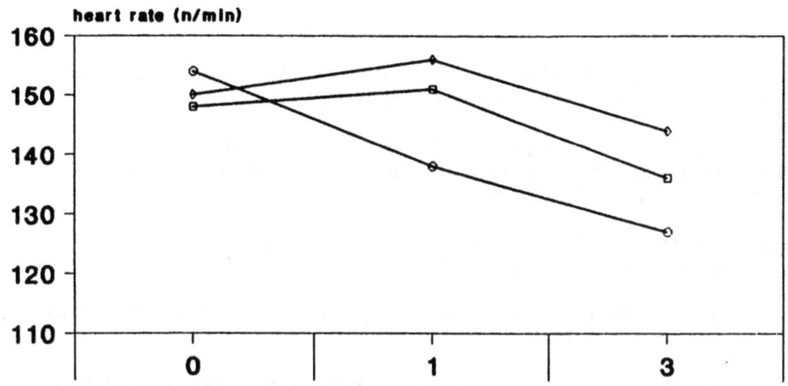

Figure 3 Heart rate increase in chick embryos after iv-injection of adrenaline (I=2µl; II=4µl; day 4 of incubation) up to 3 minutes after injection compared with physiological decrease (o)

Discussion

Most toxicity studies evaluating cardio-vascular effects utilize in vivo animal models (Anderson et al, 1990). The test system of the early chick embryo, which is presented in these experiments, has the advantages neither to use living animals, nor mammals. However, the system consists of a heart, together with the embryonic and extraembryonic blood vessel system. And, as a sign of vascular reactive ability during the first third of incubation, phagocytosis, angiogenesis and connective tissue proliferation with collagenous fibres already occur after vascular impairement (Rosenbruch, 1989).
Growth retardation and fetal death due to nicotine are well known from human studies (Farrer and Blumer, 1991). The results obtained with the chick embryo model show the comparable effects regarding the rate of undeveloped embryos and the reduction of body weight after half of the incubation period, already. In men, smoking of cigarettes leads to increased plasma levels of adrenaline (Cryer et al., 1976) and the heart rate is significantly higher after smoking as compared with the situation before (Mancia et al., 1990). A significant increase of the heart rate could also be received in chick embryos at day 5 of incubation after intravascular injection of adrenaline in the yolk sac blood vessels. Thus, the cardio-vascular reaction pattern in chick embryos after application of the test substances nicotine and adrenaline is equal to those in humans.

These results underline that

1. Intravascular injections in the extraembryonal yolk sac blood vessels of the chick embryo can be carried out with reproducible results of heart rate changes.

2. Toxicological studies in chick embryos can be evaluated including systemic effects on the test-system as well as functional cardio-vascular parameters, even during the first half of incubation.

According to our results, there are extended experimental possibilities for the application ofthis test system, "the early chick embryo". It is a suitable model for the evaluation of effects induced by drugs and environmental pollutants.

References

Anderson, P.G., S.B. Digerness, J.L. Sklar, and P.J. Boor: Use of the isolated perfused heart for evaluation of cardiac toxicity. Toxicol. Pathol. 18, 497-510, 1990

Cryer, P.E., M.W. Haymond, J.W. Santiago, S.D. Shah: Norepinephrine and epinephrine release and adrenergic mediation of smoking associated hemodynamic and metabolic events. N. Engl. J. Med. 295, 573-577, 1976

Farrar, H.C., and J.L. Blumer: Fetal effects of maternal drug exposure. Annu. Rev. Pharmacol. Toxicol. 31, 525-547, 1991

Hamburger, V., and H.L. Hamilton: A series of normal stages in the development of the chick embryo. J. Morphol. 88, 49-92, 1951

Kolesari, G.L., H.J. Schnitzler, G.M. Rajala, R.B. Lewan, and S. Kaplan: The antidysrhythmic effects of metoprolol in the epinephrine treated chick embryo. Life Sciences 42, 1159-1163, 1988

Mancia, G., A. Groppelli, R. Casadei, S. Omboni, E. Mutti, G. Parati: Cardiovascular effects of smoking. Clin. Exp. Hyper.-Theory Practice, 5, 917-929, 1990

Rosenbruch, M.: Granulation tissue in the chick embryo yolk sac blood vessel system J. Comp. Pathol. 101, 363-373, 1989

Rosenbruch, M., and A. Holst: The chick embryo yolk-sac blood vessel system as an experimental model for irritation and inflammation. Toxicol. in vitro 4, 327-331, 1990

MUTAGENICITY/CARCINOGENICITY STUDIES

Quantitation of 7-(2-Hydroxyethyl)guanine Produced by Endogenous and Environmental Ethylene Oxide in DNA of Lymphocytes

H.M. Bolt, H. Peter, M. Leutbecher, H. Ottenwälder, H.W. Goergens, and U. Föst
Institut für Arbeitsphysiologie an der Universität Dortmund, Ardeystr. 67, W-4600 Dortmund/F.R.G.

Introduction

Analysis of macromolecular adducts of ethylene oxide is a tool which allows determination of internal target doses of this carcinogen (Bolt et al., 1988). An important aspect is that ethylene is converted in the living organism to ethylene oxide, and that ethylene is produced endogenously (Filser and Bolt, 1983; Schmiedel et al., 1983).

There is a linear relation between ethylene oxide dose and alkylation of hemoglobin or DNA (Potter et al. 1989).

Watson et al. (1987) have utilized the labelization of the glycosidic bond of 7-alkylated guanine in DNA for sample preparation in the analysis of 7-(2-oxoethyl)guanine in DNA: heating the alkylated DNA (1 h at 100 °C) releases the 7-alkylated guanine from the DNA. The same procedure can be applied to 7-(2-hydroxyethyl)guanine, the main adduct of ethylene oxide. This sample preparation results in an enrichment factor of about 3 orders of magnitude for the 7-guanine adduct, compared with conventional acid or enzyme total hydrolysis of DNA, and allows determination of the basal levels of 7-(2-hydroxyethyl)guanine with GC/MS using ion impact ionization.

This new method, applied to human lymphocytes, is described in this paper.

Materials and Methods

A schematic diagram of the method is given in Fig. 1.

From 8 - 10 ml of (human or rat) blood lymphocytes are separated by centrifugation with commercial "lymphocyte separation medium", according to the standard method of Boehringer, Mannheim.

The lymphocyte suspension is then transfered into an "Eppendorf" vial (2.2 ml) and diluted with 1.5 ml 0.9% saline. After centrifugation at 100 g for 15 min the supernatant is discarded; the lymphocytes are resuspended in 2 ml 0.9% saline and again centrifuged. They are then taken into 0.5 ml of a 50 mM EDTA solution (pH 8.5), and 20 µl 10% SDS is added.

After cooling (5 min on ice) 0.5 ml water-saturated phenol is added and the whole thoroughly mixed. After separation of the phases the phenol layer is discarded, and the aqueous layer washed seve-

blood sample
↓
isolation of lymphocytes
↓
phenol extraction
↓
DNA
↓
heat treatment
($95^{\circ}C$, 1 h, neutral pH)
↓
EtOH precipitation of DNA
↓
supernatant: 7-alkylated guanine
↓
TMS derivatisation
↓
GC/MS

Figure 1: Flow scheme of the analytical method.

ral times with 0.5 ml chloroform/isoamyl alcohol (25:1). Finally, the aqueous layer is extracted twice with 0.5 ml ether and added to 50 µl 5 M NaCl solution. After cooling to $200^{\circ}C$ the DNA is precipitated with 1 ml cold ethanol. After standing overnight at $-20^{\circ}C$ the DNA is centrifuged (30 min, 9000 g); the precipitate is washed with 0.5 ml cold 70% EtOH and dissolved in 0.5 ml water. The DNA content is finally determined by photometry (260 nm); the yield of lymphocyte DNA, in our hands, varied between 40 and 67 µg DNA from 10 ml human or rat blood.

Heating for 1 h to $95^{\circ}C$ liberates the 7-alkylated guanines (Watson et al., 1987). After cooling, 1.5 ml cold EtOH is added, and the samples are stored at $-20^{\circ}C$. The remaining depurinated DNA is separated by centrifugation, and the supernatant is evaporated to dryness under a stream of N_2.

The depurination causes single-strand DNA breaks which produce ethanol-soluble oligonucleotides. These are removed on NEN-Sorb-cartridges (NEN-DuPont):

The commercial NEN-Sorb cartridges are washed with 10 ml MeOH and 20 ml H_2O. The dry DNA samples are taken up with 1 ml H_2O, under addition of 0.1 ml aqueous internal standard solution (7-methyl-guanine, 60 µg/ml), and applied to the cartridges. After percolation of 10 ml H_2O the modified bases are eluted with 10 ml methanol. This effluent is lyophilized and kept at $-20^{\circ}C$ until further processing.

For derivatization, 100 µl MSTFA (N-methyl-N-trimethylsilytri-fluoroacetamide) is added, and heated to $60^{\circ}C$ for 1 h.

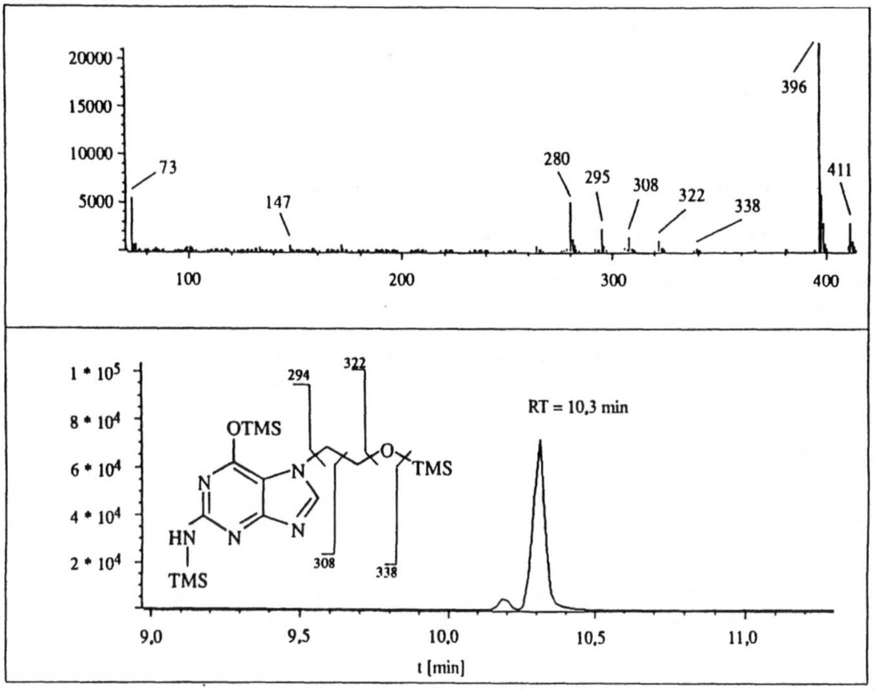

Fig. 2: GC/MS pattern of the TMS derivative of 7(2-hydroxyethyl)-guanine (electron impact ionization).
Top: fragmentation (M^+ : 411 m/z; M^+ minus CH_3 : 396 m/z); bottom: chromatographic behaviour of injected standard compound (TIC).

From this solution, 1.7 µl is injected to the "cold injection system" of a Hewlett-Packard 5890A gas chromatograph, equipped with a mass selective detector (MSD). The chromatographic conditions are: 30 m capillary column, DB-5, 0.254 mm i.d., 0.25 µm surface, carrier gas: He, 100 kPA. The samples are injected (injector temperature, 35 °C, open split), and the injector is then heated (rate: 12 °C/sec) to 300 °C. After reaching this temperature, the split is closed for 1 min to transfer the material to the column. After this time the temperature program of the column (rate 40 °C/min, final temperature 300 °C) is started.

Fig. 2 shows the chromatographic behaviour of the ethylene oxide adduct (bottom panel) and its fragmentation pattern (top panel). For quantitation of 7(2-hydroxyethyl)guanine the fragments of 396 and 280 m/z are used ("Selected Ion Monitoring" mode).

Results and Discussion

In human lymphocytes, the baseline level of the ethylene oxide adduct 7(2-hydroxyethyl)guanine was 8.49 ± 5.75 pmol/mg DNA (n = 8). The respective value for rat blood, 5.61 ± 2.99 pmol/mg DNA (n = 3), was found in the same order of magnitude.

The larger individual variability of the adduct levels in humans might be attributed to the influence of smoking because blood samples were taken both from smokers and non-smokers.

The data are consistent with previous pharmacokinetic data (Filser and Bolt, 1984) and demonstrate the presence of an "internal dose" of ethylene oxide in individuals which are not exposed occupationally. This, in consequence, is linked with a genotoxic/carcinogenic risk (Bolt and Filser, 1987; Shen et al., 1989).

References

Bolt, H.M. and J.G. Filser (1987) Kinetics and disposition in toxicology; example: carcinogenic risk estimate for ethylene. Arch. Toxicol. 60: 73-76.

Bolt, H.M., H. Peter and U. Föst (1988) Analysis of macromolecular ethylene oxide adducts. Int. Arch. Occup. Environ. Hlth. 60: 141-144.

Filser, J.G. and H.M. Bolt (1983) Exhalation of ethylene oxide by rats on exposure to ethylene. Mutation Res. 120: 57-60.

Filser, J.G. and H.M. Bolt (1984) Inhalation pharmacokinetics based on gas uptake studies. VI. Comparative evaluation of ethylene oxide and butadiene monoxide as exhaled reactive metabolites of ethylene and 1,3-butadiene in rats. Arch. Toxicol. 55: 219-223.

Potter, D., D. Blair, R. Davies, W.P. Watson and A.S. Wright (1989) The relationship between alkylation of hemoglobin and DNA in Fischer 344 rats exposed to ^{14}C-ethylene oxide. Arch. Toxicol., Suppl. 13: 254-257.

Schmiedel, G., J.G. Filser and H.M. Bolt (1983) Rat liver microsomal transformation of ethene to oxirane in vitro. Toxicol. Lett. 19: 293-297.

Shen, J., W. Kessler, B. Denk and J.G. Filser (1989) Metabolism and endogenous production of ethylene in rat and man. Arch. Toxicol., Suppl. 13: 237-239.

Watson, W.P., A.E. Crane, R. Davies, R.J. Smith and A.S. Wright (1987) A postlabelling assay for N^7-(2-oxoethyl)guanine, the principal vinyl chloride-DNA adduct. Arch. Toxicol., Suppl. 11: 89-92.

New Cell Culture System for Pulmonary Carcinogenesis and Inhalation Toxicology

Emura, M., M. Riebe-Imre, A. Ochiai, G. Singh, S.L. Katyal, M. Aufderheide, U. Mohr, J. Jacob and G. Grimmer
Institut für Experimentelle Pathologie, Medizinische Hochschule Hannover, W-3000 Hannover 61, F.R.G.; Department of Veterans Affairs, Medical Center, Pittsburgh, U.S.A. and Biochemisches Institut für Umweltcarcinogene, W-2070 Grosshansdorf, F.R.G.

Introduction

Relatively little is known about the histogenesis and the underlying mechanisms of lung cancer; in particular about the target cell type(s) of carcinogens (Schüller 1987). The same can be said about the mechanisms of epithelial regeneration in the lung after injury by toxic chemicals. One of the morphologically appreciable responses of the lung to such injury is a hyperplastic growth of bronchiolo-alveolar epithelium. There is speculation that type II pneumocytes or Clara cells may give rise to the hyperplasia. Neuroendocrine (APUD) cells and type I pneumocytes are supposed to be less capable of initiating hyperplasia because of their low activity of cell division, although hyperplasia of the former cell type is experimentally inducible (Schuller et al. 1988). Mutual transformation has even been postulated between type II pneumocytes and Clara cells in certain cases (Schüller 1987). There is another cell type which, occurring in a fetal as well as regenerating adult airway epithelium (McDowell et al. 1984; Emura 1985; Strum et al. 1990), closely resembles the small mucus granule cells (SMGC) postulated to be one of the carcinogen targets in the lung (McDowell and Trump 1983). In our cell culture study (Emura et al. 1990) a large population of cells possessing SMGC characteristics are inducible in vitro from an undifferentiated population of airway epithelial cells. Furthermore, this SMGC counterpart in culture develops into a cell type corresponding to the Clara cell in vivo in the presence of vitamin A (Emura et al. 1990; Ochiai et al. 1991). In vitro evidence of a cognate relationship between Clara cells and type II pneumocytes has also been obtained (Emura et al. 1990). In the current paper, data will be shown that certain carcinogens can be activated by the in vitro counterpart of SMGC, which is eventually transformed to show anchorage independence (AI).

Materials and Methods

Fetal Syrian hamster and human pulmonary epithelial cell lines used have already been described (Emura et al. 1990). Specific cell culture procedures for providing an in vitro counterpart of SMGC have been published previously (Emura et al. 1990; Brockmeyer et al. 1990). Identification of cell types in culture has been described elsewhere (Emura et al. 1990; Ochiai et al. 1991). Antibodies detecting antigens specific for type II pneumocytes and

Clara cells have also been published previously (Katyal et al. 1977; Singh et al. 1985). Metabolism of benzo(a)pyrene (BaP) was analysed using high performance liquid chromatography (HPLC) and that of N-diethylnitrosamine (DEN) was measured by the amount of acetaldehyde produced during incubation of cells with DEN (Conradt and Green 1982). For AI transformation, hamster and human cells were treated for 4 days and 9 weeks, respectively.

Results and Discussion

The hamster and human cells used in this study have shown capacity to differentiate in vitro into cell types resembling Clara cells and type II pneumocytes depending on the culture conditions employed (Emura et al. 1990). Detailed studies on this multipotential capacity will be published elsewhere (Ochiai et al. 1991; Emura et al. in preparation). In cells of both species no complete ciliogenesis has so far been observed (Fig. 1).

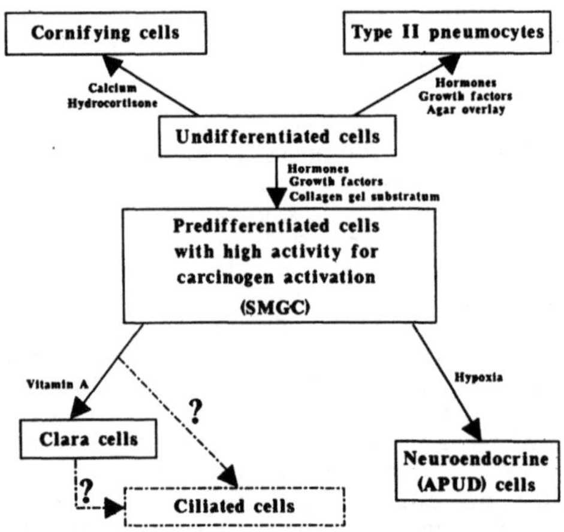

Fig. 1 Various pathways of cell differentiation from an undifferentiated type of progenitor cell

APUD cell differentiation has so far been encountered in the hamster cells (Ochiai et al. in preparation). Cornification occurs spontaneously after long-term culture, in a high calcium concentration or during tumorous growth in vivo after in vitro transformation (Emura et al. 1990). The activity of the cells to metabolize polycyclic aromatic hydrocarbons and DEN was studied on differentiation into the in vitro SMGC counterpart (Emura et al. 1990). In the 8-day incubation period, the hamster cells metabolized 97% of chrysene (CHR), while the human cells metabolized only 24% (Jacob et al. 1990). Oxidation at the 3,4 position predominates in the human cells, whereas in the hamster cells oxidation at the 1,2 position is the major metabolism.

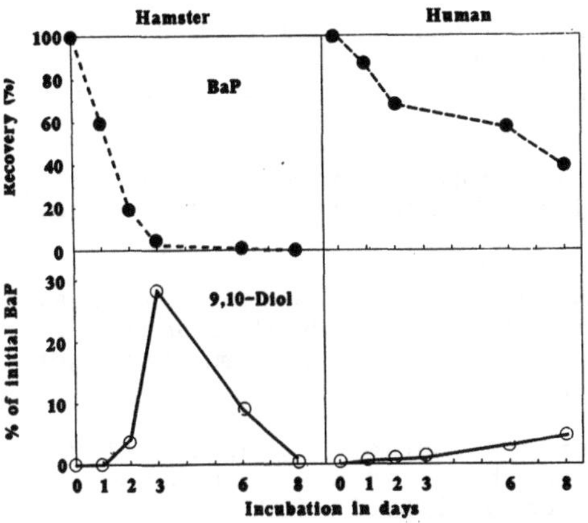

Fig. 2 Comparison of BaP metabolism by hamster and human cells in an SMGC-corresponding state

Tab. 1 AI transformation by DEN in hamster and human cells

DEN (µg/ml)	Hamster	Human
0*	0.03	0.02
150	-	0.02
450	-	0.03
500	0.24	-
1000	0.16	0.19
2000	-	0.33

* 0.03% DMSO

BaP was also metabolized more quickly by hamster cells than by human cells (Fig. 2). Maximum production of 9,10-diol of BaP occurred within a much shorter incubation period (3 days) in the hamster cells than in the human cells. A putatively proximate metabolite, 7,8-diol of BaP was produced by the hamster cells in a slightly measurable amount, while the amount produced by human cells was barely detectable. Correspondingly, AI transformation by CHR and BaP was highly inducible only in the hamster cells (~ 6 times the control by 2 µg/ml CHR, 50-100 times the control by 4-8 µg/ml BaP). In contrast, not only hamster cells but also human cells could undergo AI transformation by DEN (Table 1). The maximum metabolism of DEN as determined by the production of acetaldehyde (15 pmol/µgDNA/min) occurred on day 6 of incubation in the hamster cells. The measurement in the human cells is now in progress.

All in all, the cell culture methods developed appear to provide a potentially powerful system for analysing the regulatory mechanisms of carcinogen/toxin metabolism and intoxication, histogenesis of carcinomas and identification of the most amenable target cell types in the lung.

We thank M. Tauscher and K. Hoffmann for their technical contribution and G. Teicke for her help with the manuscript.
Partly supported by a grant from the Bundesministerium für Forschung und Technologie.

References

Brockmeyer C, Emura M, Ochiai A, Riebe M, Mohr U, Hilfrich J (1990) Secretory differentiation of human fetal bronchial epithelial cells in culture. A study by histochemistry and electron microscopy. Exp Pathol 39:11-17

Conradt P, Green U (1982) Effect of diazepam on microsomal diethylnitrosamine metabolism in gerbils. J Cancer Res Clin Oncol 104:53-61

Emura M (1985) Development of Syrian golden hamster tracheal epithelium during prenatal and immediate postnatal stages. In: Jones TC, Mohr U, Hunt RD (eds) Respiratory system. Springer Berlin Heidelberg New York pp 11-27

Emura M, Riebe M, Ochiai A, Aufderheide M, Germann P, Mohr U (1990) New functional cell-culture approach to pulmonary carcinogenesis and toxicology. J Cancer Res Clin Oncol 116:557-562

Jacob J, Grimmer G, Raab G, Emura M, Riebe M, Mohr U (1990) Metabolism of pyrene and chrysene in epithelial human bronchial and hamster lung cells. Int J Environ Anal Chem 38:221-230

Katyal SL, Estes LW, Lombardi B (1977) Method for the isolation of surfactant from homogenates and lavages of lung of adult, newborn, and fetal rats. Lab Invest 36:585-592

McDowell EM, Keenan KP, Huang M (1984) Effects of vitamin A-deprivation on hamster tracheal epithelium. A quantitative morphologic study. Virchows Arch B 45:197-219

McDowell EM, Trump BF (1983) Histogenesis of preneoplastic and neoplastic lesions in tracheobronchial epithelium. Surv Synth Pathol Res 2:235-279

Ochiai A, Emura M, Riebe-Imre M, Mohr U, Hilfrich J, Tahara E, Dungworth DL (1991) Secretory differentiation and cell type identification of a human fetal bronchial epithelial cell line (HFBE). Virchows Arch B in press

Schüller HM (1987) Experimental carcinogenesis in the peripheral lung. In: McDowell EM (ed) Lung carcinomas. Churchill Livingstone, Edinburgh London Melbourne New York pp 243-254

Schuller HM, Becker KL, Witschi H-P (1988) An animal model for neuroendocrine lung cancer. Carcinogenesis 9:293-296

Singh G, Katyal SL, Ward JM, Gottron SA, Wong-Chong M-L, Riley EJ (1985) Secretory proteins of the lung in rodents: Immunocytochemistry. J Histochem Cytochem 33:564-568

Strum JM, Singh G, Katyal SL, McDowell EM (1990) Immunochemical localization of Clara cell protein by light and electron microscopy in conducting airways of fetal and neonatal hamster lung. Anat Record 227:77-86

Stable Expression of Heterologous Microsomal Epoxide Hydrolase in BHK21 Cells: Influence on the Mutagenicity of Benzo[*a*]pyrene 4,5-Oxide

Glatt, H. R., R. Becker, A. Piée, F. Oesch and T. Friedberg
Department of Toxicology, University of Mainz, Obere Zahlbacher Strasse 67, D-6500 Mainz, FRG

Introduction

Most environmental mutagens and carcinogens require metabolic activation to electrophilic intermediates capable of reacting with cellular target structures, such as DNA. These electrophilic intermediates are in addition subject to metabolic detoxification. This metabolism is mainly controlled by enzymes whose expression is very variable. Among other things, various enzymes are inducible by environmental chemicals. Understanding the toxicology of chemicals (for example, species differences, idiosyncrasias, organotropisms) therefore requires knowledge of critical host factors. One approach towards this goal involves the use of purified enzymes in metabolism and toxicological studies (Glatt *et al.*, 1983; Glatt and Oesch, 1986). This approach, however, is laborious, the purification may alter enzymatic properties, and it may be difficult to rule out the influence of impurities in the enzyme preparation, especially when large families of structurally closely related enzymes exist, as is common in xenobiotic metabolism.
Alternatively, powerful tools are provided by gene transfer techniques. Cells in culture expressing specific heterologous cytochromes P450 are mutated by certain promutagens, whereas the parental cells are not mutated (Glatt *et al.*, 1987, 1992; Doehmer *et al.*, 1988; Dogra *et al.*, 1990; Crespi *et al.*, 1991; Gonzalez *et al.*, 1991; Thompson *et al.*, 1991). In the present study, we have applied gene transfer techniques to the major xenobiotic-metabolizing microsomal epoxide hydrolase (mEH, E. C. 3.3.2.3), to study its contribution to the detoxification of benzo[*a*]pyrene 4,5-oxide (BPO). This epoxide and other K-region arene oxides are potent mutagens in bacteria, while they are, at the most, weakly mutagenic and carcinogenic in mammalian systems (Glatt and Oesch, 1986). These differences may be ascribed to detoxification, since addition of minute amounts of purified mammalian epoxide hydrolases and glutathione transferases to bacterial test systems suppressed the mutagenicity of K-region arene oxides (Glatt *et al.*, 1983, Glatt and Oesch, 1986). The quantitative contribution of the different detoxifying enzymes was however difficult to estimate using this method, since their localization and cofactor supply were different in these cell-free systems from those in intact cells.

Materials and methods

Cells and culture media. Syrian hamster BHK21 Cl 13 cells were used for gene transfer, as the level of their own mEH is low (Glatt *et al.*, 1987). They have however substantial glutathione transferase activity (Table 1). Cells were cultured in Dulbecco's modified minimum essential medium supplemented with fetal calf serum (10 %), penicillin (100 units/mL) and streptomycin (100 µg/mL). In the selection step of the mutagenicity experiments, 50 % fresh medium and 50 % conditioned medium were used in order to increase the cloning efficiency (from < 5 % to about 60 %). Medium was conditioned for 24 h on a semiconfluent culture. The cells were grown at 37 °C in a humified atmosphere containing 5 % CO_2.

Construction of the recombinant cell line (BHK21-mEH/Mz1). A full-length cDNA clone of rat mEH, pEH 52 CK (Porter *et al.*, 1986), was kindly provided by Dr. C. Kasper (Madison, USA). The EcoRI/SalI insert encoding the entire translated mEH sequence and most of the 3'- and 5'-untranslated regions was made blunt-ended using Klenow fragment, and inserted into the expression vector pMPSV (Artelt *et al.*, 1988), linearized with BamH1 and blunt-ended with Klenow fragment. In the resulting construct (pMPEH), the expression of the mEH cDNA is controlled by the constitutive mouse myeloproliferative tumor virus promoter. pMPEH, in 20-fold excess, was co-transfected with a plasmid (LK444) conferring neomycin resistance into BHK21 cells using the calcium phosphate co-precipitate technique. Clones resistant to neomycin (1 mg/mL) were analyzed by Western blots for the presence of rat mEH, using a goat polyclonal antibody. Selected positive clones were (i) analysed for the presence of rat mEH RNA by the RNase A protection assay using radiolabelled antisense mEH transcript as the probe, (ii) immunostained with a monoclonal anti-mEH antibody and a Texas Red-conjugated goat antibody raised against mouse immunoglobulines (Diannova, Hamburg), (iii) and examined for epoxide hydrolase activity towards [^{3}H]BPO, as previously described (Glatt *et al.*, 1987).

Mutagenicity and cytotoxicity experiments. On day 1, 1.5 x 10^6 cells were seeded with 30 mL of medium in 15-cm Petri dishes. After 18 h, BPO or the solvent only (60 µL dimethylsulfoxide) was added, 3 cultures being used for each treatment. Twenty-four hours later, the exposure was terminated by a change of the medium. On day 4, the cells were detached by treating with trypsin. The cell number, expressed as a percent of the value of the solvent controls (5.2 x 10^7 cells), was used as a measure of the cytotoxicity of the treatment. The cells were then subcultured at a density of 4.5 x 10^6 cells per 15-cm Petri dish. On day 6, they were subcultured again, at a density of 10^6 per 15-cm Petri dish in medium containing ouabain (1 mM) to determine the number of mutants (2 - 3 dishes for each original culture), and at a density of 100 cells per 6-cm Petri dish in normal medium to determine the cloning efficiency (3 replicate dishes). The colonies were counted after incubating for 7 days (cloning efficiency) or 10 days (selection). Mutant frequencies were calculated.

Results

In the Western blot, probed with the anti-mEH antibody, parental BHK21 cells showed no signals, but 15 out of 20 neomycin-resistant clones analyzed were positive in the test. Clearly the strongest signal was found in clone BHK21-mEH/Mz1. The molecular weight of the immunoreactive material was indistinguishable from that of authentic rat mEH. Also immunostaining of whole cells demonstrated the presence of immunoreactive material in clone BHK21-mEH/Mz1 and its absence in BHK21 cells. The antigen appeared to be associated with a cytoplasmic network, which probably is the endoplasmatic reticulum. Different cells of the same culture showed similar staining intensities.
The highly sensitive RNase A protection assay showed that there were no sequence differences between the rat mEH antisense RNA probe and an RNA expressed in BHK21-mEH/Mz1 cells, while no hybridizing RNA was found in BHK21 cells. Epoxide hydrolase activity in the transfectant clone was about 30 times higher than in the parental

Table 1. Activities of epoxide hydrolase and glutathione transferase toward benzo[*a*]-pyrene 4,5-oxide in total homogenates of BHK21 cells, BHK21-derived cells (clone BHK21-mEH/Mz1) expressing rat microsomal epoxide hydrolase, and freshly isolated hepatocytes from adult male Sprague-Dawley rats. The radioassays described by Glatt *et al.* (1987) and Glatt and Oesch (1977) were used.

	Activity (pmol / mg protein / min)	
Cells	Epoxide hydrolase	Glutathione transferase
BHK21 Cl 13	2 - 7 [a]	2630
BHK21-mEH/Mz1	153	2770
Hepatocytes	5310	6170

[a] measured on different occasions.

line, but still substantially lower than in hepatocytes (Table 1). However, the activity in many extrahepatic tissues of the rat (Oesch *et al.*, 1977) is similar to that found in BHK21-mEH/Mz1 cells.

BPO was less toxic in BHK21-mEH/Mz1 cells than in parental BHK21 cells (Fig. 1). BPO was clearly positive in the newly developed mutagenicity test with BHK21 and BHK21-mEH/Mz1 cells. The concentration–response curve was shifted by a factor of about 4 in the transfectant line as compared to the parental line (Fig. 1).

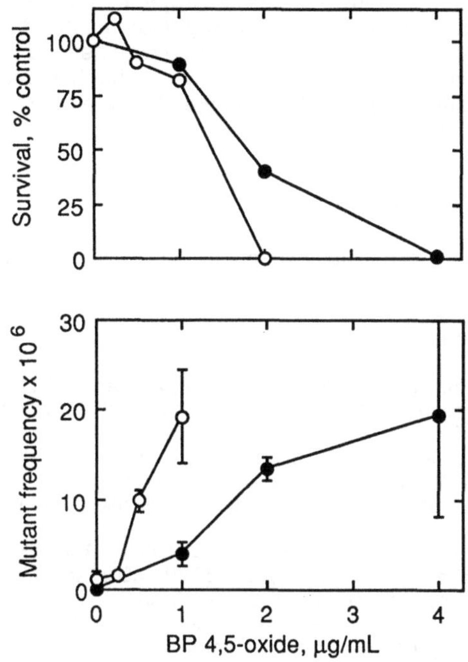

Figure 1. Cytotoxicity (above) and mutagenicity (below) of benzo[*a*]pyrene 4,5-oxide in BHK21 cells (o) and in BHK21-derived cells (clone BHK21-mEH/Mz1) expressing rat microsomal epoxide hydrolase (●). Values are means and SE of 3 cultures. The high SE of the mutant frequencies at the highest dose levels can be explained, as the cell populations were too small for an accurate determination of rare events such as mutations.

Discussion

We successfully transfected and expressed a cDNA construct, encoding rat mEH, in a cell line which had an unusually low level of mEH activity. The antibody and nucleotide probes used did not detect any signal in the parental BHK21 cells. These negative results can be explained by the weakness of the expression and potentially by species differences in the enzyme. The lack of such signals and the low activity in the parental line facilitate the genetic, biochemical and toxicological detection and characterization of the heterologous mEH in the transfectant clone. mEH RNA and protein in clone BHK21-mEH/Mz1 appeared to be identical to their authentic rat species. Furthermore, it appeared that the protein was localized in the endoplasmatic reticulum.

The increase in mEH activity had toxicological consequences in that the cytotoxicity and mutagenicity of BPO was decreased in the transfectant clone. The decrease in mutagenicity (about 4-fold) was less than the increase in mEH activity (about 30-fold). It is however important to bear in mind (i) that mEH levels in the transfectant clone were still moderate, when compared to tissues with high expression levels, such as liver, testis and kidney (Oesch *et al.*, 1977), (ii) and that the levels of the other detoxifying enzymes, the glutathione transferases, were very high in BHK21 cells (Table 1). Hence, we can conclude that even in the presence of high glutathione transferase activities, modulation of mEH on a relatively moderate level can have a marked influence on the toxicology of a substance. The experiment is now being repeated in glutathione-depleted cells.

Moreover, we have constructed a line expressing a heterologous glutathione transferase. Thus, it will be possible to elucidate the relative importance of different detoxification mechanisms in more detail.
In the present study, the active species, BPO, was added to the system externally. The concentration–time course and the distribution of the active metabolite may be very different, when it is generated continuosly in the system. In the case of benzo[*a*]pyrene, the situation is complicated as additional active metabolites may be formed. The formation of some of these metabolites requires the presence of mEH. This is true for the bay-region diol-epoxides, which appear to constitute the major DNA-binding and carcinogenic metabolites in many mammalian sytems *in vivo*. Sets of genetically engineered cells, differing in one or several defined enzyme activities, are powerful tools in elucidating the role of these highly variable host factors in the toxicology of chemicals. In the present study we demonstrated that the detoxification of mutagens can be studied using recombinant cell lines.

Acknowledgement

This work was financially supported by Deutsche Forschungsgemeinschaft (SFB302).

References

Artelt P, Morelle C, Ausmeier M, Fitzek M, Hauser H (1988) Vectors for efficient expression in mammalian fibroblastoid, myeloid and lymphoid cells via transfection or infection. Gene 68: 213-219.

Crespi CL, Gonzalez FJ, Steimel DT, Turner TR, Gelboin HV, Penman BW, Langenbach R (1991) A metabolically competent human cell line expressing five cDNAs encoding procarcinogen-activating enzymes: application to mutagenicity testing. Chem Res Toxicol 4: 566-572.

Doehmer J, Dogra S, Friedberg T, Monier S, Adesnik M, Glatt HR, Oesch F (1988) Stable expression of rat cytochrome P450IIB1 cDNA in Chinese hamster cells (V79) and metabolic activation of aflatoxin B1. Proc Natl Acad Sci USA 85: 5769-5773.

Dogra S, Doehmer J, Glatt HR, Mölders H, Siegert P, Friedberg T, Seidel A, Oesch F (1990) Stable expression of rat cytochrome P-450IA1 cDNA in V79 Chinese hamster cells and their use in mutagenicity testing. Mol Pharmacol 37: 608-613.

Glatt HR, Oesch F (1977) Inactivation of electrophilic metabolites by glutathione *S*-transferases and limitation of the system due to subcellular localization. Arch. Toxicol. 39: 87-96.

Glatt HR, Oesch F (1986) Structural and metabolic parameters governing the mutagenicity of polycyclic aromatic hydrocarbons. In: de Serres FJ (ed) Chemical Mutagens, Vol. 10. Plenum Press, New York, pp. 73-127.

Glatt HR, Friedberg T, Grover PL, Sims P, Oesch F (1983) Inactivation of a diol-epoxide and a K-region epoxide with high efficiency by glutathione transferase X. Cancer Res 43: 5713-5717.

Glatt HR, Gemperlein I, Turchi G, Heinritz H, Doehmer J, Oesch F (1987) Search for cell culture systems with diverse xenobiotic-metabolizing activities and their use in toxicological studies. Mol Toxicol 1: 313-334.

Glatt HR, Pauly K, Wölfel C, Dogra S, Seidel A, Harvey RG, Oesch F, Doehmer J (1992) Stable expression of heterologous cytochromes P450 in V79 cells: mutagenicity studies with polycyclic aromatic hydrocarbons. Polycyclic Aromatic Comp: in the press.

Gonzalez FJ, Crespi CL, Gelboin HV (1991) cDNA-expressed human cytochromes P450s: new age of molecular toxicology and human risk assessment. Mutat Res 247: 113-127.

Oesch F, Glatt HR, Schmassmann HU (1977) The apparent ubiquity of epoxide hydratase in rat organs. Biochem Pharmacol 26: 603-607.

Porter TD, Beck TW, Kasper CB (1986) Complemetary DNA and amino acid sequence of rat microsomal, xenobiotic epoxide hydrolase. Arch Biochem Biophys 248: 121-129.

Thompson LH, Wu RW, Felton JS (1991) Introduction of cytochrome P450IA2 metabolic capability into cell lines genetically matched for DNA repair proficiency / deficiency. Proc Natl Acad Sci USA 88: 3827-3831.

Detection of Suspicious Aneugenic Compounds Present in Environmental Pollutants and Their Relevance in Carcinogenicity

Hadnagy, W., C. Hader and N.H. Seemayer
Medical Institute of Environmental Hygiene at the Heinrich-Heine-University Düsseldorf, Gurlittstr. 53, D-4000 Düsseldorf, FRG

Introduction

In former studies we could demonstrate that environmental pollutants such as atmospheric particulates and particulate automotive exhaust induce genotoxic effects. They were found to induce sister chromatid exchanges and chromosomal aberrations in human and rodent tissue culture cells and to enhance the number of transformed colonies of SV40-infected Syrian hamster kidney cultures (Hadnagy et al., 1986; Hadnagy and Seemayer, 1988; Hadnagy et al., 1989; Hadnagy and Seemayer, 1990; Seemayer et al., 1988). Furthermore, it was shown that these complex chemical mixtures interfere with the mitotic spindle apparatus to produce abnormal mitotic figures and mitotic arrest in mouse macrophages (line IC-21) and Chinese hamster lung cells V79 (Hadnagy and Seemayer, 1986; Hadnagy and Seemayer, 1988; Hadnagy and Seemayer, 1990; Seemayer et al., 1989). As a consequence of this effect aneuploidy and polyploidy were induced in subsequent cell divisions. There is increasing evidence that aneuploidy and events leading to aneuploidy are mechanistically involved in malignant cell transformation and tumor formation in vivo (Oshimura and Barrett, 1986). This led to the conclusion that the carcinogenic potential of particulates from the atmosphere and particulate automotive exhaust may be the result of genotoxic and epigenetic mechanisms. Until now little is known about aneugenic (aneuploidy-inducing) compounds present in environmental pollutants. Therefore, the aim of this study was the detection of suspicious compounds contributing to the observed epigenetic effect.

Materials and Methods

As suspicious aneugenic compounds the products of incomplete combustion benzo(a)pyrene (BaP) and triethyl lead (TriEL) were chosen. Furthermore, the active mammalian metabolites of benzene hydroquinone (HQ) and benzoquinone (BQ) were included in this study. As early indicators of an aneugenic activity effects on mitotic cell division were investigated. Microscopical evaluation of mitotic arrest and induction of cell division disturbances using Chinese hamster lung cells V79 was carried out as outlined previously (Hadnagy and Seemayer, 1988). Additionally, flow cytometry studies were conducted on synchronized V79 cells treated selectively in G2/M-phase of the cell cycle to elucidate mitotic inhibition. Karyotype analysis for aneuploidy induction was performed on V79 cells with preserved cytoplasm after removal of the test substances (Hadnagy and Seemayer, 1986). Carcinogenic activity was measured with the cell transformation assay using SV40-infected Syrian hamster kidney cultures as described elsewhere (Seemayer et al., 1986).

Results and Discussion

The influence on mitotic cell division by TriEL, BaP, HQ and BQ, respectively, is shown on Fig. 1. As a parameter of mitotic arrest the increase of the mitotic index and the reduction of the proportion of anaphases and telophases to metaphases (AT/M ratio) after treatment of V79 cell cultures for 16 hours is given. As can be seen, TriEL leads to a highly increased mitotic index at a concentration of 1000 nmol accompanied by a strong reduction of the AT/M ratio. For BaP a doubling of the mitotic index was found at concentrations from 10 to 20 µg/ml as compared to the control value. Concomitantly, AT/M ratio was reduced significantly. In contrast to TrieL and Bap, HQ and BQ revealed a dose-dependent decrease of mitotic index in presence of a reduced AT/M ratio. From these results it is concluded that TriEL and BaP represent potential mitotic arrestant, whereas for the benzene derivatives HQ and BQ another mechanism for the inhibition of cell replication must be considered.

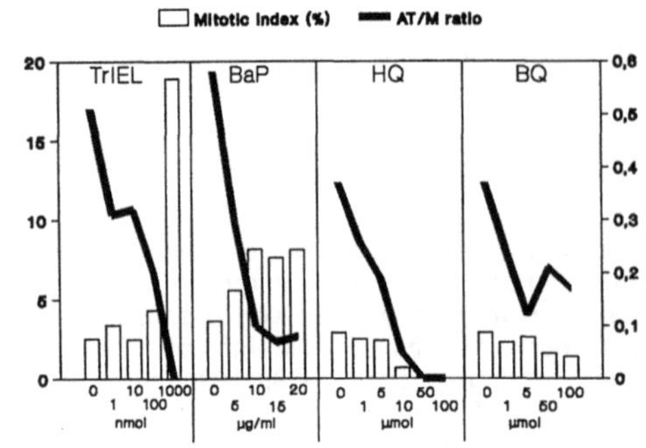

Fig.1: Mitotic arresting activity of environmental pollutants after treatment of V79 cell cultures for 16 hrs

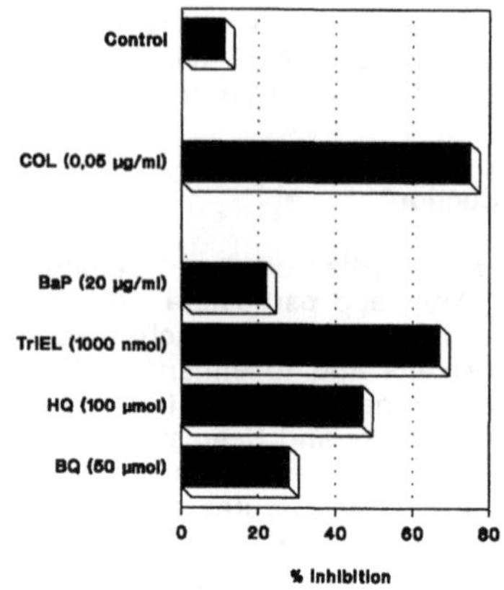

Fig.2: Inhibition of G2/M-phase of synchronized V79 cell cultures as measured by flow cytometry

For additional analysis of effects on the mitotic cell division synchronized V79 cells were treated specifically in the G2/M-phase of the cell cycle. Inhibition of G2/M-phase was measured 4 hours after treatment by flow cytometry using the fluorescence dye 2,4-diamidino-2-phenylindole (DAPI). As a positive control the known mitotic arrestant colcemid (COL) was included. In Fig. 2 the highest responses are summarized. As compared to the control all tested substances revealed an enhanced inhibition of the G2/M-phase at the given concentrations. Inhibition was highest for TriEL, reaching nearly the value of colcemid. BaP shows a twofold higher inhibition than control value. These data support the potential mitotic arresting activity of TriEL and BaP as already shown by microscopic examination of asynchrone V79 cells. Contradictive results are obtained for HQ and BQ, respectively. The increased inhibition of G2/M-phase is not in concordance with the negative response of the microscopic examination. Therefore, it is concluded that inhibition of cell cycle progression by HQ and BQ occurs in the G2-phase rather than during mitosis.

To get more information of the mitotic arresting activity the spindle-disturbing effect was investigated by assessing the appearance of mitotic abnormalities in V79 cells after exposure to the respective compounds. Abnormal mitotic figures were classified as initial C-mitoses and full C-mitoses (Schmid et al., 1989). Initial C-mitoses were characterized by chromosomes dislocated from the metaphase plate and by chromosomes which were arranged in groups or located in the center of the cell exhibiting so-called ball-metaphases. Full C-mitoses were identified by a complete scattering of contracted chromosomes in the cytoplasm similar to those induced by the known spindle poison colchicine. As shown on Fig. 3 all test substances induced initial C-mitoses in a dose-dependent manner. An increase of full C-mitoses was found with TriEL at a concentration of 1000 nmol and with BaP at a concentration of 15 and 20 µg/ml. These results reflect the appearance of mitotic arrest as indicated by an increased mitotic index and the disappearance of mitotic figures such as anaphases and telophases. Therefore, TriEL and BaP are considered to be spindle-poissoning substances with an aneugenic potential. The mitotic cell division disturbances of mostly initial C-

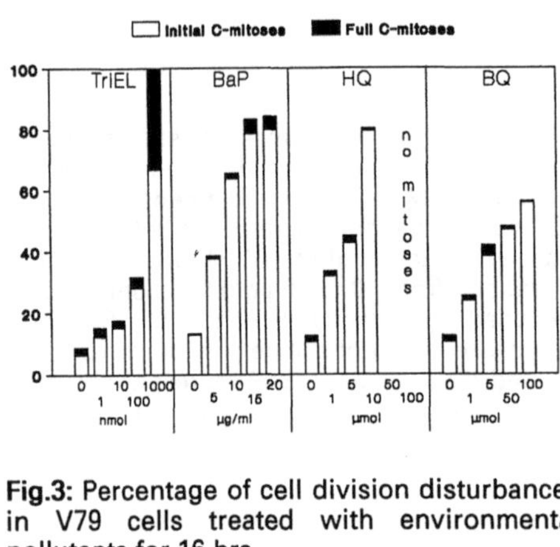

Fig.3: Percentage of cell division disturbances in V79 cells treated with environmental pollutants for 16 hrs

mitoses as found for HQ and BQ also indicate a potential aneugenic potential. These results are in good agreement with those from in vivo experiments on bone marrow cells suggesting that HQ exhibit cytotoxicity prior to mitosis resulting in abnormal metaphases (Miller and Adler, 1989). One possible explanation for this effect may be the irreversible binding to tubulin and inhibition of microtubule polymerization as had been described for HQ and BQ (Iron and Neptun, 1980; Iron and Pfeifer, 1982).

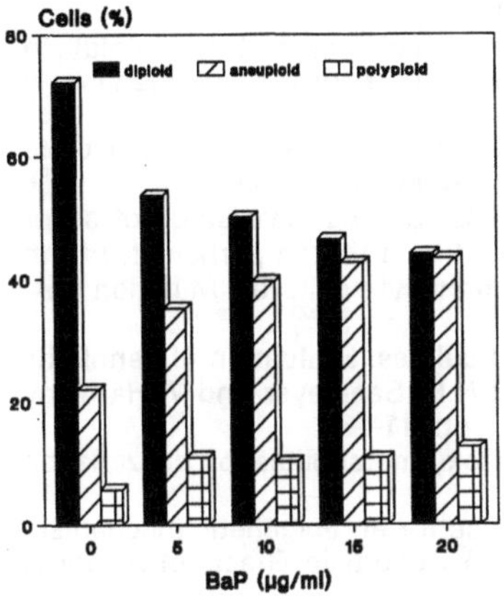

Fig.4: Induction of numerical chromosome alterations by BaP in V79 cells

BaP is considered to be one of the most important carcinogens of airborne particulates (Comittee on Biological Effects of Atmospheric pollutants, 1982), exhibiting a broad spectrum of genotoxic effects. On the other hand, also an epigenetic effect of BaP must be taken into account. As shown on Fig. 4 BaP leads to a near-diploid aneuploidy in V79 cells in a concentration dependent manner. Apparently, this effect depends on cell division disturbances as described above. The occurrence of aneuploid cells may be the result of induced initial C-mitoses. Furthermore, as compared to control an increased incidence of polyploid cells were found. To our knowledge, an aneugenic effect of BaP is only described by Benedict, 1972. In this study subcultures of Syrian hamster fetal cells were treated with the transforming dose of 10 µg/ml BaP. 72 hours after treatment 5 out of 50 metaphases were found to contain near-diploid chromosome numbers. Based on these findings it was considered that these rapidly appearing chromosomal changes may be important in cell transformation by BaP.

Our findings on in vitro cell transformation using SV 40-infected Syrian Hamster kidney cultures are presented in Fig. 5. For BaP a significantly increased number of transformed colonies was observed in the concentration range from 0.1 to 10 µg/ml. Also significant increases were registrated for TriEL (0.1 - 100 nmol), HQ (0.1 - 10 µmol) and BQ (0.01 - 1 µmol). The decreased number of transformed colonies at the highest tested concentration may be due to cytotoxic effects. Based on these results, it is assummed that above described cell division disturbances play a favorable role for the outcome of transformed colonies. Therefore, it must be considered that both genotoxic and epigenetic mechanisms contribute to the in vitro cell transformation. To what extend cell division disturbances are responsible for the cell transforming ability of TriEL, BaP, HQ and BQ has to be elucidated in further experiments. Hence, this could give an important insight in understanding cellular transformation and tumor induction.

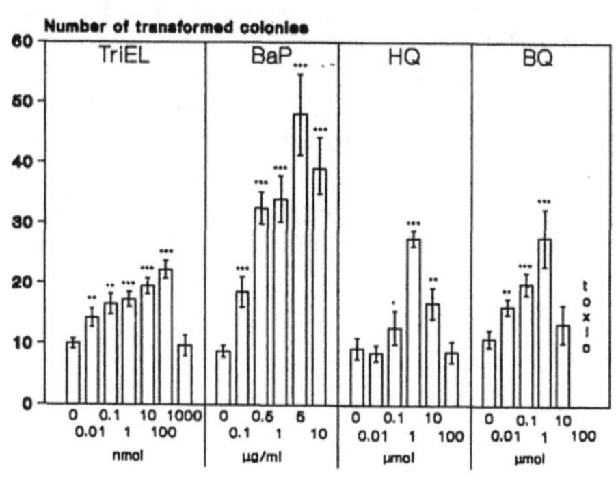

Fig.5: Number of transformed colonies in SV40-infected Syrian hamster kidney cultures treated with environmental pollutants. Bars indicate standard deviation. Significantly increased over control $*p<0.05$; $**p<0.01$; $***p<0.001$.

In conclusion, referring to the genetic and somatic consequences of aneuploidy it is recommended to include methods for the detection of aneugenic potential in routine testing of environmental agents. This may contribute to an improvement of the prediction and assessment of carcinogenic risks by environmental pollutants.

References

Benedict, W.F. (1972) Early changes in chromosomal number and structure after treatment of fetal hamster cultures with transforming doses of polycyclic hydrocarbons. J. Natl. Inst. 49: 585-590.

Committee on Biological Effects of Atmospheric Pollutants (1982) Particulate polycyclic organic matter, National Academy of Sciences, Washington.

Hadnagy, W., N.H. Seemayer and R. Tomingas (1986) Cytogenetic effects of airborne particulate matter in human lymphocytes in vitro. Mutation Res. 175: 97-101.

Hadnagy, W. and N.H. Seemayer (1986) Induction of C-type metaphases and aneuploidy in cultures of V79 cells exposed to extract of automobile exhaust particulates. Mutagenesis 1: 445-448.

Hadnagy, W. and N.H. Seemayer (1988) Cytotoxic and genotoxic effects by extract of particulate emission from a gasoline-powered engine. Environ. Mol. Mutagen. 12: 385-396.

Hadnagy, W., N.H. Seemayer, R. Tomingas and K. Ivanfy (1989) Comparative study of sister chromatid exchanges and chromosomal aberrations induced by airborne particulates from an urban and a highly industrialized location in human lymphocyte cultures. Mutation Res. 225: 27-32.

Hadnagy, W. and N.H. Seemayer (1990) Gasoline exhaust particulates: Evaluation of genotoxic and non-genotoxic effects on mammalian cell cultures. In: N.H. Seemayer and W.Hadnagy (Eds) Environmental Hygiene II, Springer Verlag Heidelberg, pp. 11-16.

Iron, R.D. and D.A. Neptun (1980) Effects of principal hydroxy-metabolites of benzene on microtubule polymerization. Arch. Toxicol. 45: 297-305.

Iron, R.D. and R.W. Pfeifer (1982) Benzene metabolites: Evidence for an epigenetic mechanism of toxicity. In: R.R. Tice, D.L. Costa and K.M. Schaich (Eds) Genotoxic Effects of Airborne Agents, Plenum Press, New York and London, pp. 241-256.

Oshimura, M and J.C. Barrett (1986) Chemically induced aneuploidy in mammalian cells: Mechanisms and biological significance in cancer. Environ. Mutagen. 8: 129-159.

Schmid, E., M. Bauchinger, K. Ziegler-Skylakakis and U. Andrae (1989) 2-Chlorobenzylidene -malonitrile (CS) causes spindle disturbances in V79 Chinese hamster cells. Mutation Res. 226: 133-136.

Seemayer, N.H., N. Manojlovic and R. Tomingas (1986) Induction of malignant cell transformation in vitro by extract and fractions of airborne particulate matter. J. Aerosol Sci. 17: 356-360.

Seemayer, N.H., W. Hadnagy and R. Tomingas (1987) The effect of automobile exhaust particulates on cell viability, plating efficiency and cell division of mammalian tissue culture cells. Sci. Total Environ. 61: 107-115.

Seemayer, N.H., N. Manojlovic, H. König and R. Tomingas (1988) Comparative investigation of carcinogenic and mutagenic activity of airborne particulate matter from polluted areas using human and rodent tissue culture cells. Ann. Occup. Hyg. 32: 247-256.

Enzyme polymorphism in human erythrocytes influences sister chromatid exchanges (SCE) induced by the fumigant methyl bromide

Hallier, E., T. Langhof, B. Gansewendt, K. Schröder, H.W. Goergens* and H. Peter
Institut für Arbeitsphysiologie an der Universität Dortmund, Ardeystraße 67, D-4600 Dortmund 1, FRG
* Arbeitsmed. Zentrum, TÜV Rheinland, Friedr.-Engels-Allee 346, D-5600 Wuppertal 2, FRG

Methyl bromide is a colourless gas with excellent qualities as an insecticide, nematocide and fungicide. It is especially useful for the fumigation of soil, due to its relatively high specific weight, which enables it to penetrate the soil very deeply. Methyl bromide is also used as a fumigant for foodstuffs, animal fodder, and various commodities. Its relevance to environmental hygiene is stressed by the fact that traces of it can be found in various foodstuffs, spices and drugs (Laug, 1941; Reeves et al., 1985). Methyl bromide is highly neurotoxic, is mutagenic and can alkylate DNA (Gansewendt et al., 1991).

Conjugation to glutathione is the major metabolic pathway for methyl bromide and the other monohalogenated methanes methyl chloride and methyl iodide. In 1971, Redford-Ellis and Gowenlock reported that glutathione conjugation of methyl chloride takes place not only in liver and kidneys, but also in human erythrocytes (Redford-Ellis and Gowenlock, 1971).

In order to investigate this metabolism, we incubated human erythrocyte cytoplasm in 22 ml head space vials with methyl chloride, methyl bromide or methyl iodide and measured the concentration of the methyl halides by gas chromatography. As a control experiment, we incubated the methyl halides with erythrocyte cytoplasm which had been boiled in order to destroy any enzyme activity.

In about 2/3 of the blood samples, a rapid elimination of methyl bromide or the other methyl halides from the gas phase was observed. These persons were called "conjugators". In about 1/3 of the population, there was no enzymatic turnover in comparison to incubations with boiled erythrocyte cytoplasm; these persons were called non-conjugators". A slight substrate disappearance was observed in the incubations with boiled erythrocyte cytoplasm due to a non-enzymatic turnover of the highly reactive substances. Simultaneously, the product S-methyl glutathione was detected by thin layer chromatography. Fig. 1 shows examples of experiments with boiled and unboiled erythrocyte cytoplasm of a conjugator and unboiled erythrocyte cytoplasm of a non-conjugator.

By now, more than 150 persons have been investigated and identified as either conjugators or non-conjugators. The two groups are completely distinct; the ability to conjugate methyl bromide or several other substrates to glutathione in erythrocytes is an either/or phenomenon. Apart from methyl bromide, methyl chloride (Peter et al. 1989) or methyl iodide (Hallier et al. 1990), the substrates ethylene oxide (Föst et al. 1991) and methylene chloride (Bolt et al. 1991) were also found to be subjected to this phenomenon. An enzymatic conjugation of these substances to glutathione in erythrocytes was observed only in humans; no such metabolism could be found in mice, rats, cattle, pigs, sheep or even rhesus monkeys (Hallier et al. 1989). The metabolism in human erythrocytes can be inhibited with sulfobromophthaleine and with chlorodinitrobenzene.

In further experiments, ^{14}C-radiolabelled methyl bromide was incubated with whole blood samples of conjugators or non-conjugators for four hours. The samples were then degassed, the different blood compartments separated and the remaining radioactivity measured by liquid scintillation counting. The distribution of bound radioactivity was found to be very different in the blood samples of the conjugators and of the non-conjugators. Whereas the conjugators had

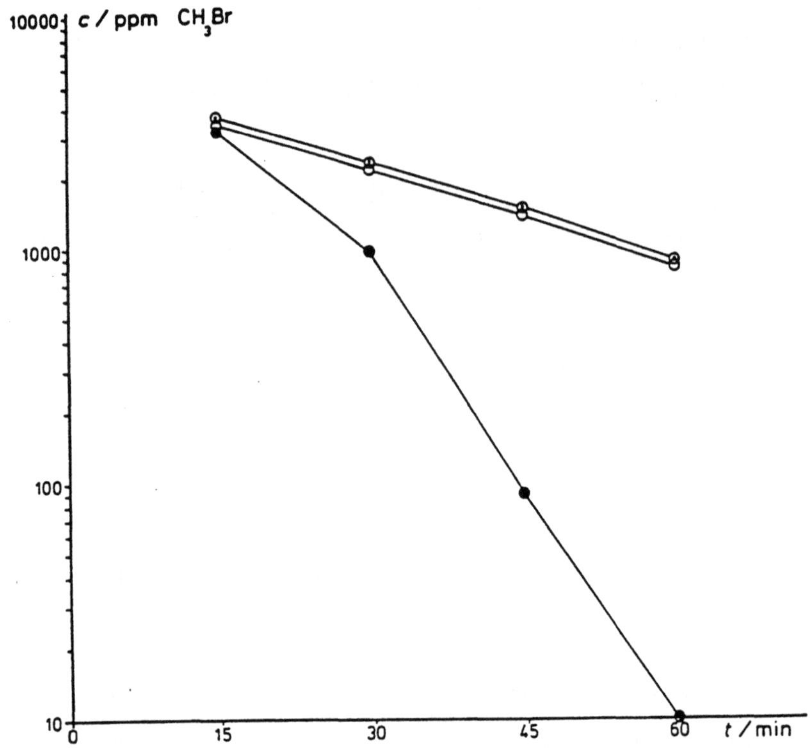

Fig. 1: Decline of methyl bromide concentration in the head space of incubation vials:
o——o control incubation with boiled erythrocyte cytoplasm
●——● incubation with erythrocyte cytoplasm of an individual conjugator
⊙——⊙ incubation with erythrocyte cytoplasm of an individual non-conjugator

most of the radioactivity in the erythrocytes (86%) and very little in blood plasma (13.8%), the non-conjugators showed the opposite picture with 36.5% in the erythrocytes and 62.9% in blood plasma. This difference in distribution implies that the enzyme polymorphism may influence the use of binding products with blood macromolecules, such as hemoglobin, for biological monitoring of a number of xenobiotics.

Finally, whole blood samples were incubated with methyl bromide for one hour. The lymphocytes were separated, cultivated in a culture medium and processed for the determination of sister chromatid exchanges according to Perry and Wolff (1974). The SCE rate increased significantly in the case of each non-conjugator, whereas incubation with methyl bromide had practically no effect on the SCE rate in the case of the conjugators, as shown in fig. 2. The difference between the two groups of conjugators and non-conjugators is especially striking when the relative increase of SCE due to methyl bromide exposure is compared, as in fig. 3. Obviously, the ability to conjugate methyl bromide to glutathione in erythrocytes seems to be a protective factor against this cytogenetic effect in lymphocytes.

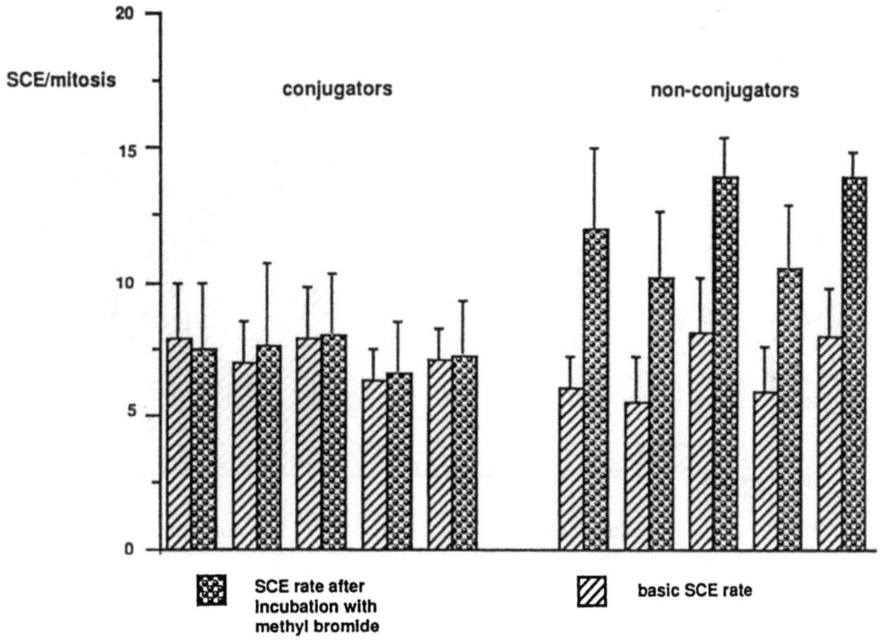

Fig. 2: SCE rate in the lymphocytes of conjugators and of non-conjugators after incubation in vitro with methyl bromide. SCE rate given per mitotic cell.

Recent results from our laboratory show that methyl bromide must be regarded as a genotoxic, directly alkylating carcinogen (Gansewendt et al. 1991). The data presented here on the induction of sister chromatid exchanges in vitro are in accordance with this view. An individually different susceptibility of humans in terms of a carcinogenic effect of methyl bromide, and possibly other related compounds, is highly probable. Furthermore, biological monitoring data is influenced by the individual metabolic disposition. The enzyme polymorphism in human erythrocytes necessitates new strategies for biological monitoring of persons occupationally exposed to methyl bromide. Interindividual differences in susceptibility to possible toxic or carcinogenic effects of the substance due to environmental pollution and residues in agricultural products must also be taken into account.

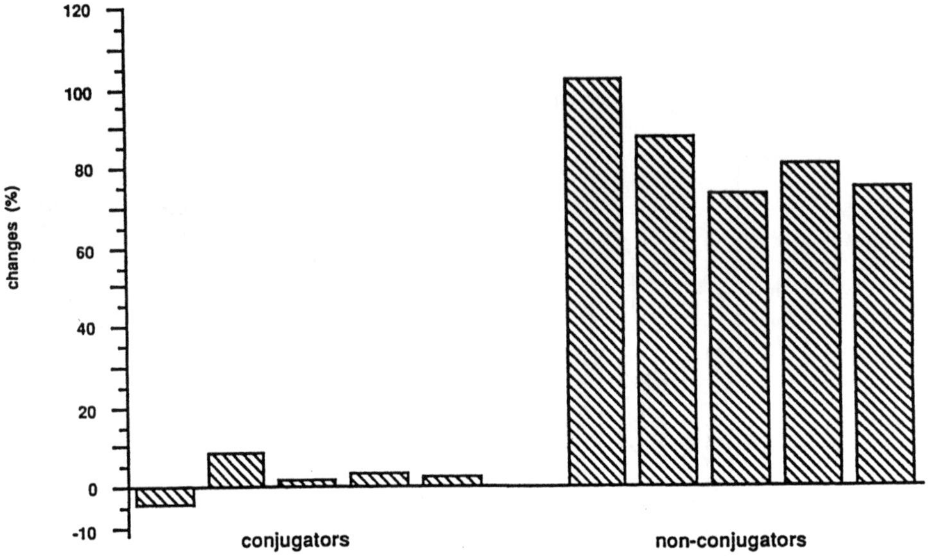

Fig. 3: Relative change of the SCE rate in the lymphocytes of conjugators and non-conjugators following incubation of whole blood samples with methyl bromide in vitro. 0 = basic individual SCE rate in lymphocytes not exposed to methyl bromide.

References

Bolt HM, Thier R, Langhof T, Hallier E, Peter H (1991) Risk assessment and species extrapolation for the carcinogenic potency of methylene chloride in humans. The 1991 EUROTOX Congress, Book of Abstracts, Maastricht, p. 38

Föst U, Hallier E, Ottenwälder H, Bolt HM, Peter H (1991) Distribution of ethylene oxide in human blood and its implications for biomonitoring. Hum Exp Toxicol 10: 25-31

Gansewendt B, Foest U, Xu D, Hallier E, Bolt HM, Peter H (1991) Formation of DNA adducts in F-344 rats after oral administration or inhalation of ^{14}C-methyl bromide. Fd Chem Toxic 29: 557-563

Hallier E, Deutschmann S, Gansewendt B, Peter H (1989) Species differences in metabolism of monohalogenated methanes. V International Congress of Toxicology, Abstracts, Taylor & Francis, Brighton, p. 39

Hallier E, Deutschmann S, Reichel C, Bolt HM, Peter H (1990) A comparative investigation of the metabolism of methyl bromide and methyl iodide in human erythrocytes. Int Arch Occup Environ Health 62: 221-225

Lang EP (1941) Bromide residues in foodstuffs. Volatile and non-volatile residues following experimental exposure to methyl bromide. Ind Eng Chem 33: 803-830

Perry P, Wolff S (1974) New Giemsa method for the differential staining of sister chromatids. Nature 251: 156-158

Peter H, Deutschmann S, Reichel C, Hallier E (1989) Metabolism of methyl chloride by human erythrocytes. Arch Toxicol 63: 351-355

Redford-Ellis M, Gowenlock AH (1971) Studies on the reaction of chloromethane with human blood. Acta Pharmacol Toxicol 30: 36-48

Reeves RG, McDaniel CA, Ford JH (1985) Organic and inorganic bromide residues in spice fumigated with methyl bromide. J Agric Food Chem 33: 780-783

Genotoxicity of Fe(III)-NTA and its Interaction with other Carcinogenic Metal Compounds

Hartwig, A., R. Schlepegrell, J. Kuhlmann, J. Benters, D. Beyersmann
University of Bremen, Department of Biology and Chemistry, Biochemistry Group, D-2800 Bremen 33, FRG

Introduction

As an essential element in humans, iron is involved in many biochemical reactions; however, it is also known to catalyse the production of highly reactive oxygen species like hydroxyl radicals and superoxide anions, which produce DNA damage in cell free systems (Inoue and Kawanishi, 1987). Since in intact cells iron is mainly bound to storage proteins like ferritin and cellular defense systems exist to detoxify free oxygen radicals, toxic reactions in vivo are prevented to a high degree (for review see Halliwell and Gutteridge, 1984).
This raises the question whether elevated levels of iron, occuring for example via occupational exposure to iron dusts might lead to adverse health effects. Even though epidemiological studies point towards a higher incidence of lung cancer among workers exposed to iron oxide, in all cases there had been a simultaneous exposure to other potentially carcinogenic agents (other metal compounds, polycyclic aromatic hydrocarbons). Animal studies provide some evidence for a cocarcinogenic potential of iron oxide, which has been shown to enhance the incidence of cancer induced by benzo(a)pyrene in hamster (reviewed by Elinder, 1986).
By using mammalian cells in culture (V79 Chinese hamster cells and human HeLa cells) we investigated whether extracellularly applied iron is able to cause DNA damage and whether there are interactions of iron with other mutagenic and/or carcinogenic metal compounds (chromium, nickel, cobalt, cadmium), where oxygen free radicals are also involved in their DNA damaging activity. To provide the bioavailability of iron, ferric Fe(III)-NTA was chosen, which has been demonstrated to be taken up by cells in culture (Planas-Bohne and Duffield, 1988; Sturrock et al., 1990). Recent investigations show furthermore, that this compound causes renal carcinomas in rats (Ebina et al., 1986); Umemura et al. (1990) found the formation of 8-hydroxyguanosine in DNA isolated from rat kidney cells after intraperitoneal administration of Fe(III)-NTA.

Material and Methods

<u>Cell culture.</u> Chinese hamster V79 cells were grown in minimal essential medium, α modified, HeLa cells in MEM, supplemented with 10 and 5 % fetal calf serum, respectively. The cells were incubated at 37°C with 5 % CO_2 in air and 100 % humidification.
<u>Chemicals.</u> To prepare Fe(III)-NTA, ferric nitrate was dissolved in 1 N HCl solution, and NTA disodium salt in bidistilled water. Both solutions were mixed to achieve a molar ratio of 1:2 of Fe:NTA. The pH was adjusted to 7.4 by the addition of $NaHCO_3$.

Colony forming ability. 300 V79 cells were allowed to attach in 60 mm-dishes for 5 h and treated with the metal compounds for 5 h in MEM α without serum. After 5 days, colonies were fixed, stained and counted.
Nucleoid Sedimentation. The procedure is based on the method of Cook and Brazell (1976) with modifications described by Yew and Johnson (1979) and is described in detail elsewhere (Hartwig und Beyersmann, 1989a).
Sister-chromatid exchange (SCE) test. After incubation of V79 cells with Fe(III)-NTA for different times, 5 μg/ml bromodeoxyuridine were added for 24 h. The further procedure of harvesting and staining is described elsewhere (Hartwig and Beyersmann, 1989b).

Results

DNA strand breaks. DNA strand breaks were determined by the nucleoid-sedimentation technique in HeLa cells. After 4 h incubation with Fe(III)-NTA, there is a dose-dependent increase of DNA strand breaks at concentrations between 100 and 1000 μM, with the highest number induced at 500 μM (Fig. 1a). These breaks were generated within one hour and their number increased within 4 h (Fig. 1b). After 24 h of incubation, however, no breaks could be detected any more (data not shown), indicating their repair.

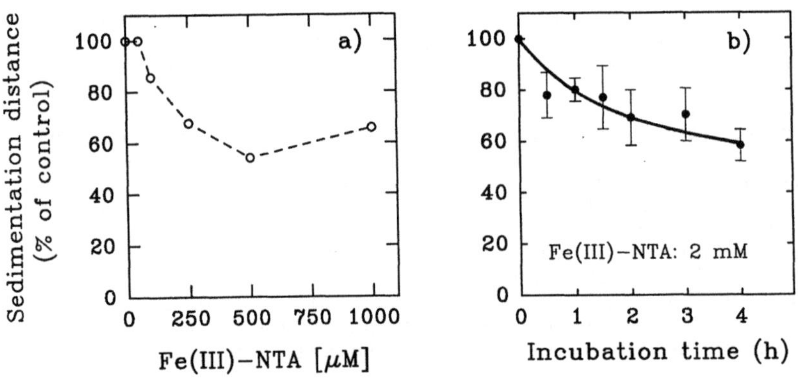

Fig. 1: Induction of DNA strand breaks by Fe(III)-NTA in HeLa cells, indicated by a decrease in sedimentation distance.

Sister-chromatid exchanges. There is a time-dependent increase in the number of sister-chromatid exchanges by 250 μM Fe(III)-NTA in V79 cells (Table 1). However, the kinetics of formation were quite different compared to DNA cleavage: the induction was low and not significant after 24 h of incubation, but increased significantly up to 72 h of incubation. No induction of SCE's were observed after incubation with 500 μM Na-NTA for the same times (data not shown).

Table 1:
Induction of sister-chromatid exchanges by 250 μM Fe(III)-NTA

Treatment time (h)	SCE's/cell ± SD	
0	8.9 ± 4.1	
24	9.7 ± 3.3	n.s. [a]
48	12.6 ± 4.3	$p < 0.01$ [a]
72	14.7 ± 4.9	$p < 0.001$ [a]

25 mitotic cells were counted for each value; n.s.: not significant; a: determined by t-test

Cytotoxicity after combined action of Fe(III)-NTA and nickel(II). As parameters for cytotoxicity both growth inhibition and colony-forming ability (CFA) were investigated after either single or combined treatments of V79 cells for 5 h with Fe(III)-NTA and nickel(II). While nickel(II) alone lowered the CFA slightly at the highest concentration tested (1 mM), and 100 μM Fe(III)-NTA were also nontoxic, there was a pronounced increase in cytotoxicity after the combined treatment (Fig. 2). Corresponding effects were observed for growth inhibition (data not shown).

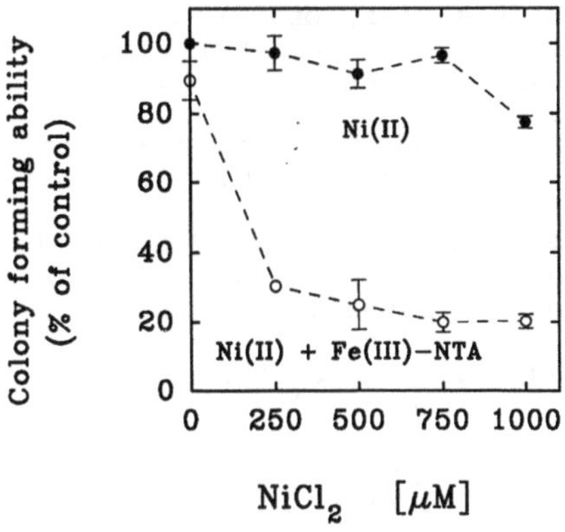

Fig. 2:
Colony-forming ability after single and combined treatment of V79 cells with $NiCl_2$ and Fe(III)-NTA.

Discussion

Fe(III)-NTA has been shown to be genotoxic in V79 and HeLa cells at two endpoints. The different kinetics observed for DNA strand break formation and induction of sister-chromatid exchanges suggest the involvement of separate mechanisms of genotoxic action of this iron complex. One possible explanation could be the generation of DNA strand breaks by lipid peroxidation, which has been shown to be evoked by Fe(III)-NTA (Hamazaki et al., 1989), while the formation of sister-chromatid exchanges could possibly require site-specific DNA damage by free radicals generated by DNA-bound iron. Interestingly, the same long incubation times were needed for the induction of mutations at the hprt-locus in V79 cells (Nakatsuka et al., 1990). Whether the genotoxic action of iron is restricted to the Fe(III)-NTA complex or also will hold for more physiological ligands has to be further elucidated.
The increase in cytotoxicity after combined treatment of V79 cells with Fe(III)-NTA and nickel(II) could be due to either direct

interactions in the formation of oxygen free radicals or else the inhibition of cellular defense systems. A partial inactivation of catalase and glutathione peroxidase as well as a reduction of the cellular glutathione content has been shown by Misra et al. (1990) in kidney and liver cells isolated from rats after i.p. administration of nickel(II).

Acknowledgement

This work was supported by grant No. 01 HK 9369 of the Bundesminister für Forschung und Technologie, Bonn.

References

Cook PR, Brazell IA (1976) Detection and repair of single-strand breaks in nuclear DNA. Nature 263: 679-682

Ebina Y, Okada S, Hamazaki S, Ogino F, Li J-I, Midorikawa O (1986) Nephrotoxicity and renal cell carcinoma after use of iron- and aluminum-nitrilotriacetate complexes in rats. J. Natl. Cancer Inst. 76: 107-113

Elinder CG (1986) Iron. In: Friberg L, Nordberg GF, Vouk VB (Eds) Handbook on the Toxicology of Metals, Elsevier, pp 276-297

Halliwell B, Gutteridge JMC (1984) Oxygen toxicity, oxygen radicals, transition metals and disease. Biochem. J. 219: 1-14

Hamazaki S, Okada S, Li J-L, Toyokuni S, Midorikawa O (1989) Oxygen reduction and lipid peroxidation by iron chelates with special reference to ferric nitrilotriacetate. Arch. Biochem. Biophys. 272: 10-17

Hartwig A, Beyersmann D (1989a) Enhancement of UV-induced mutagenesis and sister-chromatid exchanges by nickel ions in V79 cells: evidence for inhibition of DNA repair. Mutation Res. 217: 65-73

Hartwig A, Beyersmann D (1989b) Comutagenicity and inhibition of DNA repair by metal ions in mammalian cells. Biol. Trace Elem. Res. 21: 359-365

Inoue S, Kawanishi S (1987) Hydroxyl radical production and human DNA damage induced by ferric nitrilotriacetate and hydrogen peroxide. Cancer Res. 47: 6522-6527

Misra M, Rodriguez RE, Kasprzak KS (1990) Nickel induced lipid peroxidation in the rat: correlation with nickel effect on antioxidant defense systems. Toxicology 64: 1 - 17

Nakatsuka S, Tanaka dH, Namba M (1990) Mutagenic effects of ferric nitrilotriacetate (Fe-NTA) on V79 Chinese hamster cells and its inhibitory effects on cell-cell communication. Carcinogenesis 11: 257-260

Planas-Bohne F, Duffield J (1988) Factors influencing the uptake of iron and plutonium into cells. Int. J. Radiat. Biol. 53: 489-500

Sturrock A, Alexander J, Lamb J, Craven CM, Kaplan J (1990) Characterization of a transferrin-independent uptake system for iron in HeLa cells. J. Biol. Chem. 265: 3139-3145

Umemura T, Sai K, Takagi A, Hasegawa R, Kurokawa Y (1990) Formation of 8-hydroxyguanosine (8-OH-dG) in rat kidney DNA after intraperitoneal administration of ferric nitrilotriacetate (Fe-NTA). Carcinogenesis 11: 345-347

Inhibition of Gap-Junctional Intercellular Communication in V79 Cells by Airborne Particulate Matter

Heussen, G.A.H. and G.M. Alink
Department of Toxicology, Agricultural University, Bomenweg 2, 6703 HD Wageningen, The Netherlands

Introduction

For more than a decade it is known that extracts of outdoor and indoor airborne particulate matter (APM) are genotoxic. However, little attention has been paid to non-genotoxic mechanisms by which APM could play a role in the development of (lung)cancer. As there is increasing evidence that inhibition of gap junctional intercellular communication (IC) is an important mechanism in tumor promotion (for a review see Yamasaki, 1990), we investigated in the present study if extracts of APM are able to inhibit IC in V79 cells. To compare dose-effect relationships for inhibition on IC with initiating activity of extracts of APM, mutagenicity in Salmonella typhimurium TA98 was also studied.

Methods

APM was collected in the winter in Wageningen, a small rural town without important sources of industrial pollution. Sampling techniques and the degree of pollution have been described in detail previously (Van Houdt et al., 1984; 1986; 1987). After sampling filters were soxhlet extracted with methanol for 8 hours, extracts were evaporated to dryness, and the residues were dissolved in DMSO. Two indoor pollution sources were investigated. First, sampling was performed in living rooms polluted by wood combustion. Therefore two samples were taken per livingroom. One week no woodcombustion took place (control sample), while in the other week an open fire place was burning for approximately 4 hours/day (experimental sample). Secondly, sampling took place in a room (about 18 m^3) in our institute which was polluted by smoking cigarettes (60 cigarettes/week, experimental sample). A control sample was also taken in this room. At the same time outdoor sources were investigated. Therefore two extracts representing a heavily polluted sample and a relatively clean sample were prepared.

Extracts were tested for mutagenicity in the Salmonella microsome assay, as described by Ames et al. (1975), with minor modifications (Van Houdt et al., 1986).

The cytotoxicity of the extracts was determined by a cloning efficiency assay. Therefore 200 cells/dish were plated, and after the cells were attached, APM extracts were added. After incubation, cells were washed once with Hank's Balanced Salt Solution (HBSS) and new medium was added (Ham's F10 supplemented with 10% fetal bovine serum and 50 μg/ml gentamycin). Cells were cultured in a humidified atmosphere containing 5% CO_2 at 37°C. After 7 days the colonies were fixed, stained with Giemsa and scored.

For the measurement of inhibition of IC 10^6 cells were plated in 35 mm dishes. Next day a confluent monolayer was obtained and cells were exposed to extracts of APM. Before microinjection cells were washed once with HBSS and new medium was added (as above, in addition HEPES was added to a final concentration of 20 mM, pH 7.4). IC was determined after microinjection of a 10% Lucifer Yellow solution in 0.33 M lithium chloride in a single cell by means of a glass capillary tip. At least 20 cells/dish were injected with the use of a vertical microinjection system coupled to a combined light and fluorescent phase contrast microscope (Olympus IMT-2, Japan). Glass capillary tips were prepared with an automatic magnetic puller (Narishige, Tokyo, Japan) with a tip diameter of about 1 micron. Ten to fifteen minutes after the first injection the number of fluorescent (=communicating) cells was counted.

Results and Discussion

The mutagenicity data show that for both the outdoor and the indoor sources investigated, the polluted sample was more mutagenic than the corresponding control sample (Fig. 1). In all cases the effect in the presence of S9 was higher than without S9. These data agree well with previous studies conducted at our department (Van Houdt et al., 1984; 1986; 1987).

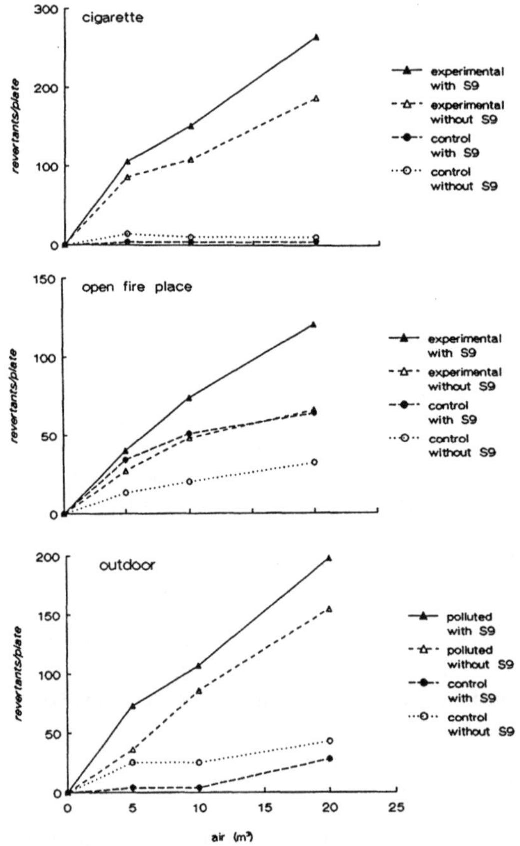

Fig. 1. Mutagenicity of extracts of airborne particulate matter in TA 98 with or without rat liver Aroclor S9. Concentrations correspond to the volume of air sampled. Data are the mean ± SD from triplicate determinations.

The cytotoxicity experiments showed that incubation for 2 hours with the open fire place sample slightly reduced cloning efficiency at the two highest doses tested. Also incubation for 24 hours was slightly cytotoxic. The cigarette smoking sample was cytotoxic only at the highest dose level tested. The outdoor sample was not cytotoxic at all (data not shown).

The data on inhibitory activity on IC of the different APM extracts are shown in Fig. 2 and 3. In all cases the polluted samples inhibited IC more than the corresponding control samples. Dose- as well as time-dependent inhibitions at non-cytotoxic concentrations were found. The data on inhibition of IC and the mutagenicity data were positively correlated ($P<0.005$, Spearman Rank Correlation).

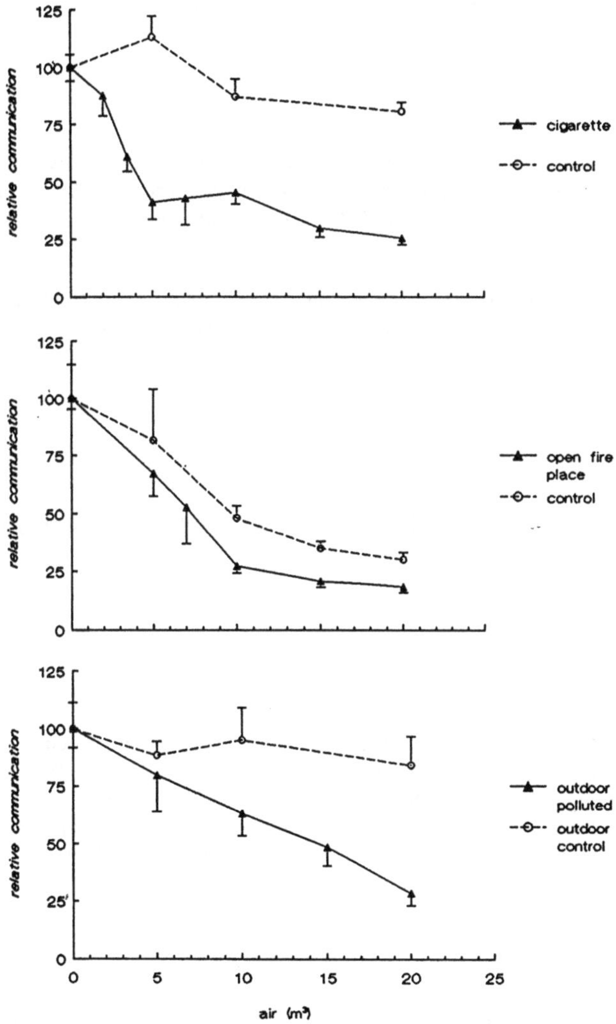

Fig. 2. Dose-dependent inhibition of intercellular communication in V79 cells by extracts of airborne particulate matter. V79 cells were incubated for 2 hours with extracts of APM. Concentrations correspond to the volume of air sampled. Results are expressed as percentage of the control and are the mean±SEM from 6-20 dishes.

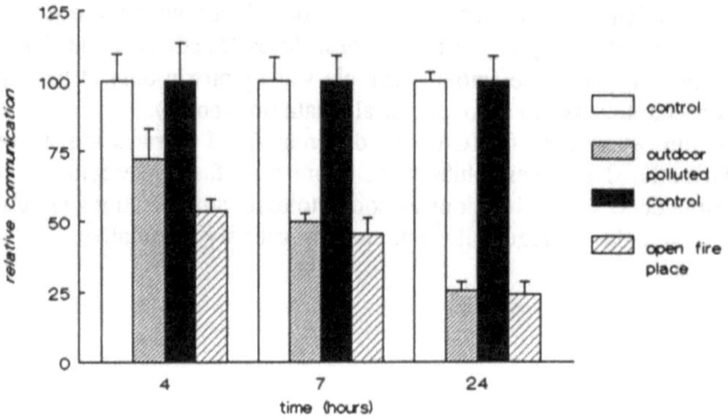

Fig. 3. Time-dependent inhibition of intercellular communication in V79 cells by extracts of airborne particulate matter. V79 cells were incubated with extracts of APM corresponding to 4 m^3 of air. Control was treated with 0.2% DMSO. Results are the mean$\pm$SEM from 6 dishes.

In conclusion, this is the first study which shows that extracts of APM, in addition to the mutagenic activity which is already known for many years, also may have a tumor promoting potency.

Acknowledgement

This study was sponsored by a grant from the Dutch Ministry of Housing, Physical Planning and The Environment.

References

Ames,B.N., McCan,J., Yamasaki,E. (1975) Methods for detecting carcinogens and mutagens with the Salmonella/mammalian microsome mutagenicity test. Mutat. Res. 31:347-364

Houdt van,J.J., Jongen,W.M.F., Alink,G.M. and Boleij,J.S.M. (1984) Mutagenic activity of airborne particles inside and outside homes. Environ. Mutagen., 6:861-869.

Houdt van,J.J., Daenen,C.M.J., Boleij,J.S.M. and Alink,G.M. (1986) Contribution of wood stoves and fire places to mutagenic activity of airborne particulate matter inside homes. Mutat. Res. 171:91-98.

Houdt van,J.J., Alink,G.M. and Boleij,J.S.M. (1987) Mutagenicity of airborne particles related to meteorological and air pollution parameters. Sci. Total Environ., 61:23-36.

Yamasaki,H. (1990) Changes of gap junctional intercellular communication during multistage carcinogenesis. In Mendelsohn,M.L. and Albertini,R.J. (Eds.) Mutation and the Environment, part D, Carcinogenesis, Wiley-Liss, Inc., New York, pp. 153-164.

THE METABOLISM OF CHRYSENE AND BENZO[A]PYRENE IN HAMSTER, RAT AND HUMAN LUNG CELLS IN CULTURE

Jacob, J.[1], G. Grimmer[1], U. Mohr[2], M. Emura[2], M. Riebe-Imre[2] and G. Raab[1]

1 Biochemical Institute for Environmental Carcinogens, W-2070 Großhansdorf (FRG)
2 Institute for Experimental Pathology, Medical School, W-3000 Hannover (FRG)

INTRODUCTION

A precondition to replace animal experiments on the carcinogenicity of ecotoxicologically relevant compounds by in vitro studies is a comparable metabolism of the latter in both systems. To this end we have studied the metabolism of chrysene and benzo[a]pyrene in hamster, rat and human lung cell cultures.

MATERIAL AND METHODS

Incubations were stopped with acetone, treated with sulfatase/glucuronidase and then extracted with ethyl acetate to give the unconverted substrate and free primary metabolites along with those metabolites originally present as sulfates and glucuronides. They were further purified by Sephadex LH 20 chromatography, converted into the OTMS-ethers and analyzed by capillary GC or HPLC (Jacob et al., 1982; Jacob and Grimmer, 1987).

RESULTS AND DISCUSSION

Chrysene metabolism in various lung cell cultures

Incubation of chrysene with mixed and epithelial hamster lung cells, human epithelial lung cells and rat bronchial and tracheal epithelial cells resulted in qualitatively similar metabolite profiles originating from oxidation at the 1,2- and 3,4-position. However, significant quantitative differences were observed for the various species with regard to both the regiospecific oxidation and the dihydrodiol formation (Fig. 1 and 2). Mixed cultures and early

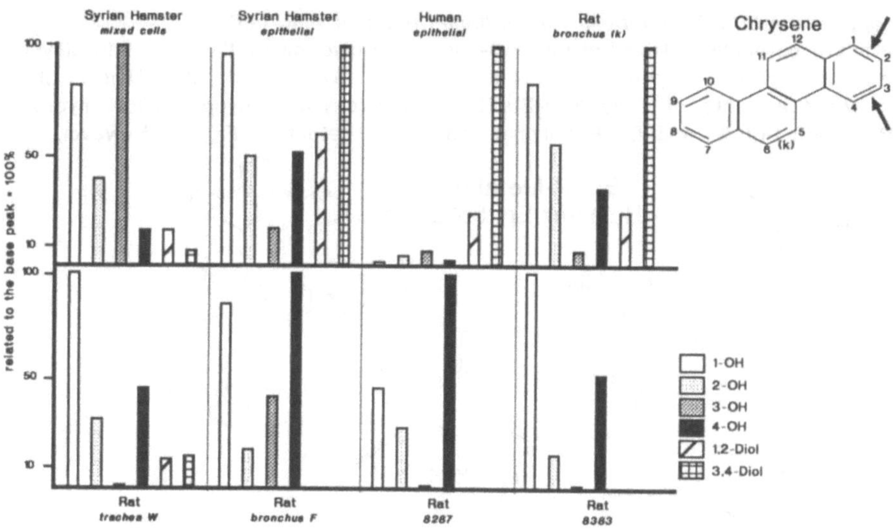

FIGURE 1 Metabolite profile of chrysene formed in lung cell cultures from human, hamster and rat (in per cent of the main metabolite).

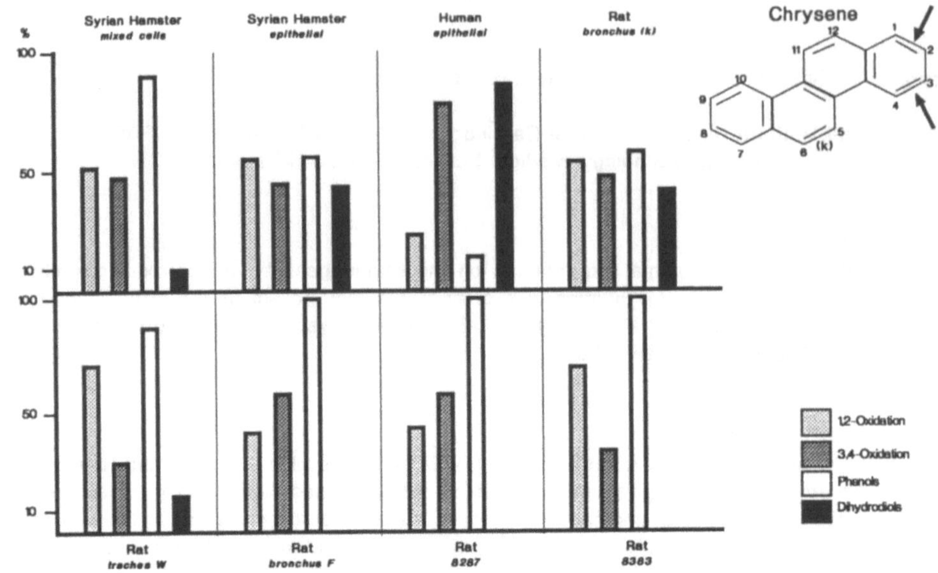

FIGURE 2 Phenol and dihydrodiol formation and regiospecific oxidation of chrysene in human, rats and hamster lung cells (in per cent of the total metabolites extractable with ethylacetate)

passages convert PAH into phenols rather than into dihydrodiols whereas epithelial cells preferentially form dihydrodiols. The amounts of the proximate carcinogen of chrysene (1,2-dihydroxy-1,2-dihydrochrysene), however, yielded comparable final concentrations in all cell cultures investigated.

Benzo[a]pyrene (B[a]P) metabolism in epithelial hamster lung cells

B[a]P is predominantly oxidized at the 9,10-position to give mainly the trans-9,10-dihydrodiol confirming previously published data (Kimura et al., 1977; Selkirk, 1978; Nemoto et al., 1979). Minor amounts of 3-hydroxy-B[a]P, 4,5-dihydroxy-4,5-dihydro-B[a]P, 7-hydroxy-B[a]P, 6-methyl-B[a]P and B[a]P-quinones were also detected (Fig. 3). However, only

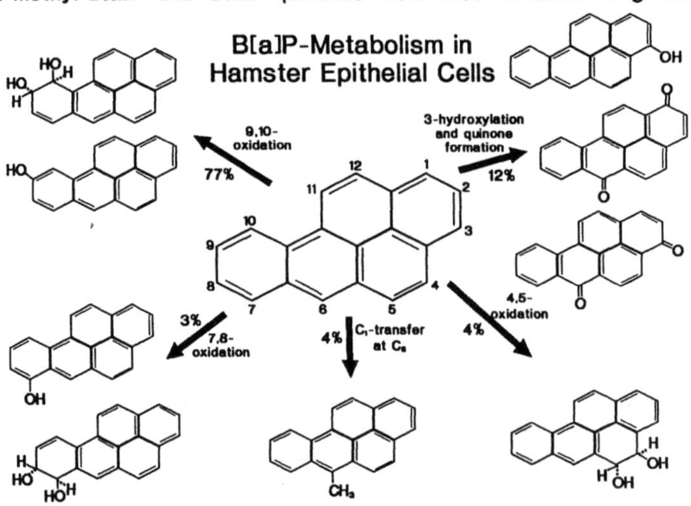

FIGURE 3 B[a]P-metabolism in hamster lung cells

traces of the 7,8-dihydrodiol, the proximate carcinogen of B[a]P, were formed.

Dependence of the metabolite profile of chrysene and B[a]P on the number of passages during cultivation

The metabolic stability of the lung cell cultures has been studied over 150 passages with chrysene and B[a]P as substrates. Results are presented in Figs. 4 and 5 in which the amounts of metabolites have been related to the cell numbers and the DNA-content. From this investigation it becomes obvious that the PAH-profile is not invariant and that the balance of phenols and dihydrodiols is altered during long-term cultivation. For both of the substrates there is a maximum of the phenol formation at the 80th passage and of the dihydrodiol formation at the 100th passage which coincides with the climax of the total turnover rate.

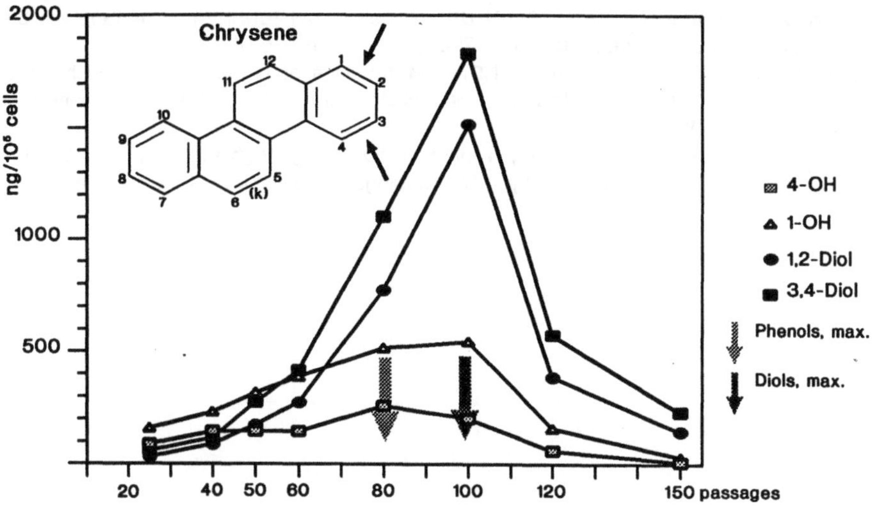

FIGURE 4 Chrysene metabolism in epithelial hamster cells in dependence on the number of passages

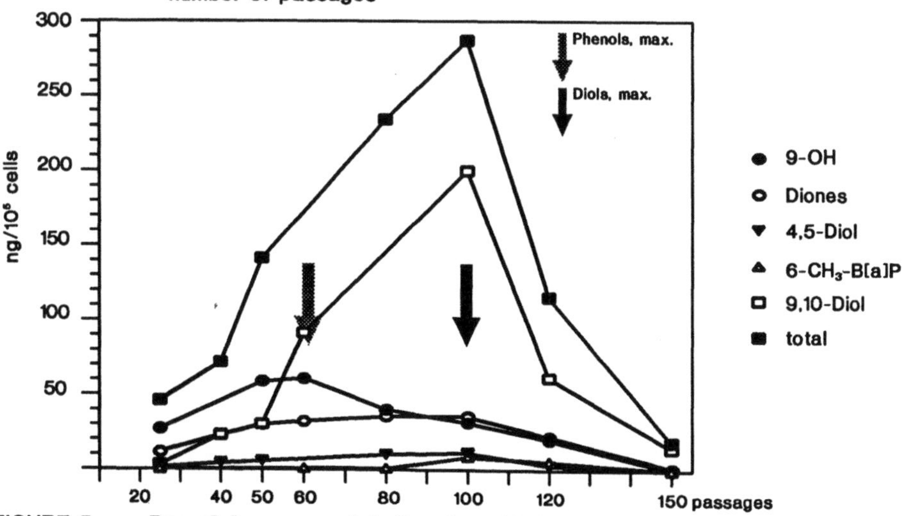

FIGURE 5 Benzo[a]pyrene metabolism in epithelial hamster cells in dependence on the number of passages

This may indicate that the epoxide hydrase is expressed more pronounced in later stages of cultivation which appears to be important for the selection of passages the metabolism of which resembles most that of complete animal test systems.

REFERENCES

Jacob J, Schmoldt A and Grimmer G (1982) Formation of carcinogenic and inactive chrysene metabolites by rat liver microsomes of various monooxygenase activities. Arch. Toxicol. 51: 255-265

Jacob J and Grimmer G (1987) Capillary gas chromatographical analysis and mass spectrometric identification of polycyclic aromatic hydrocarbon metabolites from biological materials. Rev. Anal. Chem 9: 49-89

Kimura T, Kodama M and Nagata C (1977) Differences in benzo[a]pyrene metabolism between lung and liver homogenates. Biochem Pharmacol. 26: 671-674

Nemoto N, Hirakawa T and Takayama S (1979) Metabolism of benzo[a]pyrene in hamster embryo cells. Effect of the concentration of benzo[a]pyrene on its metabolism. Chem.-Biol. Interactions, 25: 177-188

Selkirk JK (1978) Analysis of Benzo[a]pyrene Metabolism by High-Pressure Liquid Chromatography, in ' Adv. in Chromatography' (J.C. Giddings, E. Grushka, P.R. Brown, eds) Vol. 16: 1-36

ANALYSIS OF TRANSFORMING SEQUENCES IN NEOPLASTIC SYRIAN HAMSTER FETAL CELLS INDUCED BY DIVERSE CHEMICAL CARCINOGENS.

V. Notario[1], N. Popescu[2], and J.A. DiPaolo[2].
[1]Department of Radiation Medicine, Georgetown University School of Medicine, Washington, DC 20009, [2]Laboratory of Biology, NCI, Bethesda, MD 20892.

INTRODUCTION

Neoplastic transformation occurs with chemical carcinogens that interact with DNA such as carcinogenic polycyclic hydrocarbons, inorganic metals, N-Acetoxy-florenylacetamide and its derivatives and alkylating agents (1). Transformation resulting in malignancy also occurs with non-mutagenic agents including SO_2^- (sodium bisulfite). With chemicals alone or in combination with X-irradiation the incidence of transformed hamster colonies fits a one hit curve indicating that the phenomenon is inductive (2).

Analysis of 22 neoplastic cell lines, each initiated from a discrete transformed colony obtained after insult of cells with a carcinogen indicated that activation of a known oncogene was a rare event. Even though the battery of carcinogens used was extensive only N-ras point mutations were noted and that only occurred in 3 of 12 NIH 3T3 transfection positive lines, all initiated by sodium bisulfite.

Sulfur dioxide and its salts are used as preservatives in foods and medicine because of their anti-oxidant and/or anti-microbial properties. It is ubiquitous as a diesel fuel and ash pollutant. Epidemiological evidence links atmospheric sulfur dioxide with lung cancer (3) and in inhalation studies it was a cocarcinogen with benzo[a]pyrene. Bisulfite is formed from a colorless gas, SO_2^-, which hydrates upon solution to sulfurous acid and dissociates to form sulfite ions.

RESULTS AND DISCUSSION

At neutral pH sodium bisulfite is not mutagenic to either mammalian or bacterial cells but at neutral pH, it induced dose dependent, morphologic transformation of Syrian hamster fetal cells (HFC) and tumor producing lines (4); bisulfite does not induce detectable DNA damage in HFC. Alone or in combination with UV it had no effect on excision or post-replication repair; and strand breaks in parental or nascent DNA were not observed in alkaline sucrose gradients (5). Furthermore, bisulfite DNA adducts were not detected by ^{32}P-postlabeling analysis of DNA derived from HFC treated with 20 mM bisulfite (K. Randerath personal communication). However, the most that was observed was an inhibition of DNA replication probably by reducing functioning replicons. The inhibition of semi-conservative DNA synthesis by bisulfite was measured by density shift (5).

The bisulfite transformed hamster cell lines formed anchorage independent colonies in agarose and produced colonies in agarose and progressively growing subcutaneous fibrosarcomas when injected into nude (nu/nu) mice. Cytogenetic G-band analysis of metaphase and prophase chromosomes of seven neoplastic cell lines revealed structural alterations and aneuploidy (6). The chromosomal constitution of the lines varied from near diploid to near tetraploid to those containing both tetraploid and octoploid components. The number of abnormal chromosomes per line varied from 1 to 6. Structural abnormalities of chromosomes X and 1 occurred. Monosomy of 13 was the most common, found in three lines. The break point on X was at q5 which is frequently observed with other carcinogens and may result in loss of genes necessary for the maintenance of the normal phenotype.

Because of minimal effect of DNA metabolism, polypeptide differences among various bisulfite clones, a benzo[a]pyrene transformed hamster line and normal fibroblasts were analyzed by 2D-gel electrophoresis (7). Cells were pulse labeled with ^{14}C amino acids, lysed, and total cellular proteins solubilized and eventually layered in isoelectric focusing. Seven different lines were analyzed. The transformed bisulfite lines, however, differ from controls in that all neoplastic lines have four qualitative changes: two polypeptides have shifted, one is missing and a new peptide is observed (7). Transformed lines also had consistent quantitative changes greater than 2-4-fold in 21 different proteins. The qualitative peptide changes found in the bisulfite induced malignant lines were similar to those seen in a benzo(a)pyrene (BP) induced malignant line. This suggests that there is a convergence of pathways in the carcinogenesis process which is independent of the initiating agent be it DNA damaging or not.

Chemical carcinogens cause alteration of the function of normal cellular genes. Historically, NIH/3T3 cells have been used to identify oncogenic DNA sequences from transformed mouse, rat or guinea pig cells obtained after chemical carcinogens; in most cases an activated ras gene was involved (8). Thus, it is important to determine the genetic alterations that contribute to the oncogenic transformation induced by carcinogens in the in vitro Syrian hamster model. Of the 22 hamster tumor cell DNAs transfected to NIH/3T3 cells with the calcium phosphate technique (9), 12 were able to induce morphologic alterations of the NIH/3T3 cells resulting in the appearance of transformed foci in transfection assays (Table 1).

TABLE 1. TRANSFORMING ACTIVITY OF DNAs FROM NEOPLASTIC HAMSTER FETAL CELL LINES INITIATED WITH DIVERSE CARCINOGENS

Initiating Carcinogen	Transformation Frequency[1]	Transformation Efficiency[2] 1st Cycle	2nd Cycle	Activated ras
Sodium bisulfite	4/7	0.94	2.77	3/4
Benzo(a)pyrene	2/2	0.06	0.72	0/2
3-Methylcholanthrene	5/10	0.21	1.73	0/5
N-Acetoxy-florenylacetamide	1/1	0.46	4.56	0/1
Others[3]	0/4	---	---	---

[1] Results are expressed as number of cell lines initiated with the given carcinogen which generated transformed foci in transfection assays on NIH/3T3 cells over the total of cell lines tested.

[2] Efficiency is expressed as the mean number of foci per μg of DNA generated by the cell lines positive in transfection.

[3] Carcinogens included were: N'-nitro-N-nitrosoguanidine (MNNG), 7,12-dimethyl-benz(a)anthracene (DMBA) and propanesulftone.

All positive cell lines showed a marked increase (5-15 fold) in transforming efficiencies in secondary transfection experiments reaching levels within the range described for single copy dominant transforming genes in other biological systems (10). No remarkable morphologic difference was found in the appearance of the cells in foci generated by different hamster tumor DNAs. In general, foci were refractile, grew at high density and in a criss-cross pattern without specific orientation, being clearly distinguishable from spontaneous, morphologically oriented, NIH/3T3 foci. Hamster-specific MS33 (from L. Diamond) repetitive sequences were present in most NIH/3T3 transformants generated by transfection with DNAs, thus, confirming the transformed phenotype cosegregated with hamster-specific DNA repetitive sequences present within the NIH/3T3 genome and proving the existence of activated dominant transforming genes in the original chemically initiated hamster cells.

Because of the existence of activated ras in most rodent tumors, the lines were first analyzed for ras. Furthermore, recently a report has suggested that H-ras proto-oncogene activation occurs in some hamster tumor cell lines obtained from tumor explants but not in preneoplastic cells.(11) All 12 lines were examined with pan-reactive anti-ras monoclonal antibodies(12,13). This approach provides a rapid diagnostic method to screen for activated ras oncogenes. Therefore, NIH/3T3 transformants from first transfection cycles were metabolically labeled and the $p21^{ras}$ proteins immunoprecipitated and subjected to SDS-PAGE. Only the transfectants derived from some bisulfite-initiated cell lines showed altered mobilities. Similar results were obtained with monoclonal antibodies Y259 or NEI-704. In all cases the alteration was consistent with mutation events at amino acid 61. Thus, the remaining lines positive in NIH/3T3 transfection must have non-ras oncogene(s). Analysis was done utilizing 38 probes for non-ras transforming sequences including erb A, erb B, fes, fos, mos, myb, myc, N-myc, raf-1, and src (purchased from Oncor), bcr, fms, and L-myc (from Oncogene Science). Probes for blym, erb B2/neu, ets, fps, hck, int-1, kit, pim raf-2, rel, ros, yes, and p53 from the ATCC/NIH Repository of Human DNA probes. Others, abl, arg, dbl, fgr, hst, jun, mas, ret, sis and trk were generous gifts from several investigators. In all cases only restriction fragments with the mobility characteristic for NIH/3T3 proto-oncogenes could be detected (14).

Thus it must be concluded that the oncogene present in non-ras cell lines may be (i) one of the very few for which no probe is available, (ii) one of the tested oncogenes which happens to have in the hamster and the NIH/3T3 genomes indistinguishable restriction maps, or (iii) a novel oncogene. Examination of parameters such as cellular morphology, low transfection efficiency and timing of foci appearance after transfection and the structural characteristics of oncogenes for which no probes were available indicates that options (i) and (ii) are unlikely. Therefore, the results support the contention that a novel hamster gene, not yet described, may be responsible for the transformation to malignancy of most neoplastic cell lines.

CONCLUSION

Both DNA and nonDNA-damaging carcinogens induce neoplastic transformation. Furthermore transfection of NIH/3T3 with high molecular weight DNA results in transformed foci containing unique hamster specific sequences; however, with the exception of some bisulfite (a component that does not interact with DNA) lines that had activated ras oncogenes, activation was not observed with 38 other probes. Therefore the data indicated that a novel unknown oncogene may be involved.

Supported in part by grant CA49858 from the National Cancer Institute to V. Notario.

1. DiPaolo JA and Casto BC (1978) In vitro carcinogenesis with cells in early passage. Third Decennial Review Conference: Cell, Tissue, and Organ Culture. U.S. Government Printing Office, Washington, DC 20402.
2. Gart JJ, DiPaolo JA and Donovan PJ. (1979) Mathematical models and the statistical analysis of cell transformation experiments. Cancer Res. 39:5069-5075
3. Higgins ITT (1971) Effects of sulfur oxides and particulates on health. Arch. Environ. Health, 22:584-590.
4. DiPaolo JA, DeMarinis AJ and Doniger J (1981) Transformation of Syrian hamster embryo cells of sodium bisulfite, Cancer Lett., 12:203-208.
5. Doniger J, O'Neil R and DiPaolo JA (1982) Neoplastic transformation of Syrian hamster embryo cells by bisulfite is accompanied with a decrease in the number of functioning replicons. Carcinogenesis (Lond.), 3:27-32.
6. Popescu NC and DiPaolo JA (1988) Chromosome alterations in Syrian hamster cells transformed in vitro by sodium bisulfite, a nonclastogenic carcinogen. Cancer Res 48:7246-7251.
7. Wirth PJ, Doniger J. Thorgeirsson SS and DiPaolo JA (1986) Altered polypeptide expression associated with Neoplastic transformation of Syrian hamster cells by bisulfite. Cancer Res 46:390-399.
8. Balmain A and Brown K (1988) Oncogene activation in chemical carcinogenesis. Adv. Cancer Res 51:147-182.
9. Wigler M, Silverstein S, Lee LS, Pellicer A, Cheng Y and Axel R (1977) Transfer of purified herpes virus thymidine kinase gene to cultured mouse cells. Cell 11:223-232
10. Weinberg RA (1981) Use of transfection to analyze genetic information and malignant transformation. Biochim Biophys Acta 651:25-35.
11. Gilmer TM, Annab LA and Barrett JC (1988) Characterization of activated proto-oncogenes in chemically transformed Syrian hamster embryo cells. Mol Carcin 1:180-188.
12. Furth ME, Davis LJ, Fleurdelys B and Scolnick EM (1982) Monoclonal antibodies to the p21 products of the transforming gene of Harvey murine sarcoma virus and of the cellular ras gene family. J Virol 43:294-304.
13. Cooper GM and Lane MA (1983) Cellular transforming genes and oncogenesis. Biochim Biophys Acta 738:9-20.
14. Notario V, Castro R, Flessate DM, Doniger J and DiPaolo JA (1990) Frequent activation of non-ras transforming sequences in neoplastic Syrian hamster cells initiated with chemical carcinogens. Oncogene 5:1425-1430.

Carcinogenicity Testing of Bladder Carcinogens and Other Organic Compounds by the Intraperitoneal and Intravesical Route

M. Roller[1], K. Kamino[2] and M. Rosenbruch[1]
[1]Medical Institute of Environmental Hygiene at the Heinrich-Heine-Universität Düsseldorf, 4000 Düsseldorf 1, FRG
[2]Institut für Experimentelle Pathologie, Medizinische Hochschule Hannover, 3000 Hannover 61, FRG

Introduction

Up to now no simple and sensitive experimental test system is available to assess the carcinogenic properties of substances which are suspected to cause bladder cancer in humans. The usual animal model in this case is the oral application (IARC, 1987). The data show, however, that it is greatly a matter of chance to select a species which is sufficiently sensitive to a certain tested substance. Bladder carcinogens were detected in the past especially with dogs. High doses and long observation periods have to be used with this species. It can be deduced from some studies, that a total of more than one kilogram of benzidine or ß-naphthylamine had to be applied (Bonser et al., 1956; Walpole et al., 1954). Naturally this leads to many technical problems. Therefore, we started to investigate the applicability of two other test models which we expected to allow successful carcinogenicity testing with much smaller amounts of test substances.

Intravesical instillation

In the first test model the implantation of so called pellets directly into the bladder lumen was examined. This was a modification of a test system used in the past by several investigators (Ball et al., 1964; Clayson and Cooper, 1970; Harzmann et al., 1980). We replaced the surgical implantation of the pellet material by an intravesical instillation of liquid substances via the urethra of female Wistar rats. Three groups of Wistar rats received single intravesical instillations of 70 µL of paraffin wax with a melting range between 42 and 44 °C. For two groups the model carcinogens 3-methylcholanthrene or 4-aminobiphenyl had been dissolved or suspended in the molten paraffin wax. The dose of the carcinogens was 1 mg per animal. Three further groups received intravesical instillations of a fast polymerizing synthetic resin named Technovit. Two of these groups were additionally injected with intraperitoneal doses of 5 mg carcinogen. It was the aim of this double application to find out whether the physical stimulus of the artificial foreign body raises the sensitivity of the urothelium to systemically present chemical carcinogens. At autopsy the bladders were fixed in situ and extended to a length of about two centimeters by instillation of formaldehyde solution. The paraffin wax was often found as an ovate, clearly defined pellet while the Technovit plastic frequently formed irregular pieces with a relatively rough surface or broke down to small crumbs. Both materials were covered by concrements in some cases.

Histopathological examination of the dissected bladders showed alterations in all groups. Table 1 gives a summary of the diagnoses. In several bladders benign lesions were found which

Table 1. Results after intravesical (i.ves.) instillation and intraperitoneal (i.p.) injection in groups of 20 Wistar rats (lifetime observation). Rats with hyper-/metaplasia are counted irrespectively of simultaneously occuring tumours; rats with tumour are listed only once, i.e. animals with primary malignant bladder tumour are only listed in the respective column even if additionally a papilloma was found.

Treatment	Rats with hyper- and/or metaplasia	Rats with	
		Papilloma	Malignant bladder tumour
Paraffin wax i.ves.	5	3	3
Paraffin wax with 3-MC i.ves.	7	0	3
Paraffin wax with 4-AB i.ves.	6	2	1
Technovit i.ves.	4	0	1
Technovit i.ves. + 3-MC i.p.	9	3	0
Technovit i.ves. + 4-AB i.p.	12	4	3
Controls (70 untreated rats)	0	0	0

3-MC = 3-methylcholanthrene
4-AB = 4-aminobiphenyl

were classified as hyperplasia, especially papillary hyperplasia, or as squamous metaplasia. With one exception the malignant tumours were carcinomas, especially squamous cell carcinomas. In one case an undifferentiated malignant tumour of the bladder wall was diagnosed. None of all these lesions was observed among about 70 untreated control animals. Most lesions were found in the group which received instillations of Technovit and simultaneous i.p. injections of 4-aminobiphenyl. But this is not considered to be significant. The frequencies of the benign lesions as well as of the tumours were obviously lower if Technovit alone had been applied, but it seems most important that all alterations appeared relatively frequently with the paraffin wax alone. The addition of the model carcinogens to the paraffin had no observable effect on the occuring of any lesions. Therefore, it can be concluded that the observed effects are due to the presence of the foreign bodies and that no influence of the chemical carcinogens was detected.

If all animals are combined which received instillations of paraffin wax or Technovit, regardless of any additional doses, no statistically significant difference is discovered between any effects of these two materials. The frequencies of hyperplasia and/or metaplasia are 30 % with paraffin wax and 45 % with Technovit and the frequencies of tumours are 18 % and 20 %, respectively. These results confirm former data from the literature, which describe similar effects of foreign bodies on the urothelium of the bladder of laboratory rodents. This may be important if the possibility of an increased risk for humans caused by bladder stones is discussed, the applicability of the method as a test for chemical bladder carcinogens, however, is impaired by these background effects.

Intraperitoneal injection

The design of the i.p. test, as usually applied to several inorganic compounds (e.g. Pott et al., 1989), was modified to achieve an exposure of the tissue over a long period with organic compounds: the test substances were dissolved or suspended in a vehicle that could be injected in liquid form, but could be expected to form a stable depot in the animal's body and slowly release the test substances to the peritoneum. A mixture of beeswax and tricaprylin appeared to be better suited for this purpose than paraffin wax. Table 2 shows the results of a first experiment. The injected volume was 1.5 mL. The mixture usually formed defined pellets in the abdominal cavity. After two years amounts of about 500 mg could be found at autopsy. The background incidence of abdominal tumours was not significantly increased by this vehicle alone, but relatively small doses of polycyclic aromatic hydrocarbons (PAH) led to highly increased tumour incidences. For comparison the results after i.p. injection of the same amount of benzo[a]pyrene in physiological saline are given in Table 2. The difference of the effect of benzo[a]pyrene in the different vehicles becomes much more evident if latency periods are considered in addition. In Figure 1 the cumulative tumour frequency is graphed against the time of observation. It can be clearly seen that not only more tumours occur after application in beeswax-tricaprylin, but that tumours are observed much earlier. Only a low background is found with the vehicle alone, with tumours occuring relatively lately.

Table 2. Abdominal tumours (mesotheliomas and sarcomas) in female Wistar rats after intraperitoneal injection of beeswax-tricaprylin mixtures

Substance	Dose i.p. (mg)	Animals with tumour/ examined	%
Benzo[a]pyrene	5	33 / 37	89.2
3-Methylcholanthrene	5	26 / 42	61.9
4-Aminobiphenyl	5	2 / 42	4.8
Beeswax-tricaprylin 25:75	1.5 mL	3 / 41	7.3
Benzo[a]pyrene in NaCl-sol.	5	19 / 38	50.0
NaCl-sol. (historical controls)	-	11 /369	3.0

Encouraged by the positive results with the PAH we started an experiment with a number of aromatic amines which are known to cause bladder tumours in humans or after oral application in experimental animals. Preliminary results of this experiment on the basis of macroscopically visible tumours after about the half of the observation time of two and a half years indicate that the test model is less sensitive to aromatic amines than to PAH but,

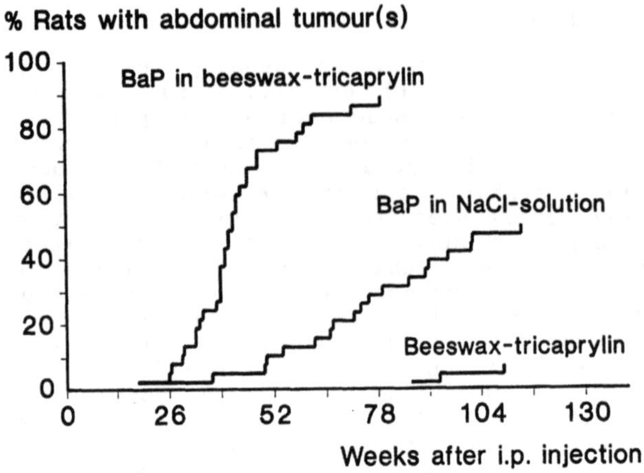

Figure 1. Time-effect relationship after i.p. injection of benzo[a]pyrene in physiological saline or beeswax-tricaprylin mixture.

in principle, is sensitive to this group of carcinogens, too. Highly significant results have been obtained with 40 mg benzidine. Final evaluation of the study will be possible with the completed histopathological diagnoses and after the full observation time of the carcinogenicity test.

References

Ball, J.K., W.E.H. Field, F.J.C. Roe and M. Walters (1964) The carcinogenic and co-carcinogenic effects of paraffin wax pellets and glass beads in the mouse bladder. Brit. J. Urol. 36: 225-237.

Bonser, G.M., D.B. Clayson, J.W. Jull and L.N. Pyrah (1956) The carcinogenic activity of 2-naphthylamine. Brit. J. Cancer 10: 539-546.

Clayson, D.B. and E.H. Cooper (1970) Cancer of the urinary tract. Adv. Cancer Res. 13: 271-381.

Harzmann, R., D. Gericke, E. Altenähr and K.-H. Bichler (1980) Induction of a transplantable urinary bladder carcinoma in dogs. Invest. Urol. 18: 24-28.

IARC Monographs: Overall Evaluations of Carcinogenicity: An Updating of IARC Monographs Volumes 1 to 42. Supplement 7. - Lyon: International Agency for Research on Cancer 1987.

Pott, F., M. Roller, U. Ziem, F.-J. Reiffer, B. Bellmann, M. Rosenbruch and F. Huth (1989) Carcinogenicity studies on natural and man-made fibres with the intraperitoneal test in rats. In: Non-occupational Exposure to Mineral Fibres. Bignon, J., J. Peto and R. Saracci, Eds. IARC Scientific Publ. No. 90: 173-179. International Agency for Research on Cancer, Lyon.

Walpole, A. L., M.H.C. Williams and D.C. Roberts (1954) Tumours of the urinary bladder in dogs after ingestion of 4-aminodiphenyl. Brit. J. industr. Med. 11: 105-109.

Cytotoxicity, Genotoxicity and Transforming Activity of 4-(methylnitrosamino)-1-(3-pyridyl)-1-butanone (NNK) in Rat Tracheal Epithelial Cells

S.Y. Zhu,[1] M.L. Cunningham,[2] T.E. Gray,[1] P. Nettesheim[1]
Laboratory of Pulmonary Pathobiology[1], Experimental Toxicology Branch[2], National Institute of Environmental Health Sciences, P.O. Box 12233 Research Triangle Park, NC, 27709, USA

Introduction

The causal relation between tobacco consumption and human cancer has been recognized for more than 50 years and NNK was considered to be the most potent carcinogenic tobacco specific nitrosamine. In present studies the cytotoxic, genotoxic and transforming effects of NNK in vivo and in vitro in rat tracheal epithelial (RTE) cells were investigated by the assays of colony forming efficiency (CFE), micronuclei formation (MN) and cell transformation frequency (TF).

Materials and Methods

NNK was obtained from Chemsyn Labs. (Lenexa, KS, USA). Methods for establishing RTE cell cultures and assays for CFE, TF and MN were described in corresponding figure legends and table footnotes.

Results

Results from the in vitro studies indicated a biphasic effect of NNK on the CFE of the cells. (Fig. 1a). Low concentrations of NNK (0.01-25μg/ml) caused up to 40% increases and high concentrations (100-200μg/ml) resulted in 50% and >90% reduction in CFE, respectively. NNK (50-200μg/ml) increased MN frequency as much as 3-fold above background (Fig. 2a) and significantly increased the TF in 4/5 (50μg/ml) and 6/8 (100μg/ml) experiments (Table 1).
The in vivo exposure of rats to NNK (150-450mg/kg) resulted in a 60-85% reduction in CFE (Fig. 1b) and a 3-5 fold increase in MN formation in RTE cells (Fig. 2b). NNK (150 and 300mg/kg) caused significant increases in TF of RTE cells from 3/3 and 2/3 rats, respectively (Table 2).

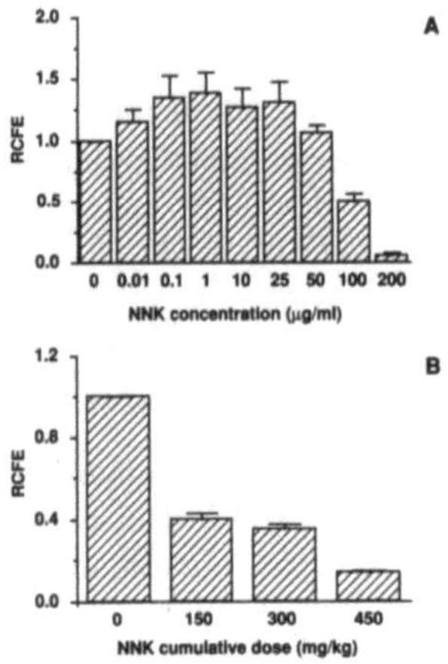

Fig. 1 Effect of NNK on CFE of RTE cells. RTE cell cultures (using F344 male rats as trachea donors) were set up with serum free medium (SFM) as described by Thomassen et al (1986). For in vitro studies (Fig. 1a), NNK exposures were initiated 24 hrs after plating (0.5-1.0x10^4 cells /dish) and continued for 7 days before the cultures were fixed and stained for colony (containing >20 cells) counting. CFE = Colonies per dish /Cells plated per dish x100%. Data based on 3-5 dishes/group, 5-12 exp. per dose. For in vivo studies (Fig.1b), rats received NNK i.p. injections in 3 consecutive days. Tracheas were removed 24 hrs after the last injection and RTE cells were isolated and plated for CFE assay as above. RCFE = relative CFE = CFE in NNK/CFE in control. Horizontal bars are ± SE (from Zhu et al. 1991, Mutation Research, in press).

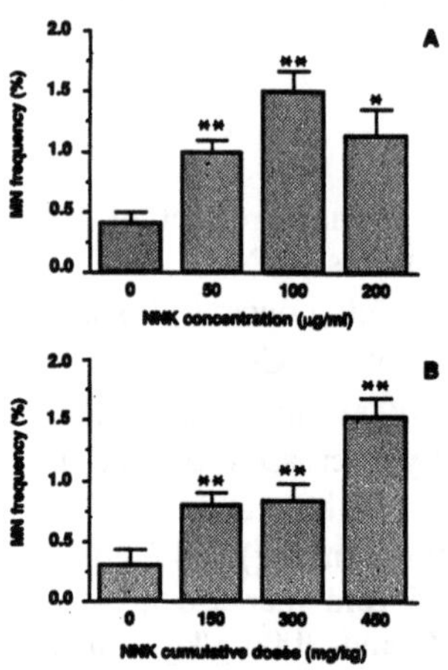

Fig. 2 Induction of micronuclei (MN) in RTE cells by NNK. For in vitro studies (Fig. 2a), 2x10^5 RTE cells/dish were plated in SFM and exposed to NNK in the next day for 24 hrs before the cytocentrifuge slides were prepared for MN evaluation after fixation and staining. Data were expressed as MN-containing cells (%) based on 1000 cells per sample and triple samples per group. For the in vivo studies (Fig. 2b), rats were treated with NNK and RTE cells were isolated as for in vivo CFE assays. Slide preparation and observation were conducted as described for the in vitro MN studies. Data based on 3 rats per NNK treatment. Compared with control, *$p<0.01$,** $p<0.001$. Horizontal bars are ± SD (from Zhu et al. 1991, Mutation Research, in press).

NNK (μg/ml)	Transformants/dish[a] Control	NNK	RTF[b]
25	0.31	1.33*	4.68
25	1.29	2.05	1.38
50	0.10	0.30	3.00
50	0.25	0.82*	3.34
50	0.31	1.05*	4.41
50	1.29	3.48**	2.71
50	1.50	4.16**	3.28
100	0.14	0.28	2.30
100	0.05	2.68**	62.21
100	0.11	0.33	2.67
100	0.31	3.47**	15.66
100	1.29	3.86**	2.45
100	0.32	1.70**	5.22
100	0.19	1.11**	5.35
100	0.15	1.29**	8.79

Table 1: In vitro transformation of RTE cells by NNK. RTE cell cultures were set up and exposed to NNK as for CFE assays and then were maintained in a selection medium for 4 weeks before the number of transformants was determined (see Kitamura et al. 1986). Transformation frequency (TF) = No. of transformants per dish/No. of colony forming cells per dish x 100%. a). Based on 20-25 dishes per group. b). RTF = relative TF = TF in NNK group / TF in control. Compared with control: * $p<0.05$, ** $p<0.001$ (modified from Zhu et al. 1991, Mutation Research, in press).

NNK (mg/kg)	Transformants/dish[a] Control	NNK	RTF[b]
150	0.57	1.71**	2.02
150	0.38	1.75**	3.70
150	0.39	1.90**	3.68
300	0.50	0.90	1.78
300	0.33	1.20**	3.63
300	0.33	1.05*	3.96

Table 2: In vivo transformation of RTE cells by NNK. Rats received NNK treatment and RTE cells were isolated and plated as for in vivo CFE assay and cultures were maintained for TF assay as in the in vitro experiments. a) and b): see table 1 for details. Compared with control: *$p<0.05$, **$p<0.005$ (modified from Zhu et al. 1991, Mutation Research, in press).

Discussion

Carcinogenesis studies in rats showed that the trachea and large bronchi were spared from developing tumors, while airways proximal (nasal cavity) and distal (bronchiolar-alveolar epithelium) to the large airways were heavily affected (for review, see Hecht and Hoffmann, 1988). We wanted to determine whether this indicates that the tracheal epithelium was completely protected from the tumorigenic effects of NNK, or whether the animals were perhaps dying from competing tumors before they developed tracheo-bronchial neoplasms. The RTE cell culture system is well suited to assess the effects of chemicals on tracheal epithelium (both in vitro and in vivo, for review, see Nettesheim and Barrett, 1984; and Nettesheim and Marchok, 1983). Several endpoints, including mitogenicity, cytotoxicity, clastogenicity and transforming activity can be measured in the same cell system.

Our studies show that NNK was toxic and clastogenic and caused transformation of RTE cells regardless of whether the compound was injected i.p. reaching the target tissue via the circulation or whether it was applied directly to the cells in culture. One interpretation is that RTE cells have sufficient drug metabolizing activity to generate cytotoxic and genotoxic and transforming products from NNK. Another interpretation is that the parent compound itself is toxic, clastogenic and has transforming activity. Which one is correct is still to be determined.

In conclusion, the present studies suggested that the rat tracheal epithelium is a target for the cytotoxic, genotoxic and transforming effects of NNK even though no rat tracheal tumors were reported following NNK exposure.

References

Hecht, S.S. and D. Hoffmann (1988) Tobacco-specific nitrosamines, an important group of carcinogens in tobacco and tobacco smoke. Carcinogenesis, 9, 875-884.

Kitamura, H., D.J. Fitzgerald, Lung-An Li, J.C. Barrett and P. Nettesheim (1986) Morphological characterization of transformed colonies in rat tracheal epithelial cell cultures exposed to carcinogen. Cancer Res., 46, 4631-4641.

Nettesheim, P. and J.C. Barrett (1984) Tracheal epithelial cell transformation: A model system for studies on neoplastic progression. CRC Critical Reviews in Toxicology,12, 215-240.

Nettesheim, P. and A. Marchok (1983) Neoplastic development in airway epithelium. In: Advances in Cancer Research, vol.39, G. Klein and S. Weinhouse, eds. Academic Press, pp. 53-54.

Thomassen, D.G., U. Saffiotti and M.E. Kaighn (1986) Clonal proliferation of rat tracheal epithelial cells in serum-free medium and their responses to hormones, growth factors and carcinogens. Carcinogenesis, 7, 2033-2039.

INHALATION STUDIES

Toxicological Criteria for Comparison of Smoke Screens by an Inhalative Experimental Animal Study. An Example: Toxic Zinc Oxide/ Hexachloroethane Smoke

U. Glaser and F. J. Mönig
Fraunhofer Institut für Umweltchemie und Ökotoxikologie,
D-5948 Schmallenberg-Grafschaft

Introduction

This inhalation study with Wistar rats aims to present an investigation concept that helps authorities and industry in the replacement of toxic smoke screens among which several mixtures of ZnO/ hexachloroethane (HC-) smoke had been preferentially used for decades in military and fire-service training.

Material and Methods

The test protocol followed the OECD-guidelines and included an acute inhalation study (a single 4 hrs exposure period in a 36 m^3 smoke chamber and 24 hrs and 14 days observation periods) as well as a subacute inhalation test (28 days with daily 4 hrs exposure periods in 225 l flow-through inhalation chambers that were connected with the smoke chamber). The mean temperature in the smoke and inhalation chambers was $\leq$ 22°C, the mean relative humidity was restricted to about 50 percent.

Results and Discussion

In the acute inhalation test the beginning particle concentration amounted to 1000 mg/m^3 (Fig. 1). This smoke was not refreshed while decreasing in concentration for group HC-1 (mean 570 mg/m^3) or renewed by igniting a further 13 g smoke pot volume after 2 hrs for group HC-2 (mean 772 mg/m^3) or for group HC-3 (mean 1620 mg/m^3) by ignition of 26 g for every hour. The particle size distribution was within the 5 μm respirable fraction range, the mass median diameter grew with time and fell due to new ignition (Fig. 1).

Lethality was among all 10 male rats of group HC-3 seen within 3 days, and amounted to 20 % each for the HC-1 and HC-2 exposure groups. Body weight gain was highly reduced among the HC-1 and HC-2 subgroups of 5 rats, which were observed for 24 hrs and 14 days periods in comparison to unexposed controls. Hematology found significantly elevated protions of blood leucocytes (Tab. 1).

Exposure effects were restricted to lung and upper airways and could be best quantified by bronchioalveolar lavage (BAL) analyses. The mean total BAL cell counts had increased within 24 hrs after HC-smoke exposure by factors > 5 due to raised proportions of leucocytes. The mean albumin contents in the cell-free BAL fluid, an indicator of lung edema, were elevated by factors > 9 and the mean LDH release into BAL fluid, a cytotoxicity indicator, had increased by factors > 5 vs. controls. But these effects were nearly normalized during the 14 days observation period (Tab. 1). However, histopathological examination still found pneumonia and few incidences of edema as well as alveolar accumulation of black-brown pigment in the rats of the HC-1 group subsequently observed for 14 days (Tab. 3).

In the subacute inhalation test exposures were by factors > 10 lower amounting to 18 mg/m³, 35 mg/m³, and 77 mg/m³ for the groups HC-4, HC-5, and HC-6, respectively (Fig. 2).
Exposures did not affect mortality among the five male and female rats in each group, but had significantly and dose-dependently reduced the mean body weight gain, and increased the mean portion of blood leucocytes during the 28 days period (Tab. 2). However, all the rats were of normal appearance and behaviour.

Again exposure effects, which were slightly more expressed in the male rats, were restricted to the lung tissue. The mean lung weight had dose-dependently increased after the subacute inhalation period. Mean total BAL cell counts were elevated due to raised leucocytes in the BAL fluid by factors of about 2 at lowest (HC-4) and > 4 at highest (HC-6) smoke concentration. The mean total BAL protein content (factors > 5 vs. controls) were dose-dependently increased similar to the acute test at higher HC-concentration. However, though being significantly elevated the mean BAL albumin and LDH values were lower than measured for the acute test groups (Tab. 2).
Histopathology found neither pneumonia, nor proliferation and fibrosis, but inflammatory cell infiltrates, histiocytosis and accumulation of black-brown pigment as well as exudate in the lung tissue (Tab. 3).

Results show the extreme respiratory toxicity of the HC-smoke even at relatively low concentration after repeated exposure of rats, posing a risk for passengers and animals incidentialy inhaling these smoke screens. In addition ecotoxicological tests, also performed according to OECD-guidelines in our institute found severe damage of important environmental species like fish, daphnia etc. Therefore, HC-smoke should be banned, since alternatives are available as seen in parallel inhalation tests, where other smoke screens reaveled to be much lower toxic than HC-smoke.

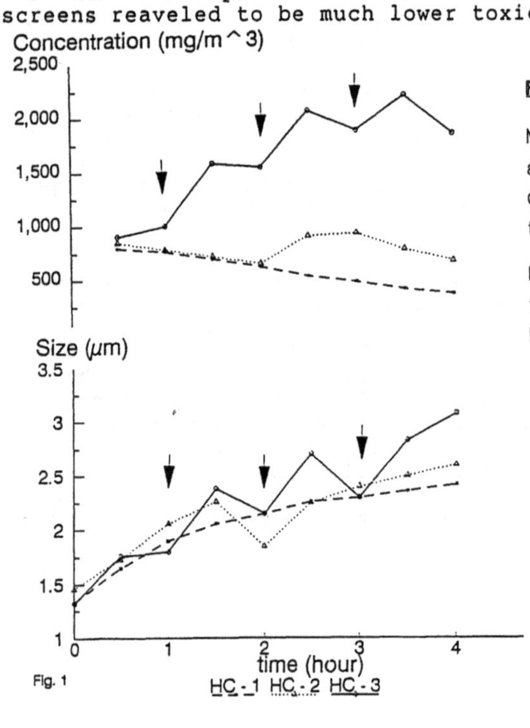

Fig.1: ACUTE INHALATION TEST

Mean particle concentration (mg/cbm) and particle size (mass median aerod. diameter, μm) for the groups exposed to HC-smoke (↓ new ignition).

Lethality amounted to 20%, 20% and 100% in the groups exposed to HC-1, HC-2 and HC-3, respectively.

Tab. 1: Acute inhalation test: Mean body weight (B.W.) gain and lung weight, hematology, as well as cytological and biochemical results in the bronchioalveolar lavage (BAL) fluid of 5 male Wistar rats each group exposed to HC-smoke and thereafter observed for 24 hours and 14 days.

Concentration	(mg/m^3)		Controls	HC - 1 570	HC - 2 772
B.W. gain	(g)	24 h	3.2 ± 2.6	- 23.3 ± 1.3*	- 33.2 ± 5.3*
		14 d	25.4 ± 6.8	- 7.0 ± 37.6	- 7.3 ± 35.9
RBS count	($10^{12}/l$)	24 h	7.3 ± 0.3	8.5 ± 0.4*	8.4 ± 1.0
		14 d	7.8 ± 0.4	8.5 ± 0.9	8.0 ± 0.5
WBC count	($10^9/l$)	24 h	6.5 ± 1.3	4.7 ± 1.9	4.8 ± 1.4
		14 d	5.9 ± 0.9	8.3 ± 2.2	6.4 ± 2.1
Blood leucocytes	(%)	24 h	7.4 ± 1.8	23.8 ± 7.2*	42.6 ± 14.5*
		14 d	12.0 ± 6.6	14.2 ± 5.5	18.7 ± 9.6
Lung weight	(g)	24 h	0.66 ± 0.03	1.08 ± 0.17	1.11 ± 0.34
(lobus sinister)		14 d	0.77 ± 0.05	0.99 ± 0.04	0.81 ± 0.01
BAL cell count	($10^5/ml$)	24 h	0.7 ± 0.2	5.5 ± 2.0*	3.7 ± 1.7*
		14 d	1.2 ± 0.5	2.4 ± 1.5	2.4 ± 0.3*
BAL leucocytes	(%)	24 h	0.4 ± 0.4	66.4 ± 15.7*	74.5 ± 13.0*
		14 d	0.1 ± 0.2	1.0 ± 1.4	0.3 ± 0.3
BAL protein	(mg/l)	24 h	158 ± 16	1453 ± 124*	1597 ± 686*
		14 d	254 ± 37	440 ± 180	327 ± 77
BAL albumin	(mg/l)	24 h	95 ± 14	1152 ± 123*	902 ± 311*
		14 d	125 ± 20	198 ± 73	211 ± 49*
BAL LDH	(U/l)	24 h	24.4 ± 2.1	146.1 ± 40.5*	123.8 ± 60.5*
		14 d	36.2 ± 3.5	48.2 ± 17.4	44.3 ± 11.0

Tab. 2: Subacute inhalation test: Mean body weight (B.W.) gain and lung weight, hematology, as well as cytological and biochemical results in the bronchioalveolar lavage (BAL) fluid of 5 Wistar rats each group and sex daily exposed to HC-smoke

Concentration	(mg/m^3)	sex	Controls	HC - 4 18.0	HC - 5 35.3	HC - 6 76.6
B.W. gain	(g)	m	74.8 ± 7.7	58.1 ± 7.1*	52.8 ± 5.4*	47.9 ± 10.9*
		f	24.1 ± 7.1	17.2 ± 4.4	16.7 ± 6.5	16.5 ± 8.2
Bood leucocytes	(%)	m	7.8 ± 4.1	9.2 ± 3.7	10.8 ± 4.7	14.8 ± 5.6
		f	7.2 ± 4.0	10.4 ± 3.4	10.8 ± 5.2	17.8 ± 5.3*
Lung weight	(g)	m	0.69 ± 0.02	0.92 ± 0.05*	0.95 ± 0.05*	1.04 ± 0.06*
(lobus sinister)		f	0.61 ± 0.02	0.76 ± 0.03*	0.78 ± 0.11*	0.97 ± 0.10*
BAL cell count	($10^5/ml$)	m	0.9 ± 0.3	2.4 ± 0.3*	3.0 ± 0.3*	6.0 ± 1.0*
		f	0.9 ± 0.2	1.7 ± 0.2*	2.7 ± 0.8*	4.0 ± 0.6*
BAL leucocytes	(%)	m	0.3 ± 0.4	37.2 ± 10.4*	42.0 ± 12.5*	66.5 ± 7.9*
		f	0.7 ± 0.3	24.8 ± 3.0*	40.5 ± 9.1*	70.7 ± 4.8*
BAL protein	(mg/l)	m	215 ± 66	1094 ± 340*	1225 ± 76*	1314 ± 174
		f	163 ± 25	716 ± 179*	776 ± 97*	1486 ± 286*
BAL albumin	(mg/l)	m	72 ± 27	359 ± 104*	370 ± 61*	399 ± 138*
		f	59 ± 8	205 ± 68*	237 ± 40*	368 ± 125*
BAL LDH	(U/l)	m	25.8 ± 5.4	65.8 ± 5.5*	71.0 ± 10.4*	110.1 ± 14.9*
		f	21.4 ± 3.8	55.4 ± 10.5*	58.6 ± 8.3*	97.2 ± 12.9*

*$p \leq 0.05$: comparison of HC-groups vs. controls by Student's t-test or Mann-Whitney U-test

Tab. 3: Histopathological findings in Wistar rats after acute and subacute inhalation of HC-smoke

	ACUTE INHALATION TEST				SUBACUTE INHALATION TEST			
	Controls		HC-1		Controls		HC-4	
	24 h	14 d	24 h	14 d	m	f	m	f
Bronchiopneumonia	0/5	0/5	2/5	0/5	0/5	0/5	0/5	0/5
Pneumonia, interst.	0/5	0/5	0/5	5/5	0/5	0/5	0/5	0/5
Inflam. cell infiltrates	0/5	0/5	1/5	5/5	0/5	0/5	3/5	3/5
Histiocytosis alveol.	0/5	0/5	0/5	3/5	0/5	0/5	2/5	4/5
Edema alveol.	0/5	0/5	5/5	1/5	0/5	0/5	0/5	0/5
Edema interst.	0/5	0/5	1/5	2/5	0/5	0/5	3/5	3/5
Pigment alveol.	0/5	0/5	4/5	1/5	0/5	0/5	4/5	4/5
Hyperaemia	1/5	4/5	4/5	1/5	5/5	2/5	2/5	2/5
Hemorrhage	0/5	2/5	2/5	2/5	1/5	0/5	1/5	0/5
Atelectase	5/5	5/5	1/5	0/5	2/5	5/5	5/5	5/5

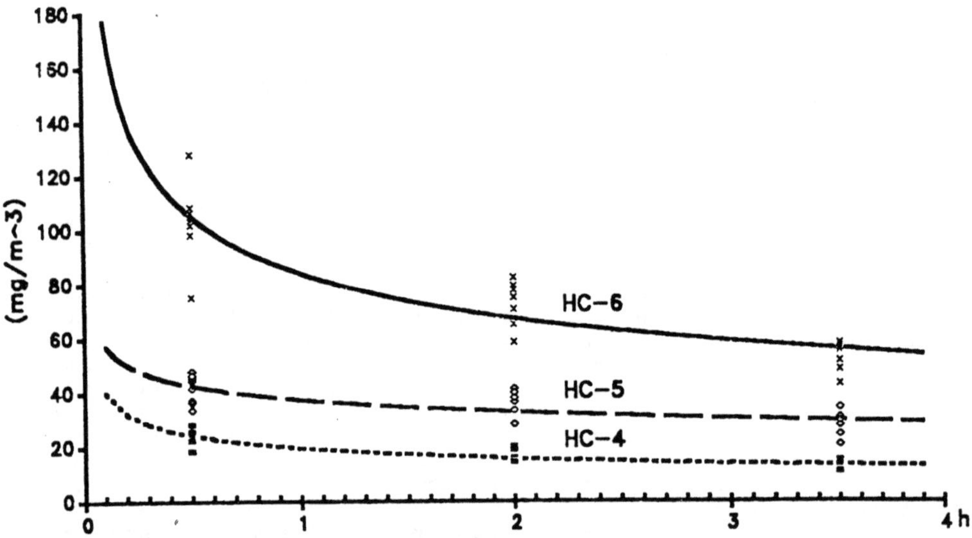

FIG 2 : Subacute inhalation test: Mean particle concentrations of HC-smoke during the daily inhalation periods (■ HC-4: 18.0 mg/m~3, ○ HC-5: 35.3 mg/m~3, × HC-6: 76,7 mg/m~3)

RESPONSE OF THE CANINE RESPIRATORY TRACT DURING LONG-TERM EXPOSURE TO A SULFUR(IV) AEROSOL AT LOW CONCENTRATION

J.Heyder, I.Beck-Speier, G.A.Ferron, P.Heilmann, E.Karg, W.G.Kreyling, A.G.Lenz, K.Maier, H.Schulz, S.Takenaka, T.Tuch

GSF-Forschungszentrum für Umwelt und Gesundheit GmbH
(Center for Environmental and Medical Research)
Projekt Inhalation, D-8042 Neuherberg, Federal Republic of Germany

INTRODUCTION

A few air pollutants, in particular certain gases, have been recognized as being noxious. The noxious effects are generally observed in acute or subchronic animal and human exposure studies with agents at concentrations exceeding by far average ambient concentrations, or as a result of observations during environmental episodes with high levels of air pollution. However, the effects resulting from inhalation of these agents over long periods of time at or close to ambient concentrations have only been studied rarely.

Sulfur dioxide (SO_2) is one of these noxious agents, whose effects at atmospheric concentrations are still unknown. Its effects are due to the reactive, IV-valent sulfur. It has been shown that the effects may be enhanced when aerosol particles act as carriers taking SO_2 to the lower portion of the respiratory system which it would not reach in its gaseous state because of complete absorption on the walls of the upper airways. Equally, the interaction between SO_2 and atmospheric particles can lead to the direct formation of particulate S(IV) as sulfites.

There are a number of reasons why sulfite is a candidate for a pneumotoxin. Bisulfite is highly reactive with biomolecules. Sulfite also causes intracellular ATP depletion. Moreover the lungs are not as well protected from sulfites as other organs like liver and kidney since less sulfite oxidase is available in the lungs. This enzyme carries out the oxidation of sulfite to sulfate, thereby detoxifying S(IV). Sulfites can also cause bronchoconstriction, especially when inhaled at acidic pH.

The exposure study described in this paper was conducted in order to determine the possible deleterious effects on the canine respiratory system at an early stage resulting from continuous long-term inhalation of a S(IV) aerosol at a level below that at which acute toxicity would be expected. Respirable sulfite particles were used as a surrogate for particle-bound inorganic S(IV)-compounds at a concentration equivalent to 0.6 mg SO_2 m^{-3}. This is a level of SO_2 which triggers an early smog warning in the FRG.

STUDY DESIGN

This long-term exposure study with beagle dogs was designed as a longitudinal study, characterized by repeated observations of individual respondents. To establish baseline data for lung parameters and determine their dependence on confounding factors, each dog was observed on a number of occasions while housed in a whole body chamber ventilated with clean air (control period of the study). In the exposure period of the study, the chambers were ventilated with a S(IV) aerosol and each parameter was again measured repeatedly for each dog. Thus each individual served as its own control.

MATERIALS AND METHODS

The beagle dog was choosen for this study since its lungs are considered to resemble human lungs and are large enough to allow serial lung lavages as a tool for investigating biochemical and cellular responses of the lungs to air pollutants. Eight dogs were exposed to airborne S(IV), three dogs served as sentinels.

Four large whole body chambers were available for housing 12 dogs at minimum restraint and allowing controlled delivery of particulates to the respiratory tract of the animals. This exposure system was designed as a dynamic system with continuous horizontal, unidirectional, low turbulent displacement airflow and continuous aerosol injection into the airflow. The exposure environment was maintained at a comfortable temperature, relative humidity and air velocity.

S(VI)-free sodium metabisulfite was dissolved in degased water at a concentration of 3.5%. The solution was adjusted to pH 7 with NaOH and nebulized to yield dry particles with a mass median aerodynamic diameter of about 0.5 μm and a geometric standard deviation of about 2. Despite rapid drying of the nebulized droplets, gaseous S(IV) was released and S(IV) was partially oxidized to S(VI). Thus the final exposure atmosphere was composed of gaseous S(IV) (26%), particulate S(IV) (63%) and particulate S(VI) (11%). This exposure atmosphere was a sulfur(IV) aerosol slightly contaminated with sulfur(VI). About 50% of the S(IV) is converted into sodium bisulfite and 50% into sodium sulfite upon contact with the airway surfaces.

Functional pulmonary response of the animals to inhaled S(IV) aerosols was tested by measuring respiratory lung functions, by examining lung lavage fluid and by studying clearance of radiolabeled test particles from the lung periphery. Serial segmental lung lavages were performed at 7-week intervals in order to obtain epithelium lining fluid from the lungs for analysis of cellular contents and cell function and biochemical indicators of lung injury. Radiolabeled, moderately soluble, monodisperse $^{57}Co_3O_4$ particles served to study clearance of material deposited in the lung periphery by translocation comprised of intracellular dissolution in alveolar macrophages and transfer of dissolved material into the circulation. Respiratory lung function tests were performed in order to assess exposure related changes in lung mechanics and alveolar-capillary gas transfer. The structural responses of the airways and peripheral airspaces of the animals to the inhaled S(IV) aerosol was studied by semiquantitative morphology and quantitative morphometry using light, scanning and transmission electron microscopy.

RESULTS AND DISCUSSION

The protein and albumin content of the lavage fluid increased after about 200 days of exposure, accompanied by a lowered oxidative status of the proteins. This lowered oxidative status points to a reduced cellular release of oxidants or an increased antioxidant capacity in the epithelial lining fluid. In addition, there was increased release of the lysosomal enzyme ß-N-acetylglucosaminidase from phagocytes. The coincidence in increase of protein content and ß-N-acetylglucosaminidase activity may be an indication for the participation of macrophages and leukocytes in enhanced protein transudation into the lungs. Alveolar macrophages showed a reduced phagocytic capacity and a reduced production of oxygen-derived free radicals indicating an impaired bacterial defence capacity.

The intracellular dissolution capacity was affected in such a way that a significant enhancement was observed in the clearance of these particles from the lung periphery. Respiratory lung function tests revealed a reduction in specific lung compliance and membrane diffusing capacity due to the S(IV) exposure.

Morphological examinations revealed disturbance of ciliated cell development in the trachea. Structural alterations in the lung periphery were indicated by airspace enlargement resulting in a significant reduction in the calculated pulmonary diffusing capacity of the lungs.

CONCLUSIONS

The observations suggest: 1. involvement of leukocytes in producing vascular leakages of protein-rich plasma in the alveolar microcirculation. 2. reduction in the release of oxidants from leukocytes or enhancement of the antioxidant capacity of the lungs. 3. impairment of the cellular defence against bacteria as indicated by a decline in phygocytic capacity and a decreased formation of oxygene radicals. 4. enhancement of the macrophage mediated clearance of inorganic particles.

The early functional response of the lungs to exposure was accompanied by a structural response resulting in a reduction in the surface area available for gas exchange, which corralated with the observation of a reduced diffusing capacity.

The alterations in lung parameters observed in this study are likely to reflect an initial stage of impairment which could increase after longer periods of exposure. The effects of S(IV) on the respiratory system may well be enhanced in the human environment by the presence of other air pollutants and by additional stresses like respiratory infections and for people at risk of being easily affected, such as children, sick and elderly persons. Thus the results obtained from this long-term exposure study represent the minimum change after a relatively short period of chronic exposure to low S(IV) concentrations and indicate the possibility of a health hazard associated with ambient gaseous and particulate sulfur(IV).

Acute Ozone-Induced Change in Lung Permeability of Genetically Susceptible and Resistant Mice is Differentially Mediated by Cytochrome P-450

Kleeberger, S.R., M.A. Trush, and L.Y. Zhang
Department of Environmental Health Sciences, The School of Hygiene and Public Health, The Johns Hopkins University, Baltimore MD, USA 21205

Introduction

Ozone (O_3) has become a common air pollutant in urban areas and in the workplace. Because O_3 is highly reactive and potentially toxic, it has received increasing attention regarding effects on pulmonary function. However, the precise mechanisms which lead to O_3-induced lung injury are still not completely understood. A role for cytochrome P-450 enzymes in the development of oxidant-induced lung injury has recently been suggested. Rietjens et al. (1988) found that prolonged exposure to 0.8 ppm O_3 caused qualitative and quantitative changes in pulmonary cytochrome P-450-linked enzyme characteristics in whole lung and within specific lung cells of rats. Takahashi and Miura (1990) showed that prolonged exposure to 0.2 ppm O_3 caused increased content of specific isoforms of cytochrome P-450 in lungs of rats. Although these studies suggest P-450 inducibility by O_3 may be important, a physiologic or pathologic correlate to the activity of these enzymes has not been demonstrated.

In a previous study, we described a reversible model of O_3-induced airway inflammation in inbred mice (Kleeberger et al., 1990). Two parameters of inflammation were characterized by bronchoalveolar lavage (BAL): influx of polymorphonuclear leukocytes (PMNs) and total protein (a marker of epithelial permeability). We demonstrated that 3 hr exposure to 2 ppm O_3 induced PMN influx that was maximal 6 hr post-exposure, and was controlled by a single gene at a locus we have named Inf. The exposure also caused significant increase in BAL protein that was maximal 24 hr after O_3. The genetic control of this hallmark of oxidant injury is not clear. The phenotypes of the C57BL/6J (B6, termed SUSCEPTIBLE) and C3H/HeJ (C3, termed RESISTANT) strains were easily distinguished by the difference in magnitude of their inflammatory responses to O_3. In the present study we hypothesized that acute O_3-induced lung inflammation in B6 and C3 mice is mediated by the cytochrome P-450 system, and that differential P-450 activity may account for between-strain variation in O_3-induced inflammation. To test this hypothesis, we 1) inhibited and 2) induced cytochrome P-450 systems of B6 and C3 mice prior to their exposure to 2 ppm O_3, and determined the effects of these treatments on O_3-induced airway inflammation.

Materials and Methods

B6 and C3 mice were purchased from Jackson Laboratories (Bar Harbor, ME). To inhibit cytochrome P-450, mice of each strain were injected intraperitoneally (IP; 0.2 ml) with piperonyl butoxide (PBO, 800 mg/kg) or SKF-525A (25 mg/kg) 1 hr before 3 hr exposure to O_3 or filtered air. Vehicle control animals were injected with olive oil or sterile water and similarly exposed. To induce cytochrome P-450, mice of both stains were injected IP with 3-methylcholanthrene (3-MC; 30 mg/kg daily, 5 days) or vehicle (olive oil) before exposure to O_3 or filtered air. O_3 generation (2 ppm) and exposure protocol were performed as described previously (Kleeberger et al., 1990). Effects of the drug treatments on O_3-induced inflammation were assessed by BAL 6 and 24 hr after exposure. The BAL procedure is described in detail elsewhere (Kleeberger et al., 1990). The effects of each drug on O_3-induced inflammation were assessed statistically by analysis of variance (ANOVA). Student-Newman-Keuls (SNK) a posteriori procedure was used to compare means. Significance was accepted at $P<0.05$.

Results and Discussion

The hypothesis that cytochrome P-450 enzymes contribute to acute O_3-induced inflammation was tested initially by pre-treating susceptible B6 mice with structurally and mechanistically distinct cytochrome P-450 inhibitors, piperonyl butoxide or SKF-525A. Both inhibitors significantly attenuated the maximum O_3-induced change in epithelial permeability (24 hr post-exposure; Table). Compared to the change in vehicle controls, piperonyl butoxide and SKF-525A significantly attenuated the maximum O_3-induced change in total BAL protein by 48% and 38%, respectively, over air-exposed animals ($P<0.025$). To further test the hypothesis, B6 mice were treated with 3-MC to induce cytochrome P-450 enzyme activity. Compared to vehicle controls, 3-MC significantly increased by 23% the change in mean total BAL protein elicited by O_3 ($P<0.05$; Table).

Table. Effects of piperonyl butoxide (PBO), SKF-525A, and 3-methylcholanthrene (3-MC) on maximum acute O_3-induced changes in %PMN influx (6 hr post-O_3) and BAL protein concentration (24 hr post-O_3) of B6 and C3 mice. Means±SEM are presented. Statistical comparison of treatment vs vehicle: **, $P<0.025$; *, $P<0.05$.

		Max Δ%PMNs		Max ΔProtein (μg/ml BAL return)	
Strain	Treatment	Air	Ozone	Air	Ozone
B6	PB	0.5±0.2	18.5±4.2	121±05	***278±26
	Vehicle	0.8±0.1	20.4±5.1	134±06	435±21
C3	PBO	0.5±0.2	4.7±1.5	84±04	192±29
	Vehicle	0.7±0.1	6.3±1.6	94±10	209±19
B6	SKF	0.6±0.2	13.5±0.9	79±04	**265±28
	Vehicle	0.9±0.3	13.3±1.5	70±07	369±27
C3	SKF	0.7±0.1	3.0±0.6	69±23	169±15
	Vehicle	1.8±0.8	4.3±1.6	59±14	130±17
B6	3-MC	1.4±1.0	14.0±0.5	140±04	*396±22
	Vehicle	0.7±0.2	11.2±1.1	132±11	341±20
C3	3-MC	1.7±0.8	6.5±1.5	96±04	234±26
	Vehicle	2.2±0.8	8.0±3.6	102±13	267±43

This increase approaches the maximum obtainable acute O_3-induced change in mice (Kleeberger et al., 1990). Results from these experiments with P-450 inhibitors and an inducer strongly suggest that acute O_3-induced change in airway permeability of susceptible B6 mice is related, in part, to the activity of cytochrome P-450 enzymes. The mechanism(s) by which these enzymes contribute to altered permeability is not clear. It is now recognized that several hundred separate P-450 isoforms exist in nature (Parke et al., 1991), and the biological roles of most have not been elucidated. However, at least two mechanisms may be proposed which could account for P-450 involvement in O_3-induced permeability changes. One possibility is that P-450 enzymes metabolize O_3-generated substrates into products with greater physiologic activity or cytotoxicity than the precursor. Ozone has been demonstrated to induce the release of a variety of products, including prostanoids, fatty acids, and lipid hydroperoxides (see Figure), and each is a substrate for an isoform of cytochrome P-450 (Porter and Coon, 1991). In some instances the biologic functions of O_3-derived products are well-characterized. The activities of their P-450-mediated metabolites are less clear, but it is conceivable that they could contribute to epithelial membrane dysfunction and subsequent permeability changes. A second possibility is that "activated oxygen" and hydrogen peroxide which are generated by P-450-catalyzed reactions with O_3-generated substrates may contribute to the epithelial injury of acute O_3 exposure. The importance of hydrogen peroxides and oxygen metabolites has been demonstrated in a number of models of oxidant lung injury, and they may also contribute to the lung injury described in our model. It is important to note that the involvement of cytochrome P-450 does not account for the entire change in lung permeability induced by acute O_3 exposure. The "non-P-450" factors may include direct effects of O_3 on membrane lipids, and the release of low molecular weight mediators which may cause vasodilatation and increased vascular permeability.

Interestingly, neither the inhibitors nor the inducer altered the O_3-induced change in mean lavageable PMNs of B6 mice (Table). Results therefore suggest that cytochrome P-450 activity does not contribute to PMN influx induced by acute O_3 exposure. As depicted in the Figure, results also indicate that the events which lead to altered airway permeability and PMN influx are not mutually dependent. This conclusion is consistent

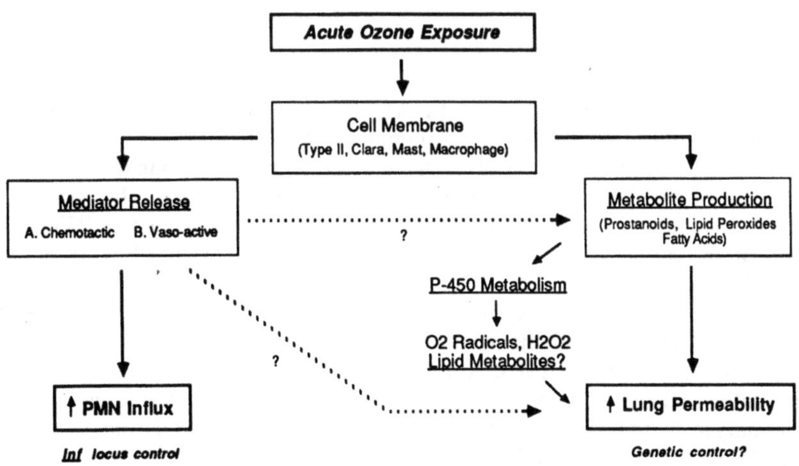

Figure. Schematic of proposed events of inflammation that follow acute ozone exposure.

with that of Kleeberger and Hudak (in press) who found that B6 mice which were depleted of PMNs, or whose PMNs were inactivated or inhibited from migration, were not protected against acute O_3-induced increase in epithelial permeability.

In contrast to the marked effects in B6 mice, pharmacological modulation of cytochrome P-450 enzymes did not alter events of O_3-induced inflammation in the resistant C3 strain. Compared to vehicle controls, the change in total BAL protein elicited by acute O_3 exposure was not significantly altered by treatment with either piperonyl butoxide or SKF-525A (Table). The influx of lavageable PMNs was likewise unaffected by either of the treatments (Table). Interestingly, induction of P-450 enzymes with 3-MC did not affect either the O_3-induced change in PMNs or total BAL protein (Table). Results therefore suggest that the cytochrome P-450 system does not have a major role in the acute inflammatory response in C3 mice. Assuming that P-450 enzymes of B6 and C3 mice are similarly inducible by 3-MC, these results may also suggest that the "defect" or mechanism which confers differential susceptibility to O_3-induced inflammation in these two strains of mice may occur prior to the participation of P-450. We hypothesize that P-450 enzymes are unimportant in development of lung inflammation in C3 mice because O_3 exposure does not elicit equivalent production of substrates in these animals compared to B6 mice. As depicted in the Figure, it may be suggested that differential susceptibility is manifested at the cell membrane, perhaps related to antioxidant defense at the site of O_3 interaction with membrane lipid. This would account for differential production of lipid metabolites, etc., and their subsequent processing by P-450 enzymes. An alternative hypothesis is that O_3 generates qualitatively and quantitatively similar substrates that are metabolized identically by B6 and C3 mice, but the metabolites exert differential pro-inflammatory effects. This may be due to inter-strain differences in handling of P-450 derived substrates, or subsequent events which are initiated by these products.

Acknowledgements

Research was supported by EPA R-816557, CIAR 90-27, and ES 03819.

References

Kleeberger SR, Bassett DJP, Jakab GJ, Levitt RC (1990) A genetic model for evaluation of susceptibility to ozone-induced inflammation. Am J Physiol 258:L313-L320.

Kleeberger SR, Hudak BB (1991) Acute ozone-induced change in airway permeability: the role of infiltrating leukocytes. J Appl Physiol (In Press).

Parke DV, Ioannides C, Lewis DFV (1991) The role of the cytochromes P450 in the detoxication and activation of drugs and other chemicals. Can J Physiol Pharmacol 69:537-549.

Porter TD, Coon MJ (1991) Cytochrome P-450. Multiplicity of isoforms, substrates, and catalytic and regulatory mechanisms. J Biol Chem 266:13469-13472.

Rietjens IMCM, Dormans JAMA, Rombout PJA, van Bree L (1988) Qualitative and quantitative changes in cytochrome P-450-dependent xenobiotic metabolism in pulmonary microsomes and isolated Clara cell populations derived from ozone-exposed rats. J Toxicol Environ Health 24:515-531.

Takahashi Y, Miura T (1990) Responses of cytochrome P-450 isozymes of rat lung to in vivo exposure to ozone. Tox Lett 54:327-335.

Photocorrosion of Cadmium Sulfide Suspensions Applied in Animal Inhalation Studies

König, H.P., U. Heinrich, H. Kock and L. Peters
Fraunhofer Institute of Toxicology and Aerosol Research
Nikolai Fuchs Straße 1, D-3000 Hannover 61

Introduction

In 1989 aerosols of the oxide (CdO), the sulfate ($CdSO_4$) and the sulfide (CdS) of Cadmium had been classified by the german Commission for the Investigation of Health Hazards of Chemical Compounds in the Work Area (MAK Commission) as substances with a carcinogenic potential for humans. Lung tumors had been observed by exposing rats to these chemical species (Glaser et al., 1990); no carcinogenic effect occured in hamsters or mice (Heinrich et al., 1989).

The aerosol generation for the inhalation exposure of the animal started from a aqueous suspension of pigment CdS E (Bayer AG, Krefeld) by using a jet nebulizer. It is well known that CdO as well as CdS are insoluble in water (Page et al., 1986; Stoeppler, 1984). Later, when the final results of the inhalation experiments were published, we were told by SCM Chemicals, Baltimore, Ohio, USA, that CdS may be insoluble in water only when kept in the dark. Under influence of light a considerable amount of cadmium sulfide might dissolve by photocorrosion interactions.

As the carcinogenic potency of the various Cd compounds depends on their solubility in the lung or on the bioavailability of the Cd^{++} ion generally, it was of great importance to know whether some of the CdS was already dissolved before deposition of the Cd-aerosol in the lung. If this happened, then at least part of the carcinogenic effect observed in rats after exposure to the CdS aerosol has to be attributed to the Cd^{++} ions, present in the aqueous suspension or the generated aerosol, respectively. To answer this question, we started a three-month CdS aerosol generation study, using the same equipment as was employed in the long-term animal inhalation studies.

Materials and Methods

A stainless steel tank (450 x 450 mm) containing 5 l of the CdS suspension (0.38 g/l CdS, 20° C, pH 6) used for aerosol generation was covered with a transparent plastic plate. Light intensity underneath this cover amounted to 50 Lux, depending on the contamination of this plate with solid CdS. The laboratory light was

automatically switched off every day after 12 hours according to the light cycle employed with the animal study. The CdS aerosol concentration was kept at approximately 90 µg Cd/m^3. For aerosol generation a CdS pigment was used (Cadmiumgelb E 1, Bayer AG, Krefeld) with a purity of 98.7% and loads of 0.14% Zn and 0.8% $BaSO_4$. In additional experiments on photocorrosion of CdS in saline and ethanolic suspensions and aqueous suspensions under shield gases, we applied CdS with a purity of 99.999% (Aldrich, Steinheim).
The CdS suspensions and the filter samples of the resulting aerosol were analysed for total and soluble Cd^{++} by atomic absorption spectrometry (AAS) and for anions like SO_4^{--} by ion chromatography (More details are given in: König et al., 1991).

Results

Conditions, according to the animal study, resulted in a decomposition rate of 0.3mg CdS/l/h at a light intensity of 50 Lux. Within 4 weeks a maximum of about 40% soluble Cd^{++}out of total cadmium was formed (Fig.). After this period, we obtained almost constant decay. Under constant experiment conditions like light intensity, replenished suspension volume and aerosol generation rate the ratio of Cd^{++}/total Cd kept constant, too. The ratio of Cd^{++}/total Cd found in the dry CdS aerosol produced was sligthly lower. This might be related to the technique

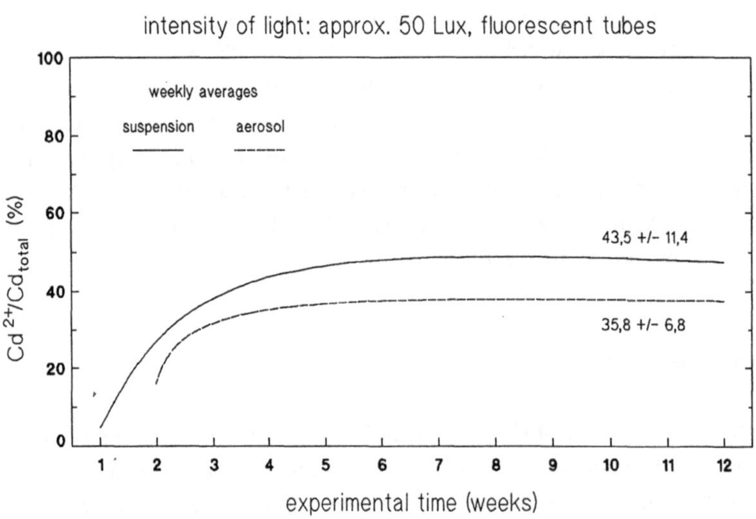

Fig.: Cd concentrations found in CdS suspension (set point concentration: 0.38 g CdS/l) and in the aerosol produced. Aerosol generation conditions similar to animal inhalation study (Glaser et al., 1990).

of nebulation with selection of smaller particle sizes and backflush of the larger ones.
CdS suspensions in saline solutions (3.33 and 0.83 g Cd/l) had been applied in intratracheal instillation experiments with rats (Pott et al., 1987). These samples had been illuminated for 24 h with 800 Lux from fluorescent tubes; only for the analysis the samples had been stirred to obtain a homogeneous suspension, in the meantime the suspension was allowed to settle down. Even immediatly after preparing the CdS suspensions some Cd^{++} ions could be detected.
CdS at the same concentration level in ethanol showed a greater stability. Suspensions kept in the dark, remained stable; lowering the pH-value down to pH 3 did not change the photochemical decay rate significantly (König et al., 1991). Also under shield gases (Ar, N_2) the photooxidation of CdS in aqueous suspensions was observed.

Discussion

Aqueous suspensions of CdS exposed to daylight or incandescent light are not stable with respect to the photooxidation to $CdSO_4$. With respect to the CdS inhalation study with rats, we have to assume that the CdS aerosol generated from a suspension as described above to produce a 90 µg Cd/m³ aerosol concentration, consisted of approx. 40% $CdSO_4$ (36 µg Cd/m³) and 60% CdS (54 µg Cd/m³). Therefore the observed lung tumor rate in rats (80%, average of males and females taken together) cannot be attributed solely to the effect of CdS.
It is known that, after inhalation exposure of rats to 30 µg Cd/m³ as $CdCl_2$ or 30 µg Cd/m³ as CdO dust, a lung tumor rate of 75% and 72%, respectively, was observed (Glaser et al., 1990). For this reason, the exposure to 36 µg Cd/m³ as $CdSO_4$ could well account for the total lung tumor rate of 80% observed after inhalation of the above-mentioned mixture of CdS and $CdSO_4$.
On the other side, inhalation experiments with rats and monkeys to determine the pulmonary retention of $CdCl_2$, CdO and CdS clearly showed the bioavailability of CdS retained in the lungs (Oberdörster and Cox, 1989). CdS in these experiments was suspended in ethanol and the exposure to the aerosol lasted for 10 minutes only. Based on our data, CdS in ethanol is much more stable than CdS in water, and therefore we can assume that almost no $CdSO_4$ was inhaled by these animals. Also, tumors occured in rats after intratracheal instillation and intraperitoneal injection of CdS (Pott et al., 1987).

From this the following can be deduced:

* Because of the high carcinogenic potency of Cd^{++} ions in the lungs also very small amounts of of CdS dissolved in the lungs could lead to a

tumor response (the exposure to only 12.5 µg/m³ $CdCl_2$ already resulted in a significantly increased lung tumor rate, Takenaka et al., 1983), and therefore inhalable CdS has to be regarded as a probable human carcinogen.

* Because of the relatively long biological half-time of inhaled particles like CdS in the human lung compared to the rat lung the carcinogenic risk of CdS dust for humans could be higher than expected on the basis of the rat data.

References

Glaser, U., D. Hochrainer, F.J. Otto, H. Oldiges (1990) Carcinogenity and Toxicity of four Cadmium Compounds inhaled by rats. Toxicol. Environm. Chem. 27: 153-162

Heinrich, U., L. Peters, H. Ernst, S. Rittinghausen, C. Dasenbrock, H. König (1989) Investigation on the carcinogenic effects of various cadmium compounds after inhalation exposure in hamsters and mice. Exp. Pathol. 37: 253-258.

König, H.P., U. Heinrich, H. Kock, L. Peters (1991) Effect of Photocorrosion on Cadmium Sulfide Suspensions applied in Animal Inhalation Studies with CdS Particles. Arch. Environm. Cont. Toxic. (in press).

Page, A.L., M.M. El-Amamy, A.C. Chang (1986) Cadmium in the Environment and its Entry into Terrestrial. In: Foulkes, E.C. (ed.) Cadmium. Springer Verlag, Berlin: 33-74.

Pott, F., U. Ziem, F.J. Reiffer, F. Huth, H. Ernst, U. Mohr (1987) Carcinogenicity Studies on Fibres, Metal Compounds, and some Dusts in Rats. Exp. Pathol 32: 129-152.

Oberdörster, G., C. Cox (1989) Kinetics of inhaled $CdCl_2$, CdO and CdS in rats and monkeys. In: Hiscock, S.A., R.A. Volpe (ed.) Ed. Proc. Sixth Int. Cadmium Conf. Paris 19-21 April: 147-154.

Stoeppler, M. (1984) Cadmium. In: Merian, E. (ed.) Metalle in der Umwelt. Verlag Chemie, Weinheim: 375-408

Takenaka, S., H. Oldiges, H. König, D. Hochrainer, G. Oberdörster (1983) Carcinogenicity of Cadmium Chloride Aerosols in W rats. J. Nation. Cancer Inst. 70: 367-373.

Risk Estimation of Human Lung Cancer from Environmental Exposure to Diesel Exhaust: Extrapolation from Particle Induced Experimental Lung Tumors?

Oberdörster, G.[1], W.P. Pepelko[2], C.P. Yu[3] and C. Chen[2]
[1]Environmental Health Sciences Center, University of Rochester, Rochester, NY, USA
[2]U.S. Environmental Protection Agency, Washington, DC, USA
[3]Department of Mechanical & Aerospace Engineering, SUNY, Buffalo, NY, USA

The U.S.Environmental Protection Agency is presently preparing a Health Assessment Document of inhaled diesel engine emissions which will also address the question of an assessment of its carcinogenic potency. The evidence of a carcinogenic potential of inhaled diesel exhaust from epidemiological studies is limited whereas it is sufficient from long-term rat inhalation studies (IARC, 1989). Table 1 lists several of those rat inhalation studies, and the question arises whether the result of these studies could be used for assessing a carcinogenic risk for human exposure.

Table 1

CHRONIC DIESEL EXHAUST INHALATION STUDIES IN RATS RESULTING IN LUNG TUMORS IN AT LEAST ONE EXPOSURE CONCENTRATION

Authors	Exposure Conc. mg/m^3	Duration Hrs/Day	Days/Wk	Years
Brightwell *et al.*, 1986	0.7 - 6.6	16	5	2.0
Ishinishi *et al.*, 1986	0.5 - 3.7	16	6	2.5
Heinrich *et al.*, 1986	4.0	19	5	2.5
Iwai *et al.*, 1986	4.9	8	7	2.0
Mauderly *et al.*, 1987	0.35 - 7.0	7	5	2.5
Takaki *et al.*, 1989	0.1 - 3.7	16	6	2.5

(for summary of these studies, see Ishinishi *et al.* 1986)

Since total diesel exhaust consists of a vapor phase and a particle phase, with the latter composed of adsorbed organics and a carbon core, the observed induction of lung tumors in rats after exposure to diesel exhaust could in principle be due to either of the components in these phases or to a combination thereof since carcinogenic active compounds are present in all of them. Thus, the vapor phase contains benzene and aldehydes which are either known or probable human carcinogens. However, concentrations of these carcinogenic compounds in diesel exhaust is quite low on the order of 0.1 ppm (benzene) to 1.3 ppm (total aldehydes) (Williams and Chock, 1981) µg/m^3, and rat exposures to relatively high concentrations of filtered diesel exhaust, i.e., vapor phase components only, did not result in detectable increases in lung tumors (Heinrich *et al.*, 1986a). With respect to organic compounds adsorbed onto particles, a variety of tests has shown that organics extracted from the particles have significant mutagenic activity (Lewtas, 1983). They contain known carcinogens such as BaP and nitroaromatics, and organic extracts were shown to induce lung tumors following intratracheal instillation (Grimmer *et al.*, 1987), dermal papillomas following skin painting (Nesnow *et al.*, 1982) and subcutaneous tumors following injections (Kunitake *et al.*, 1986). Furthermore, exposure to diesel exhaust containing particle adsorbed organics resulted in the induction of DNA adducts (Bond *et al.*, 1990) which were found in lung regions where tumors occurred. While all this speaks in favor of a carcinogenic potential of the organic compounds in diesel exhaust and might justify their use for assessing a carcinogenic risk other evidence speaks against this: The concentration of known carcinogenic organics is quite low. For example, Tong and Karasek (1984) reported a BaP concentration of less than 0.1 µg/mg of particulate diesel exhaust; and long-term experiments with inhaled pyrolyzed pitch condensate which lacks an inorganic carbon core yet contains about three orders of magnitude greater concentrations of BaP than present in diesel exhaust resulted in a tumor incidence as seen after exposure to diesel exhaust (Heinrich *et al.*, 1986b).

In contrast, recent evidence has been reported independently by two investigator teams that long-term inhalation of carbon black particles by rats induced lung tumors in rats at about the same exposure concentrations as diesel exhaust (Mauderly *et al.*, 1991; Heinrich *et al.*, 1991). The carbon black particles used in these inhalation studies were essentially free of organics. Also, repeated intratracheal instillation of carbon black led to the same lung tumor incidence in rats as intratracheal instillations of diesel particles (Kawabata *et al.*, 1986; Pott *et al.*, 1991). DNA adducts in the lung have also been found to be induced by inhalation of pure carbon black similar to inhalation of diesel exhaust particles containing organics. Moreover, a tumorigenic particle effect is not restricted to carbon black alone but was also observed after long-term inhalation of high concentrations of TiO_2 and coal dust (Lee *et al.*, 1986; Martin *et al.*, 1977). Effects observed in these studies have generally been referred to as particle "overload" phenomena, *i.e.*, effects related to a high particulate burden in the lung such as impairment of alveolar macrophage (AM) mediated particle clearance, fibrotic changes and lung tumors. It appears that very different lung particulate mass burdens for the different particle types are necessary to induce the different effects. For example, while about 12-15 mg of toner or diesel particles retained in a rat lung after 24 or 30 months of exposure resulted in an impairment of AM-mediated particle clearance, only the diesel particles induced a significant lung tumor incidence. Moreover, more than 600 mg of retained TiO_2 (particle size ~0.2-0.3 µm) in the rat's lung was necessary to induce a significant lung tumor incidence compared to about 15 mg of diesel exhaust or carbon black. These data seem to indicate that particle-induced lung tumors do not occur at the same levels of lung "overload" and one might, therefore, question whether diesel- induced lung tumors are due to a particle overload effect alone. However, it is conceivable that the retained particle mass *per se* is not the appropriate parameter to correlate with lung tumors. Similar to overload related retardation of AM-mediated particle clearance, which has been shown to correlate better with the retained particle volume rather than the mass (Morrow, 1988; Muhle *et al.*, 1990), particle-induced lung tumors may also correlate better with a different particle parameter. Indeed, a comparison of relationships of different particle dosimetric parameters with lung tumor incidences reveals that only the retained particle surface area correlates reasonably well with lung tumor incidence (Fig. 1). Preliminary results with carbon black by Mauderly *et al.* (1991) would also fit well in this dose-response relationship. Thus, when discussing particle overload induced effects in the lung, one has to consider the appropriate dose parameter for a specific effect. The particle surface area-lung tumor incidence plot could even be interpreted as showing a threshold (in terms of retained particle surface area) below which no tumor response would be observed. Certainly, more data are necessary to confirm the existence of such a threshold. It follows from the foregoing discussion that (1) the particulate matter in diesel exhaust rather than particle adsorbed organics or vapor phase components can explain the induction of lung tumors in the long-term inhalation studies; (2) the induction of lung tumors after diesel exhaust exposure may be caused by an unspecific particle overload effect. Therefore, an assessment of carcinogenic risk could be based on the lung burden of particulate matter. This conclusion does not preclude the possibility that the particle adsorbed organics or vapor phase components have some carcinogenic or promotional effects. However, these effects are likely to be very small and difficult to integrate in a quantitative risk assessment as presented in the following section. Fig. 2 sows a schematic overview of hypothetical pathogenic sequences of diesel exhaust induced effects.

A dosimetric extrapolation model was used for assessing lung tumor risk (Yu *et al.*, 1990). The model takes into account species differences (rat *vs.* man) in both particle deposition and particle clearance including particle transport to lung associated lymph nodes and species differences in ventilatory parameters and lung parameters. It was also considered that high particulate lung burdens as they occurred in the rat studies inhibit particle clearance. A unit cancer risk estimate (upper 95% bound) was calculated with a linearized multistage model using data of Mauderly *et al.* (1987) from rats chronically exposed to diesel exhaust at 3 different concentrations. Additionally, the model incorporated available data for elapsed time-to-tumor occurrence in the rat lung. Fig. 3 shows the

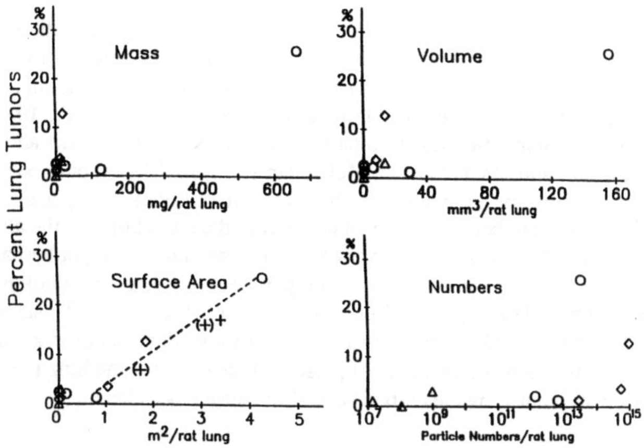

Figure 1 Retained mass, volume, surface area and numbers of particles in correlation to lung tumor incidences in rats after long-term exposure to TiO2 (O, Lee *et al.*, 1986), diesel exhaust (◊, Mauderly *et al.*, 1987) and toner (Δ, Muhle et al., 1990). + indicates result of inhalation study with carbon black (Heinrich *et al.*, 1991) and (+) indicates preliminary result of Mauderly *et al.* (1991) with carbon black.

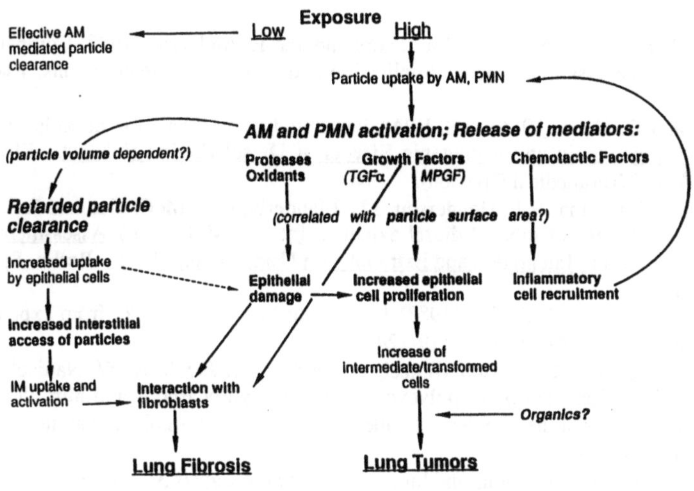

Figure 2. Hypothetical Pathogenic Sequence for Pulmonary Effects of Inhaled Diesel Exhaust

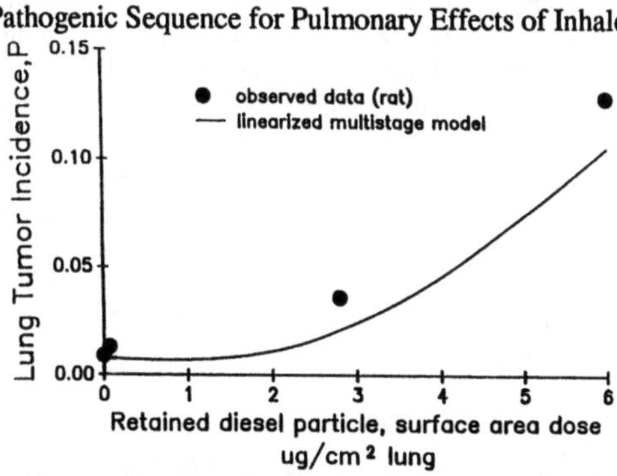

Figure 3. Lung tumor incidence after chronic diesel exhaust exposure, Mauderly *et al.*, 1987. The linearized multistage model was adapted to a "time-to-tumor" model.

result; the estimated lifetime risk of lung cancer from exposure to 1 μg/m^3 diesel particulate matter based on the study of Mauderly *et al.* (1987) is 1 x 10^{-5}. This is the upper 95% boundary, *i.e.*, the risk maybe lower than this estimate while it is not likely to be higher. While this estimate compares favorably with previous estimates based on comparative potency (Albert *et al.*, 1983; Cuddihy *et al.*, 1983; Harris, 1983) and other estimates from cancer bioassays in rats (Albert and Chen, 1986; Pott and Heinrich, 1988; Smith and Stayner, 1990; McClellan *et al.*, 1989), it should be kept in mind that it is based on several assumptions made for the clearance model, *e.g.*, a simplified retention compartment not differentiating between retained material in the interstitial *vs.* the alveolar space. One basic assumption is also that the same retained particle dose per unit rat lung or human lung will lead to the same effects. This assumption can be challenged since no data are available to support this assumption. Thus, the calculated upper 95% unit lung cancer risk of 1 x 10^{-5} should be viewed as a rough estimate only which needs to be improved when more data become available both on the dosimetry as well as on mechanistic aspects. At present, it appears that the lung tumor incidence in the rat studies can be fully explained by the carbon core of the diesel particles.

Acknowledgement

This paper was supported by NIH grants ESO4872 andES01247.

References

Albert, R.E., J. Lewtas,S. Nesnow, T.W. Thorslund and E. Anderson (1983) Comparative potency method for cancer risk assessment: Application to diesel particulate emissions. *Risk Analysis* **3**: 101-117.

Albert, R.E. and C. Chen (1986) U.S.-EPA diesel studies on inhalation hazards. In: Ishinishi *et al.*(eds.) Carcinogenic and Mutagenic Effects of Diesel Engine Exhaust. Elsevier Science Publishers B.V. (Biomedical Division).

Bond, J.A., J.R. Harkema, R.F. Henderson, J.L. Mauderly, R.O. McClellan and R.K. Wolff (1989) Molecular dosimetry of inhaled diesel exhaust. In: U. Mohr (ed.) Assessment of Inhalation Hazards Integration: Integration and Extrapolation Using Diverse Data. Berlin, Springer-Verlag, pp. 315-324.

Cuddihy, R.G. and r.O. McClellan (1983) Evaluating lung cancer risks from exposures to diesel engine exhaust. *Risk Analysis* **3**: 119-124.

Grimmer, G., H. Brune, R. Deutsch-Wenzel, G. Dettbarn, J. Jacob, K.-W. Naujack, U. Mohr and H. Ernst (1987) Contribution of polycyclic aromatic hydrocarbons and nitro-derivatives to the carcinogenic impact of diesel exhaust condensate evaluated by implantation into the lungs of rats. *Cancer Letters* **37**: 173-180.

Harris, J.E. (1983) Diesel emissions and lung cancer. *Risk Analysis* **3**: 83-100.

Heinrich, U., H. Muhle, S. Takenaka, H. Ernst, R. Fuhst, U. Mohr, F. Pott and W. Stöber (1986a) Chronic effects on the respiratory tract of hamsters, mice and rats after long-term inhalation of high concentrations of filtered and unfiltered diesel engine emissions. *J. Appl. Toxicol.* **6**: 383-395.

Heinrich, U., F. Pott, U.Mohr, R. Fuhst and J. Donig (1986b) Lung tumours in rats and mice after inhalation of PAH-rich emissions. *Exp. Pathol.* **29**: 29-34.

Heinrich, U., D. Dungworth, F. Pott, A. Schulte, L. Peters, C. Dasenbrock, K. Levsen, W. Koch and O. Creutzenberg (1991) The carcinogenic effects of carbon particles and tar/pitch condensation aerosol after inhalation exposure of rats. Presented at: Seventh International Symposium on Inhaled Particles, Edinburgh, Scotland, September, 1991.

IARC (International Agency for Research on Cancer) (1987) IARC Monographs on the Evaluation of the Carcinogenic Risk of Chemicals to Humans. Vol. 46, Engine Exhausts and Nitroarenes. Lyon.

Ishinishi, N., A. Koizumi, R. O. McClellan and W. Stöber, eds. (1986) Carcinogenic and Mutagenic Effects of Diesel Engine Exhaust. Elsevier Science Publishers B.V. (Biomedical Division).

Kawabata, Y., K. Iwai, T. Udagawa, K. Tukagoshi, and K. Higuchi (1986) Effects of diesel soot on unscheduled DNA synthesis of tracheal epithelium and lung tumor formation. In: Ishinishi, N., K. Koizumi, R.O. McClellan and W. Stöber (eds.) Carcinogenic and Mutagenic Effects of Diesel Engine Exhaust, Elsevier Science Publishers, Amsterdam, New York, Oxford, pp. 213-222.

Kunitake, E., K. Shimamura, H. Katayama, K. Takemoto, A. Yamamoto, A. Hisanaga, S. Ohyama, and N. Ishinishi (1986) Studies concerning carcinogenesis of diesel particulate extracts following intratracheal instillation, subcutaneous injection, or skin application. In: N. Ishinishi *et al.* (eds.) Carcinogenic and mutagenic effects of diesel engine exhaust. Amsterdam. Elsevier, pp. 361-380.

Lewtas, J. (1983) Evaluation of the mutagenicity and carcinogenicity of motor vehicle emissions in short-term bioassays. *Environmental Health Perspectives* **47**: 141-152.

Lee, K.P., N.W. Henry III, H.J. Trochimowicz and C.F. Reinhardt (1986) Pulmonary response to impaired lung clearance in rats following excessive TiO_2 dust deposition. *Environ. Res.* **41**: 144-167.

Martin, J.C., H. Daniel, and L. LeBouffant (1977) Short- and long-term experimental study of the toxicity of coal-mine dust and of some of its constituents. In: Inhaled Particles IV, Part I, ed. W. H. Walton, p. 361-370, Pergamon Press, Oxford.

Mauderly, J.L., R. K. Jones, W.C. Griffith, R. F. Henderson and R.O. McClellan (1987) Diesel exhaust is a pulmonary carcinogen in rats exposed chronically by inhalation. *Fund. & Appl. Tox.* **9**: 208-221.

Mauderly, J. L., M.B. Snipes, E.B. Barr, W.E. Bechtold, R.F. Henderson, C.E. Mitchell, K.J. Nikula and D.G. Thomassen (1991) Influence of particle-associated organic compounds on carcinogenicity of diesel exhaust. Presented at: Health Effects Institute, Annual Conference, April, 1991, Colorado Springs.

McClellan, R.O., R.G. Cuddihy, W.C. Griffith and J. L. Mauderly (1989) Integrating diverse data sets to assess the risks of airborne pollutants. In: U. Mohr (ed.) Assessment of Inhalation Hazards Integration: Integration and Extrapolation Using Diverse Data. Berlin, Springer-Verlag, pp. 3-22.

Morrow, P.E. (1988) Possible mechanisms to explain dust overloading of the lungs. *Fund. & Appl. Tox.*, **10**: 369-384.

Muhle, H., O. Creutzenberg, B. Bellman, U. Heinrich and R. Mermelstein (1990) Dust overloading of lungs: Investigations of various materials, species differences and irreversibility of effects. *J. Aerosol Med.* **3**: Suppl. 1, S-111 - S128.

Nesnow, S., C. Evans, A. Stead, J. Creason, T.J. Slaga and L.L. Triplett (1982) Skin carcinogenesis studies of emission extracts. In: J. Lewtas (ed.) Toxicological Effects of Emissions from Diesel Engines. Amsterdam, Elsevier, pp. 295-320.

Pott, F. and U. Heinrich (1988) New findings on the carcinogenic effect of diesel engine exhaust. *Z. Gesamte Hygiene* **34**: 686-689.

Pott, F., D.L.Dungworth, U. Heinrich, H. Muhle, K. Kamino, P.-G. Germann, M. Roller, and R.M. Rippe (1991) Lung tumors in rats after intratracheal instillation of dusts. Presented at VII International Symposium on Inhaled Particles, Edinburgh, September, 1991.

Smith, R.A. and L. Stayner (1990) An exploratory assessment of the risk of lung cancer associated with exposure to diesel exhaust based on a study with rats. A final report, Division of Standards Development and Technology Transfer, NIOSH, Cincinnati.

Tong, H.Y. and F.W.Karasek (1984) Quantitation of polycyclic aromatic hydrocarbons in diesel exhaust particulate matter by high-performance liquid chromatography fractionation and high resolution gas chromatography. *Anal. Chem.* **56**: 2129-2134.

Williams, R.L. and D.P. Chock (1981) Characterization of diesel particulate exposure. *Environment International* **5**: 199-209.

Yu, C.P., K.J. Yoon and Y.K. Chen (1990) Retention modeling of diesel exhaust particles in rats and humans. J. Aerosol Med., in press.

Short term chromate inhalation in a Computer-assisted inhalation chamber: immediate toxicity and late cancer development

Popper, H.H., G. Wiespainer, E. Leingartner, W. Weybora, M. Ratschek
Institute of Pathology, University of Graz, Medical School, Institute of Biomedical Engineering, Technical University of Graz, Graz, Austria

Introduction

Chromates have been shown by many authors to be highly toxic to mammalian cells and animals. In cell cultures using pneumocytes, fibroblasts and alveolar macrophages we have demonstrated toxic reactions (Popper et al., 1990). Also genotoxic effects have been shown using the micronucleus assay. To further extend our data we investigated an inhalation experiment using Fisher 344 rats. In these experiments we looked for toxic and carcinogenic effects at the onset of inhalation and for the late outcome.

Material and methods

Disease free male and female Fisher 344 rats were used for whole body inhalation experiments. All were 4-5 months old and were separated into groups of 12 animals each, consisting of 6 females and 6 males, respectively (early phase reactions and late phase reactions) . For each concentration and substance one group of rats was used.

Inhalation was performed for 7 hours/day, 5 days/week, 4 weeks, weekends and holidays were spared. Within the inhalation chamber the rats were given water ad libitum. After inhalation the rats were placed in normal cages and were feed a normal diet. As chromate compounds we used potassium dichromate (K2Cr2O7) and barium chromate (BaCrO4; only used for comparison of early toxic reactions) in two concentrations (HC 10 mg/m^3 and LC 0,5 mg/m^3) and as chromium salt chromium-aluminium-sulfate (CrAlSO4; 10 mg/m^3 and 0,5 mg/m^3). Two types of inhalation experiments were done: for the study of acute toxic reactions one female and one male rat were killed at 3, 6, 10, 28 days during inhalation and in the 2nd and 5th months (when inhalation was already finished). For late toxic and carcinogenic effects K2Cr2O7 and CrAlSO4 exposures were done as described above. After one month of inhalation the animals were placed in normal clean cages and feed a normal diet. The rats were not killed until they looked ill or they died from chromate induced disease. Controls were killed at bimonthly intervals between 17th to 26th months. Animals were killed with an overdose of CO2. One group of rats served as negative control in each experiment.

Designs of inhalation procedure:

K2Cr2O7, BaCrO4 or CrAlSO4, respectively were placed into an electric mill. Milled chromate or chromium salt dust was transferred into the inhalation chamber by insufflation of clean prefiltered air. Only particles smaller than 10µm could reach the chambers: This particle size was achieved by reducing the air stream within the mill (Fig.1). Within the chambers the concentration of particles was held constant by a computer connected to a laser particle counter. Particle counts were correlated to the desired concentration by chemical analysis (titration of chromates or chromium salts respectively and weight analysis). The evaluated particle concentration was held constant by an Apple MacIntosh Computer®. A minimum and maximum concentration was set, and by connection to the magnetic mill the computer could control supply of chromates or chromium salt dust (Fig. 2).

Histology

After autopsy sections from all organs were taken and fixed in neutral buffered formaldehyde solution. Nose and paranasal sinuses, sternum and femur were decalcified with EDTA. All tissue blocks were processed as routinely. Sections were stained with H&E, PAS and Elastica van Gieson. Fresh frozen tissue samples were used for chromium analysis.

Chemical analysis

Tissue samples from all organs were ashed with liquid concentrated HN03, dried under N2 stream and chromium content was analyzed by spectrophotometry. Values are given as µg Cr / g wet tissue.

Results

Pathological changes in rats inhaling chromates could be divided into early phase reactions (i.e. changes within the 4 weeks of ongoing inhalation and 2 weeks after cessation of inhalation). Afterwards most rats recovered and stayed alive and well until the 20th week. A late phase reaction was found after 17 months from the beginning of chromate inhalation.

Only few animals of the HC dichromate group were alive after the 3rd month. Therefore late phase changes were mainly seen in animals from the LC dichromate group.

Fig 1: Schema of production of chromate or chromium salt dust

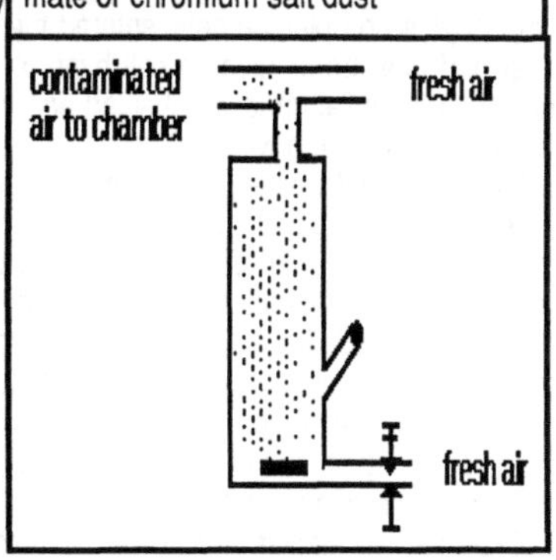

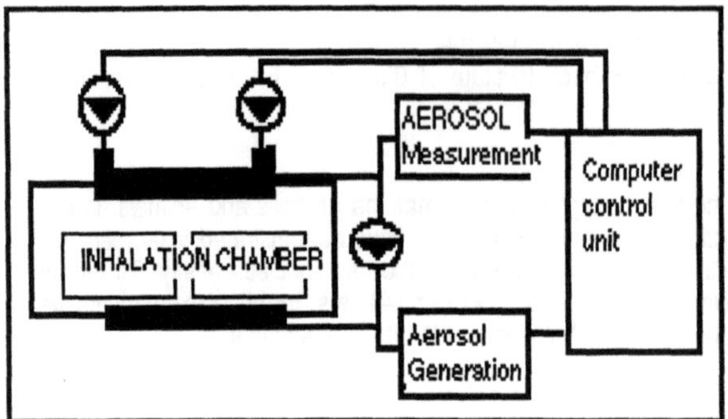

Fig. 2: Schema of computer controlled inhalation chamber; aerosol measurement was done by a laser particle counter; 2 cages were placed into each chamber; males and females were separated by these 2 cages

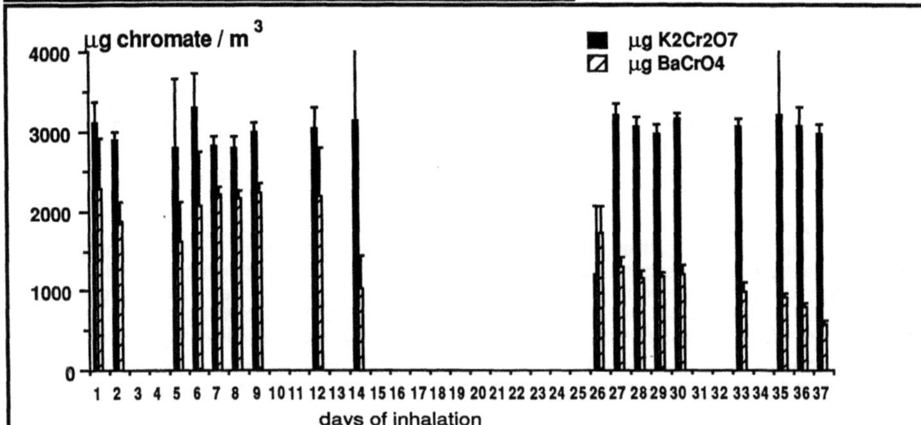

Fig. 3: Comparison of dichromate and bariumchromate inhalation; the given values are calculated chromium values (molecular weight as part of 10 mg total salt) corresponding to the 10 mg/ m^3 high dose of each compound; zero values are weekends and/or holidays

Respiratory tract

Early phase reactions:

Acute inflammatory reactions in the respiratory tract were more pronounced in the HC dichromate groups than in the LC groups, but the reactions were similar. Dichromate caused a more severe inflammatory reaction compared to BaCrO4.

In the upper respiratory tract an acute inflammation in the nose and nasal sinuses started within 3 days of dichromate inhalation. The inflammation was exclusively confined to the respiratory epithelium, leaving the olfactory epithelium completely unchanged. The cell composition was first predominated by neutrophils, after 6 days also macrophages, lymphocytes and plasma cells entered the mucosa. The inflammation resulted in ulceration of the epithelial layer and later on in reserve cell hyperplasia and squamous cell metaplasia. After cessation of chromate inhalation the inflammation subsided and vanished from the 7th week on. Hyperplasia and metaplasia were no longer visible from week 10 on.

In the larynx and trachea only a mild transient inflammatory response could be found, the same changes were seen in the bronchi and bronchioli. In contrast in the peripheral lung parenchyma there was a patchy interstitial infiltrate with damage of type I pneumocytes, hyperplasia of type II pneumocytes and an acute neutrophilic interstitial infiltrate. This patchy interstitial pneumonia persisted throughout the inhalation period of 4 weeks. In addition to neutrophils which were present throughout, also lymphocytes and monocytic cells entered the interstitium from day 6 on. Pneumocyte type II hyperplasia was a prominent feature until the 5.-6. week when inflammation in the lungs subsided. After the 9th week no inflammatory changes were seen in the alveolar region, but a mild inflammation was found in the bronchi and bronchioli, starting at the 6th week and lasting throughout to the 20th week.

BaCrO4 caused similar inflammatory changes in the respiratory tract, but the infiltrate was not as dense as with dichromate.

CrAlSO4 caused only a mild inflammation in the bronchi, bronchioli and alveolar region without hyperplasias or metaplasias. The inflammation within the nasal cavities was similar to that induced by dichromates. In all control animals no inflammation and no other pathologic changes were found in the respiratory tract.

Late phase reactions:

In 3 of 12 animals from the HC dichromate group squamous cell carcinomas were found. Carcinomas arose from the nasal and/or oral cavity, in one animal with an extension to the skin of the snout including an infiltration of the basal epidermal layer and ulceration (fig. 4). All squamous cell carcinomas were well differentiated, with > 20% keratinization of the tumor area, mild mitotic activity, lymphatic vessel invasion, but without distant metastases. In addition a mild chronic sinusitis and rhinitis was present in all animals from the LC and HC dichromate and CrAlSO4 groups. Larynx and trachea were normal in all animals. In the lungs severe bronchiectases were found in 12/12 and 6/12 animals from the HC and LC dichromate experiments, respectively. In the CrAlSO4 experiments bronchiectases were found in 3/12 (HC) and 1/12 (LC), respectively (fig. 4). This bronchiectases were accompanied by a recurrent purulent bronchitis with ulceration. In some cases a squamous cell metaplasia started from remnants of intact bronchial mucosa. In many cases bronchiectases and bronchitis were accompanied by purulent peribronchial pneumonias. 9/48 animals died with septic shock derived from these purulent bronchitis/ bronchiectases. No bronchiectases were found in control animals.

Gastrointestinal tract

Early phase reactions:

No pathological changes were seen in the oral mucosa, the pharynx and the esophagus in all groups of animals. Animals from the HC and LC dichromate groups suffered from chronic gastritis, which was more pronounced in the HC group and less pronounced in the LC group. In 1 animal (HC dichromate group) a gastric ulcer was found, directly located at the border of the forestomach and the glandular stomach, erosions were found in 3 animals.

Within the first 15 days 10/12 animals in the HC dichromate group died with paralytic ileus. There was no inflammation of the mucosa wether in the small intestine nor the colon. But a toxic degeneration of the nerval ganglia of the submucosal and muscular plexus was found in all these rats The nuclei showed condensed chromatin fragments and a vacuolar degenerated cytoplasm. No similar changes could be found in the CrAlSO4, BaCrO4 and LC dichromate groups.

Late phase reactions:

No pathological changes occurred after 3 months in the gastrointestinal tract. Control animals were free of pathological changes throughout the experiments.

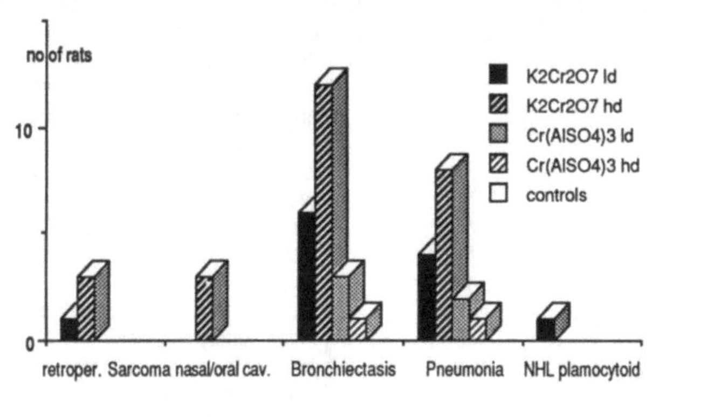

Fig. 4: Number of tumors or bronchiectases in inhalation experiments; retroperitoneal sarcomas were either malignant fibous histiocytomas or fibrosarcomas; nasal/oral cav. = squamous cell carcinomas arising from nasal cavity, extending to oral cavity and in 1 case also snout ; included are pneumonias: these are peribronchial types corresponding to bronchiectases; NHL= non Hodgkin lymphoma

Liver and spleen

Early phase reactions:

In 3/16 and 1/16 animals from the HC and LC dichromate groups, respectively, a granulomatous hepatitis was found. The granulomas consisted of neutrophils, eosinophils, lymphocytes and plasma cells and were located within the centroacinar region. The lesions persisted for approximately 3 weeks. Afterwards no inflammatory changes were found. No inflammation was found within the liver in CrAlSO4 groups and all control animals and mild inflammatory changes in BaCrO4 groups. In the spleen a depletion of lymphocytes from the trabecular areas and a pronounced siderosis was found in all animals from the HC and in 4/12 animals from the LC dichromate groups. Similar changes were seen in 5/12 and 1/12 HC and LC inhalation experiments with BaCrO4. No pathological effects were caused by CrAlSO4 inhalation.

Late phase reactions:

No pathological changes could be found in the liver in all groups. In the spleen in one animal with HC dichromate inhalation a non-Hogdkin lymphoma of plasmocytoid subtype was found (fig. 4). No other late change effects could be seen in all other animals, no pathological changes were found in CrAlSO4 inhalation.

Other organ systems

Soft tissues: In 3/12 rats from the HC dichromate group and 1/12 rats from the LC group malignant fibrous histiocytomas (MFH) or a fibrosarcoma developed from retroperitoneal soft tissues. MFH is the most common mesenchymal tumor in humans with uncertain etiology. We did not recognize visible changes in the early inhalation period and therefore did not include retroperitoneal tissues in our investigation. Also no chemical analysis for chromium was done using retroperitoneal tissue.

No pathological changes were found in the central nervous system including the spinal cord, the hemopoietic system (tested were bone marrow of femur and sternum), the endocrine system (thyroid and parathyroid glands, pancreas, adrenal glands), the reproductive system (ovary, uterus, fallopian tubes, vagina, testis, epidydimis, spermatic cords, prostate) and surprisingly the urinary system (kidneys, ureters, bladder). This was especially interesting because toxic changes have been reported from the kidneys. But neither in the HC nor the LC dichromate or BaCrO4 groups we found toxic changes: no glomerular disease, no toxic tubulopathies, no inflammation or metaplasia in the urinary bladder. CrAlSO4 did not cause toxic or tumorigenic changes. No spontaneous tumors occurred in the control animals.

Chromium and chromate analysis:

Chromium could be found in all organ system including the central nervous system. The chromium concentrations were highest in the lungs with peak values during the first week (fig. 5). Afterwards the concentration was lowered but never reached levels from control animals.

Discussion

From experimental and clinical investigations chromates are thought to be carcinogenic for humans (Alcedo, Wetterhahn, 1990, Costa, 1991, Mancuso, 1975). We recently have shown that dichromates are highly toxic for different lung cells. In our present investigation we further extend our knowledge about chromium toxicity and carcinogenicity by in vivo inhalation experiments. Four different organ systems are mainly affected by chromates, esp. the water soluble dichromates: the respiratory tract, the gastrointestinal tract, the spleen and liver and the retroperitoneal soft tissues. Interestingly carcinogenicity is solely confined to the upper respiratory and digestive tract (nasal and oral cavities), the retroperitoneal soft tissues and the spleen, whereas the lower respiratory tract is the site of severe inflammatory changes.

In contrast to different nickel compounds which primarily affect the olfactory epithelium of the nose (Benson et al., 1988) chromates act on the respiratory epithelium and caused a recurrent ulcerous rhinitis and sinusitis associated with regeneration, hyperplasia and

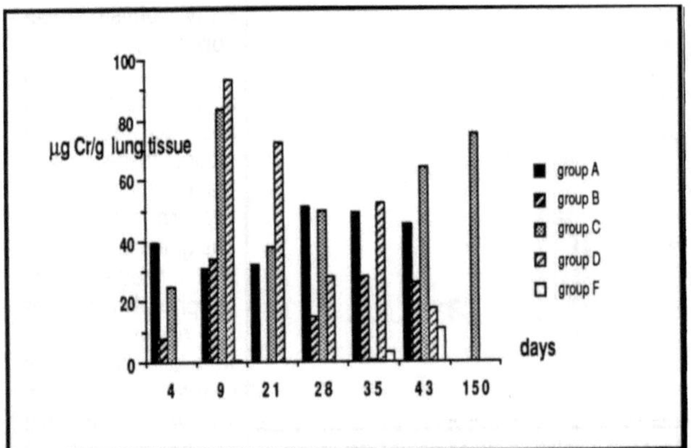

Fig. 5: Example of chromium content in dichromate inhalation experiments; groups A and B are LC (males and females), C and D are HC, F are controls; given are the chromium values per gramm lung weight; each column represents one animal

squamous cell metaplasia of the epithelium. As late effect a significant number of squamous cell carcinomas arose from the nasal cavity, infiltrated the oral one and the snout. This carcinomas could be attributed to the inhaled dichromates with great certainty (no spontaneous tumors in the controls). Mechanisms of dichromate carcinogenicity might be associated with the release of oxygen radicals (Popper, Woldrich, 1991) inducing DNA strand breaks and DNA-protein complexes (Costa, 1991). In addition chronic recurrent inflammation with attempts of regeneration might be a "triggering" factor.

The effects of dichromates in the gastrointestinal tract were surprising. We were aware of the problem of chromate ingestion by the rats cleaning their fur. This oral uptake of chromates could not be prevented in this whole body inhalation set up. Therefore we anticipated gastritis in the high dose dichromate groups; but this was also seen in the low dose ones. The toxic degeneration of nerval ganglia was not expected. This could be either an effect of dichromate itself or of circulating toxic metabolites induced by dichromates. It was more probable to assume a direct toxic effect of dichromate, because beyond the 20th day no more rats died although they still inhaled and ingested dichromate. There might be an adaptation mechanism present.. From the knowledge of the toxic effects on the small and large intestine the inflammatory lesions in the liver were expected, because dichromate would be transported from the intestine into the liver. These granulomatous lesions are typically found under these conditions.

The spleen lesions were unexpected, because no abnormalities were found in the hemopoietic system. There was no anemia, no abnormalities in erythrocytes, no pathologic changes in the bone marrow. Therefore we can just speculate about probable disturbances in the ferritin/transferrin system. The one case of NHL in the spleen can be attributed with some certainty to the dichromate effect itself. No spontaneous NHL are normally found in Fisher rats.

The most striking event was the significant number of MFH and fibrosarcoma arising from the retroperitoneal soft tissues. The etiology of MFH in humans is completely unknown. This finding of MFH in our animals might be a first step into appreciation of the etiology of soft tissue sarcomas.

Another striking effect of dichromates was the "induction" of bronchiectases in the lung. There exist a long lasting debate on the etiology of bronchiectases in humans. Most facts are related to tobacco smoke. Different factors have be accused to cause bronchiectases. The most reliable one is an imbalance of proteases and antiproteases. On the other hand many metals have been found in tobacco and oxides thereof in smoke main stream inhaled by humans. Chromium is one of those metals. Therefore one might think that there could be a causal relationship of dichromate induced bronchiectases in our experimental system and bronchiectases caused by metal oxides in the tobacco smoke.

Acknowledgements

The investigation was supported by the Austrian Research Fund (FFWF) Project P6065, which is gratefully acknowledged.

References

Alcedo JA and Wetterhahn KE: Chromium toxicity and carcinogenesis. In: Int.Rev.Exp.Pathol. 31, eds. GW Richter, K Solez, P Aisen, G Cohen, p85-108, Academic Press, San Diego, New York, Boston, London, Sydney, Tokyo, Toronto, 1990

Benson JM, Burt DG, Carpenter RL, Eidson AF, Hahn FF, Haley PJ, Hanson RL, Hobbs CH, Pickrell JA, Dunnick JK: Comparative inhalation toxicity of nickel sulfate to F344/N rats and B6C3F1 mice exposed for twelve days. Fundam.Appl.Toxicol. 10, 164-178, 1988

Costa M: DNA protein complexes induced by chromate and other carcinogens. Environ.Health Persp. 92, 45-52, 1991

Mancuso TF: Considerations of chromium as an industrial carcinogen. In: Proceedings of the international conference on heavy metals in the environment; ed. TC Hutchinson, p343-356, Institute for Environmental Studies, Toronto, 1975

Popper HH, Grygar E, Ratschek M, Weybora W, Ingolic E, Wiespeiner G, Wawschinek O. Cytotoxicity of chromium compounds: Comparison of in vitro and in vivo tests. Environmental Hygiene, Vo. 2, 73-76, 1990, Springer Verlag, Berlin, Heidelberg, New York

Popper HH, Woldrich A: Oxygen radical formation - a probable mechanism for chromate toxicity. In: Histo- and Cytochemistry as a Tool in Environmental Toxicology, eds. W. Graumann, J. Drukker. Prog. Histochem.Cytochem. 23, 220-226, 1991

Studies on the exhalation of volatile N-Nitrosamines following systemic application to Sprague-Dawley rats

P. Schmezer, B.M. Schmezer, B.L. Pool-Zobel* and R.G. Klein
Institute for Toxicology and Chemotherapy, German Cancer Research Center, INF 280, W-6900 Heidelberg; *Institute for Hygiene and Toxicology, Federal Research Center for Nutrition, Engesserstr. 20, W-7500 Karlsruhe, F.R.G.

Introduction

Specific volatile N-nitrosamines (e.g. N-nitrosodimethylamine, NDMA) have been detected in different sources of the human environment, such as in tobacco smoke (Tricker and Preussmann, 1989) or in several work places (rubber, leather, chemical industries; ECETOC, 1990). In these exposure situations, amounts of up to several hundred micrograms per cubic metre have been found. For this reason, we initiated a long-term inhalation study with NDMA in Sprague-Dawley rats using exposure schedules in which the lowest concentration was comparable to peak concentrations at those work places (Klein et al. 1991). It is feasible to make risk assessments on the basis of such inhalation experiments with animals at dose levels relevant to the human situation. For this purpose, it is furthermore important to know the actual absorption rate in the respiratory tract. Therefore, in addition to the long-term assay, several short-term experiments were performed to study the exhalation rate of NDMA and some other volatile N-nitrosamines in rats. In earlier inhalation experiments with endotracheally intubated rats, we had shown that some volatile nitrosamines including NDMA are not completely absorbed in the respiratory tract and several % of the nitrosamines inhaled for a few minutes were exhaled again within the subsequent hour (Klein and Schmezer, 1984). From the kinetic analysis of these data, exhalation rates as high as 30% could be extrapolated for NDMA and NPYR during the first breaths immediately after inhalation. This was an unexpected observation and it could be due to the local deposition of the compounds in the respiratory tract of the animals and their subsequent exhalation. To elucidate this matter, we performed additional exhalation studies following systemic application of the nitrosamines by intraveneous injection.

Material and Methods

Animals, N-nitrosamines, inhalation-, and exhalation-procedures were as described earlier (Klein and Schmezer, 1984). As shown in Figure 1, endotracheally intubated rats received the nitrosamines by i.v. injection; the animals were then transferred to a thermostatted chamber, to avoid condensation of the moisture associated with expired air. All exhaled air was collected into nitrosamine-specific absorption tubes (ThermoSorb), then the nitrosamines were eluted from these tubes, and finally quantified by a Thermal Energy Analyzer.

Results and Discussion

In contrast to the inhalation/exhalation results of our earlier studies, the results in Table 1 show that for any of the investigated nitrosamines less than 1% of the doses applied i.v. were exhaled unchanged. This was the case when the injections were given as bolus. Table 2 shows the results for the data when the injections were given as 10-minute infusions, a situation which should be more comparable to the conditions in the short-term inhalation studies. But the results reveal the same picture as for the bolus injections: For any compound less than 1% of the applied doses were exhaled unchanged. We can conclude from this that the higher exhalation rates, observed earlier after inhalative application, are probably due to the local deposition of the compounds in the respiratory tract of the animals and show the subsequent exhalation of the deposited material. Parallel to these investigations, we measured the half-lives of the different nitrosamines in the blood under the same conditions as for the presented exhalation experiments. As an example, the blood-levels for NDMA are shown in Figure 2: We measured several points over a time period of 1 hour after i.v. injection. The exponential regression analysis revealed a half-life for NDMA of about 25 minutes. The values for NDEA and NMOR were 20 and 38 minutes, respectively. If we look at the half-lives as well as other parameters such as volatility or solubility, we can see that the observed exhalation rates were positively correlated with the lipophilia (Druckrey et al., 1967) of the nitrosamines but not with the volatility (Klein 1982) or with the elimination from the blood of the rats. Since the human exposure situation is usually more complex than with only one substance it was of importance to analyse the effects of additional exposures which possibly arise. Therefore, the long-term inhalation study with NDMA, mentioned above, was also aimed at classifying the additional effect of the common air pollutants SO_2 on the NDMA-induced carcinogenesis. In this context, we examined if the exhalation rates are influenced by a pretreatment with SO_2. The results are shown in Table 3. We used two different treatment groups: one group of rats received 6ppm SO_2 over 3 weeks before we performed the exhalation studies, a second received 54ppm SO_2 over 12 weeks. In both cases, the additional SO_2 inhalation did not influence the NDMA exhalation rates.

References

Druckrey,H. Preussmann,R. Ivankovic,S. and Schmähl,D. (1967) Organotrope carcinogene Wirkungen bei 65 verschiedenen N-Nitrosoverbindungen an BD-Ratten. Zeitschr. Krebsforsch 69: 103-201.

ECETOC (1990) Human exposure to N-nitrosamines, their effects, and a risk assessment for N-nitrosodiethanolamine in personal care products. ECETOC Technical Report No 41, Bruxelles, Belgium.

Klein,R.G. (1982) Calculations and measurements on the volatility of N-nitrosamines and their aqueous solutions. Toxicology 23: 135-147.

Klein,R.G. and Schmezer,P. (1984) Quantitative measurement of the exhalation rate of volatile N-nitrosamines in inhalation experiments with anaesthetized Sprague-Dawley rats. In: O'Neil,I.K. Von Borstel,R.C. Miller,C.T. Long,J. and Bartsch,H. (eds.), N-nitroso Compounds: Occurence and Biological Relevance to Human Cancer, IARC Sci. Publ. 57: 513-517.

Klein,R.G. Janowsky,I. Pool-Zobel,B.L. Schmezer,P. Hermann,R. Amelung,F. Spiegelhalder,B. and Zeller,W.J. (1991) Effects of long-term inhalation of N-nitrosodimethylamine in rats. IARC Sci. Publ. 105: 322-328.

Tricker,A.R. and Preussmann,R. (1989) Preformed nitrosamines in smokeless tobacco. In: Tobacco and cancer, perspectives in preventive research, A.P. Maskens (ed.), Elsevier Publ., 35-47.

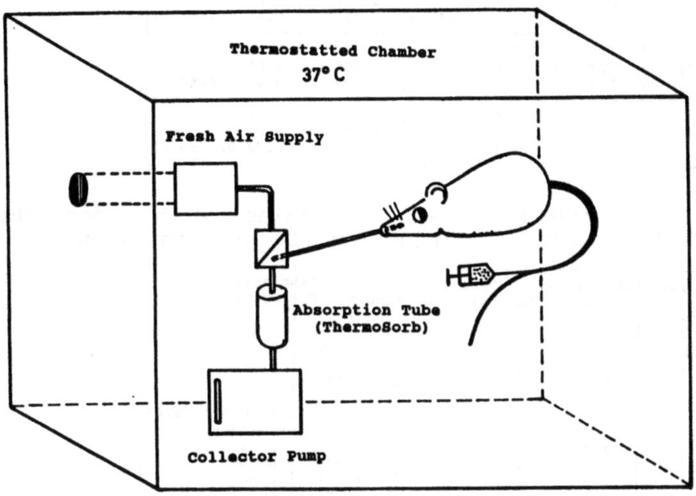

Fig. 1: Exhalation Diagram

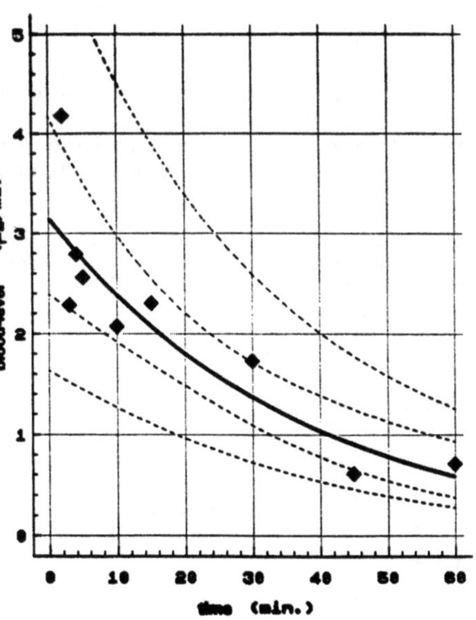

Fig. 2: Regression Analysis; NDMA; Half-live = 24.9 min.

Tab. 1: Exhalation (60 min.) of nitrosamines with endotracheally intubated rats following i.v. injection (bolus)

Compound (mg in 1ml phys.NaCl)	Animal-weight (g)	Dose (mg/kg)	x±SD	Exhalation amount of NNO (µg)	rate in %	x ± SD
NDMA (0.63)	280	2.3		0.272	0.04	
	280	2.3		0.338	0.05	
	269	2.3		0.238	0.04	
	273	2.3	**2.3±0**	0.269	0.04	**0.04±0.01**
NDEA (0.55)	258	2.1		2.611	0.48	
	270	2.0		1.986	0.36	
	285	1.9		1.222	0.22	
	258	2.1	**2.0±0.1**	2.136	0.39	**0.36±0.11**
NMOR (0.52)	270	1.9		0.044	0.01	
	317	1.6		0.035	0.01	
	278	1.9	**1.8±0.2**	0.039	0.01	**0.01±0**
NPYR (0.85)	254	3.4		0.110	0.01	
	248	3.4		0.111	0.01	
	275	3.1	**3.3±0.2**	0.148	0.02	**0.01±0.01**

Tab. 2: Exhalation (60 min.) of nitrosamines with endotracheally intubated rats following i.v. injection (10-min. infusion)

Compound (mg in 1ml phys.NaCl)	Animal-weight (g)	Dose (mg/kg)	x±SD	Exhalation amount of NNO (µg)	rate in %	x ± SD
NDMA (0.63)	230	2.7		0.333	0.05	
	230	2.7		0.350	0.06	
	230	2.7		0.575	0.09	
	225	2.8		0.352	0.09	
	230	2.7	**2.7±0.1**	0.385	0.06	**0.07±0.02**
NDEA (0.55)	295	1.9		1.449	0.26	
	330	1.7		3.070	0.56	
	232	2.4		4.711	0.86	
	229	2.4		4.479	0.81	
	270	2.0		2.654	0.48	
	254	2.2	**2.1±0.3**	1.352	0.25	**0.54±0.26**
NMOR (0.52)	282	1.8		0.022	0.004	
	295	1.8		0.071	0.01	
	200	2.6		0.079	0.02	
	215	2.4		0.061	0.01	
	249	2.1		0.048	0.01	
	236	2.2	**2.2±0.3**	0.066	0.01	**0.01±0.01**

Tab. 3: Exhalation (60 min.) of nitrosamines with endotracheally intubated rats following i.v. injection (10-min. infusion) and SO_2-inhalation

Compound (mg in 1ml phys.NaCl)	SO_2 Inhal.	Animal-weight (g)	Dose (mg/kg)	x±SD	Exhalation amount of NNO (µg)	rate in %	x ± SD
NDMA (0.63)	3 weeks 6 ppm	230	2.7		0.249	0.04	
		254	2.5		0.299	0.05	
		248	2.5	**2.6±0.1**	0.144	0.02	**0.04±0.02**
	12 weeks 54 ppm	290	2.2		0.555	0.09	
		280	2.3		0.118	0.02	
		308	2.1	**2.2±0.1**	0.492	0.08	**0.06±0.04**
NDEA (0.55)	3 weeks 6 ppm	230	2.4		3.158	0.57	
		215	2.6		4.006	0.73	
		235	2.3	**2.4±0.2**	4.007	0.73	**0.68±0.09**
	12 weeks 54 ppm	262	2.1		1.282	0.23	
		290	1.9		1.216	0.22	
		336	1.6	**1.9±0.3**	1.073	0.20	**0.22±0.02**
NMOR (0.52)	3 weeks 6 ppm	235	2.2		0.055	0.01	
		243	2.1		0.065	0.01	
		249	2.1	**2.1±0.1**	0.083	0.02	**0.01±0.01**

THE BEHAVIOUR OF INSPIRED AEROSOL PULSES IN MEN

Ch.F. Schiller-Scotland, J. Gebhart*, R. Siekmeier*

Institut für Biophysik der J.W.Goethe-Universität
Paul-Ehrlich-Str.20, D-6000 Frankfurt a.M.
*GSF Forschungszentrum für Umwelt und Gesundheit, D-Frankfurt a.M.

INTRODUCTION

Up to now the behaviour of inert aerosol particles within the human respiratory tract has primarily been studied in young healthy people. The application of bronchoconstrictive or brochodilatory pharmaca to this group results in a corresponding variation of aerosol derived effective airway dimensions (EADs). Each subject has been measured at least twice, once without pharmacon and then after drug application. By this way, each subject has been his own reference. It turned out that variations in EADs are most pronounced in proximal airways; the effect decreases continuously towards the lung periphery until it disappears. For the antagonistic drugs carbachol (bronchoconstrictor) and oxitropiumbromide ((Ventilat^R) bronchodilator) the sites of action are in almost identical volumetric lung depths V_{LD}s. The percental changes in EADs are equivalent for the applied doses of pharmaca (Schiller-Scotland (1990)).

In the present study these pharmaca experiments are repeated to study the effect of bronchoconstriction and bronchodilation on inspired aerosol pulses. In many respects small aerosol particles behave like a non-diffusing gas which trace almost exactly the mass movement of air in which particles are suspended. As a consequence the volumetric dispersion of an aerosol pulse describes the convective mixing of air during a breathing cycle. Changes of aerosol pulses during a respiratory cycle may therefore indicate functional disturbances of pulmonary ventilation. Such disorders are also characteristic for severe smokers. Therefore the behaviour of inspired pulses has been regarded in middle-aged healthy nonsmokers and smokers.

EXPERIMENTAL METHOD

By means of an inhalation apparatus which allows continuous monitoring of respiratory volumes and of aerosol concentration close to the mouthpiece changes of aerosol pulses during a respiratory cycle can be studied. Particle concentration is measured with a laser photometer and respiratory flow rate by a pneumotachograph.

Experiments are performed by use of a nonhygroscopic monodisperse di(2ethylhexyl)sebacate aerosol, condensed on NaCl nuclei. The particle size was $\approx 0{,}8\,\mu m$ (σ_g=1,1).

Dispersion of a 70 cm^3 pulse of monodisperse aerosol has been investigated applying the following maneuver: By breathing particle free air, airborne particles are removed from the respiratory tract. Starting from functional residual capacity (FRC) the subject inspires 1000 cm^3 (air and aerosol) with a flow rate of 250 cm^3s^{-1}. By means of a valve system the aerosol bolus is injected into the inspired air at a preset step. After end inhalation the subject immediately expires with the same flow rate to residual volume (RV). Within one series of measurements the aerosol pulse is placed at different volumes corresponding to different V_{LD}s.

The dispersion of the expired pulse, which is a function of convective gas mixing between inspired air and pulmonary trapped air is characterized by the bolus half-width ($H = V_{h1}+V_{h2}$). The loss of particles during the breathing maneuver is described by the deposition DE. (Fig. 1).

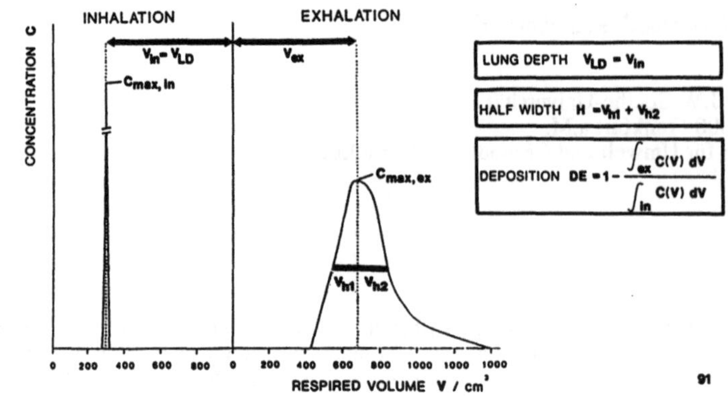

Fig. 1:

RESULTS AND DISCUSSION

In the pharmaca studies the behaviour of aerosol pulses were investigated in 8 healthy nonsmokers (5 men, 3 women, 20-30 yrs.). For each pharmacon the subjects were measured twice: Before and after inhalation of 1,75 mg carbachol (CARB), and before and after inhalation of 0,2 mg oxitropiumbromide (OXI), respectively. Carbachol was inhaled to achieve bronchoconstriction, Oxitropiumbromide is known as a bronchodilator. Both pharmaca affect cholinergic receptors but there was a time interval of several days between both pharmaca applications.

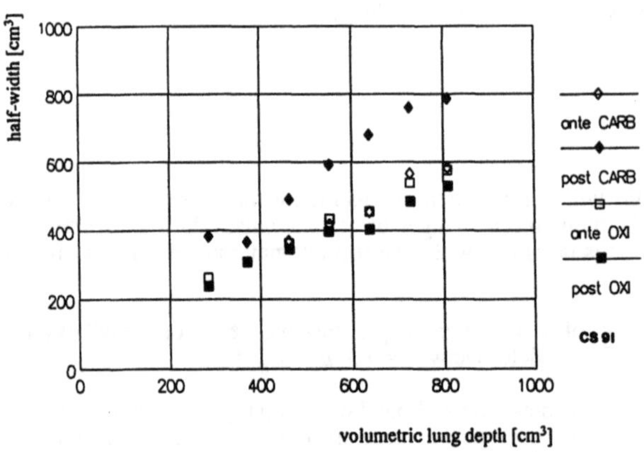

Fig. 2: Dispersion (half-width) of respired aerosol pulses in expired air as function of volumetric lung depths; data represent means of 8 nonsmokers

Before pharmacon application H (Fig. 2) and DE (Fig. 3) show no significant intrapersonal variability for both experimental runs. H as well as DE increase with increasing V_{LD}. The inhaled aerosol pulse preserves its identity during a breathing cycle in a remarkable manner, following the principle of "last in first out and vice versa". After carbachol inhalation dispersion expressed by H is significantly higher (Fig. 2) than the reference values ($p<0.01$, Wilcoxon). DE (Fig. 3) now also increases with increasing V_{LD}, especially after carbachol inhalation. In comparison an equivalent oxitropiumbromide induced bronchodilation only slightly indicates corresponding effects. While bronchoconstriction seems to interfere with convective gas transport, an additional bronchodilation does not induce a significant variation.

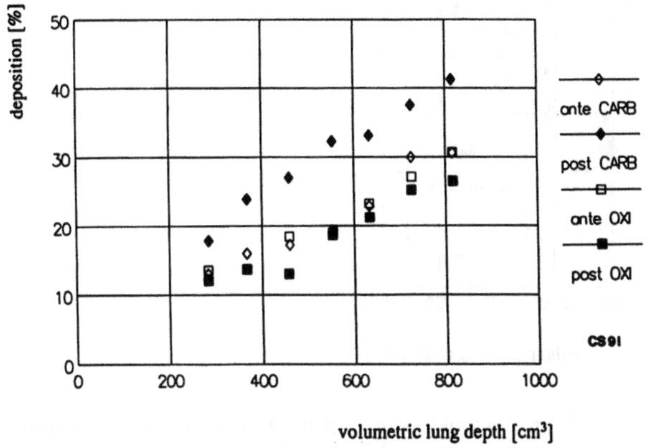

Fig. 3: Deposition of 0,8 μm-particles from aerosol pulses inspired in different volumetric lung depths; data represent means of 8 nonsmokers

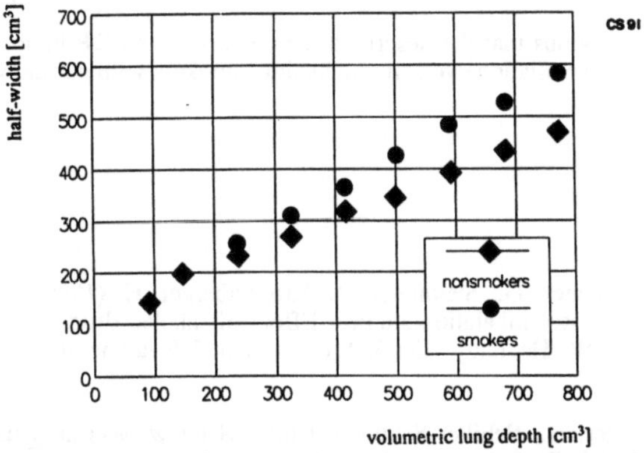

Fig. 4: Dispersion (half-width) of respired aerosol pulses in expired air as function of volumetric lung depths; data represent means of 32 nonsmokers and 32 smokers

In a second study, where no pharmaca were applicated, 32 nonsmokers (NS:50,0±6,0 yrs.) and 32 healthy smokers (S: age 49,3±6,3 yrs., 45,6 pack yrs.) matched for age, body size and weight volunteered in the examination. For all subjects the values of conventional pulmonary function tests are within the normal limits.

It turns out that the absolute values of the bolus half-width as well as the differences between both groups (Fig. 4) increase with V_{LD}. For $V_{LD} < 200$ cm³ there are no significant differences for the H-values. Differences get significant for $V_{LD} \approx 240$ cm³ (see Table) and are highly significant for $V_{LD} \geq 500$ cm³.

Results for the WILCOXON matched pairs signed ranks test:

HW_{240}	HW_{330}	HW_{420}	HW_{590}	HW_{770}
$p<0.06$	$p<0.01$	$p<0.01$	$p<0.001$	$p<0.001$

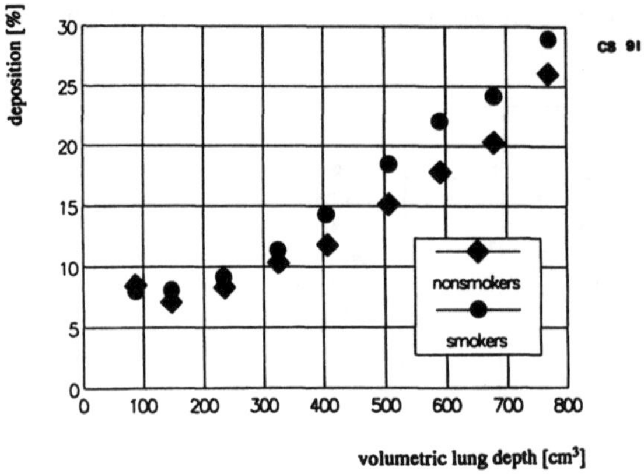

Fig. 5: Deposition of 0,8 μm-particles from aerosol pulses inspired in different volumetric lung depths; data represent means of 32 nonsmokers and 32 smokers

DE (Fig. 5), which also increases with V_{LD} shows no significant differences between S and NS for V_{LD} <400 cm^3. Only for V_{LD}> 400 cm^3 there are small differences (Wilcoxon p<0.1).

From these experimental findings it is obvious that the described parameters H and DE indicate the existance of small disorders in pulmonary ventilation even among healthy smokers with normal values in pulmonary function tests.

REFERENCES:

Schiller-Scotland, Ch.F., Gebhart, J., Scheuch, G., Siekmeier, R., Kronenberger, H. (1990): Human airway diameters derived from aerosol inhalation studies: Effect of inhaled drugs on airway dimensions. In: N.H. Seemayer, W. Hadnagy (Eds.): Environmental Hygiene II, Springer-Verlag, Berlin, pp. 229-232.

Siekmeier, R., Gebhart, J., Kronenberger, H., Schiller-Scotland, Ch.F., Schultze-Werninghaus, G., Heyder, J., Meier-Sydow, J. (1990): Einfluß pharmakoninduzierter Bronchokonstriktion bzw. Bronchodilatation auf die Dispersion inhalierter Aerosolpulse. Atemwegs-Lungenkrankheiten 16, 289-291.

Siekmeier, R., Gebhart, J., Kronenberger, H., Schiller-Scotland, Ch.F., Schmidt, U. (1991): Behaviour of inspired pulses of inert aerosol in middle-aged healthy smokers and non-smokers. J. Aerosol Medicine 4, Supplement 1, p. 43.

Acknowledgement - This work was partially supported by the CEC under Contract No. PL 89 00 82.

The Behavior of Inhaled Spherical Iron Oxide Particles in Human Lungs: Magnetometric Studies

Stahlhofen, W. and W. Möller,
GSF - Forschungszentrum für Umwelt und Gesundheit GmbH, Institut für Biophysikalische Strahlenforschung, Paul-Ehrlich-Straße 20, 6000 Frankfurt am Main 70, Germany

Introduction

The measurement of the remanent magnetic field of the human lung contains a lot of information about the behavior of magnetic material, either in the lung or in the macrophages. Iron oxide can be used as a test aerosol for studies of particle clearance and macrophage activity.

Magnetopneumography (MPG) is a useful tool to characterize the defense mechanisms of the lung. After voluntary inhalation of ferrimagnetic particles, the retention of the particles can be estimated. During the first day after deposition, the particles are engulfed by alveolar macrophages. The decay of the remanent magnetic field (rmf) of the lung after a magnetization process, called relaxation, reflects the mobility of the alveolar macrophages. This decay is influenced by the activity of the macrophages and by the hydrodynamic (viscoelastic) properties of the surrounding medium. With a secondary magnetization process (Cohen et al., 1984) this properties can be independently studied. In this models the macrophage activity was characterized as a randomization process with an intra-cellular energy E_Z (Nemoto, 1982). The shape of the particles has an influence on the estimated values for viscosity and cell energy. If the particles consist of smaller aggregates, then during mag-netization, chains might be build up with the induction of a changing hydrodynamic behavior and changing magnetic properties of the particles. In vivo measurements of intracellular viscosity and macrophage activity with the use of solid spherical ferrimagnetic particles are discussed in this study.

Material and Methods

Spherical particles were produced by nebulizing a Fe_2O_3-colloid with a spinning top aerosol gener-ator (Möller et al., 1990). After the addition of hydrogen gas (2 vol.-%) the particles were reduced to Fe_3O_4 (magnetite) in a furnace at 800 °C. After the reduction stage, O_2 was added in order to get the composition of normal air. This aerosol was breathed directly, until 0.5 - 1 mg were deposited in the human lungs. The aerodynamic diameter of the particles was 2.8 μm (geometric standard devia-tion < 1.1) and allowed the deposition mainly in the alveolar region of the lung. The deposition of 0.5 mg requires about 10^8 particles, so nearly one particle is deposited per alveoli. The particles should behave like single magnetic spheres in the lung.

For magnetopneumographic measurements the person lay in prone position on a bed that could be moved between the magnetizing and the sensing system (Stahlhofen et al., 1988). Magnetization could be done in the localized and in the homogeneous field technique, respectively. In localized field technique, two magnetizing coils were available with 8.66 cm diameter for long-term magneti-zation and 10 cm diameter for pulse magnetization, both having 1 T central flux density. For homo-geneous field technique, a single coil with 40 cm diameter and a maximum central flux density of 0.1 T was used. Magnetopneumographic measurements were done with a superconducting 2nd order SQUID-gradiometer sensing system with 10 cm diameter of the sensing coils and 15 cm baselength. In order to achieve sufficient sensitivity the whole system was placed within a magneti-cally shielded room. This allowed a lower detection limit below 50 μg of magnetite in the whole lung.

After inhalation the particles were magnetized in a strong external magnetic field (primary magneti-zation). The relaxation phenomenon can be observed in a decrease of the remanent magnetic field after the magnetizing field was switched off (Stahlhofen et al., 1990a). The main process that destroys the aligned magnetic dipoles was found to be the activity of the macrophages. This misalignment happens within the inner cell structure of the macrophages. The particles deposited within the alveolar region are engulfed very fast by macrophages and stored into phagolysosomes. When using spherical particles the membranes of the phagolysosomes was found to be tightly adhered to the surface of the particles (Stahlhofen et al., 1990b). So it was assumed that the whole phagolysosome containing a magnetic sphere should be rotated. One hour after magnetization the lung becomes again nonmagnetic. In this state all remanent magnetic dipoles are randomly oriented.

Primary magnetization of the particles was done in homogeneous field technique with 70 mT. After completed relaxation all remanent dipoles were randomly oriented, making the lung nonmagnetic. The secondary magnetization curves were then recorded for different field strengths in the range from 0.22 mT to 4.4 mT in homogeneous field technique. In a weak magnetic field (< 5 mT) where no change of the magnetic domain structure occurs, a torque acts on magnetic spheres and tends to align the magnetic dipoles parallel to the direction of the magnetic field. In a viscous environment the hydrodynamic retarding torque balances the magnetic torque, describing the time dependence of magnetic field increase. The rate of increase gives information on the apparent viscosity of the cytoskeleton of the macrophages. With the use of spherical magnetic particles with 1.3 μm diameter in the human lungs the apparent viscosity of the cytoplasm was determined. This measurements are influenced by the activity of the macrophages and by elastic properties of the cytoplasm.

Results and Discussion

The course of increase of magnetization of the lung is described with the balance between magnetic torque and the resistance of the surrounding medium due to its viscosity. This problem was solved for the initially random orientation of the remanent dipole particles (Stahlhofen et al., 1990c). Recorded secondary magnetization curves of one subject are plotted in fig. 1 with three different strengths of the weak magnetizing field. A fit of the expected course of the Newtonian viscosity model could be done only during the initial phase. The rate of increasing magnetization of the lung allows to estimate the apparent viscosity of the surrounding medium. The mean value of estimated viscosity of this subject is shown in tab. 1 together with the results for two other non-smoking subjects.

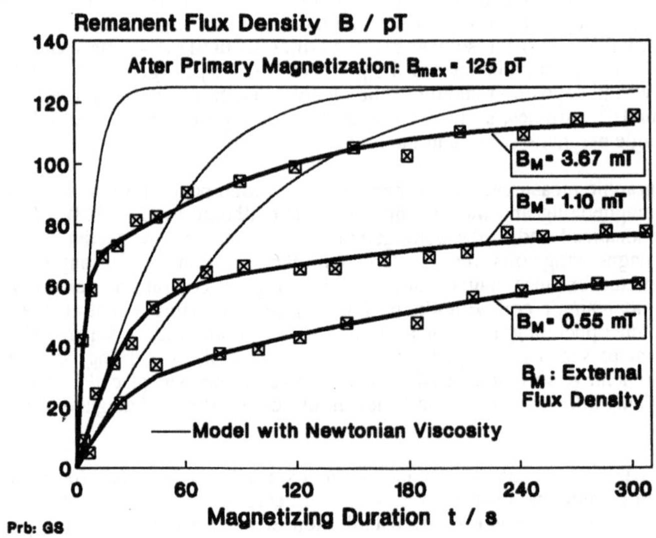

Fig. 1: Secondary magnetization curves with spherical magnetite particles for different external field strengths B_M.

For increasing duration of the secondary magnetization process the intracellular activity of the macrophages influences the ordering process in the magnetic field, yielding in a lower value for the stationary remanent magnetic field of the lung compared to primary magnetization. The equilibrium rmf of the lung increased with increasing magnetizing field strength indicating a behavior like in paramagnetism (Nemoto, 1982). The application of the model of paramagnetism (Langevin function) allowed to characterize the activity of the alveolar macrophages as a randomization process with an energy of $E_Z \approx 6 \cdot 10^{-18}$ J (Tab. 1).

SUBJECTS:	Prb: FH	Prb: GS	Prb: GR
APPARENT VISCOSITY η / Pa·s (MEAN±SD)	88.8±48	99.2±25	69.7±19
RANDOMIZING ENERGY E_z / J	$6.8 \cdot 10^{-18}$	$5.8 \cdot 10^{-18}$	$7.6 \cdot 10^{-18}$

Table 1: Predicted values for randomizing energy and viscosity from a fit of the above models to the measured behavior.

Cell energy E_Z is much larger than thermal energy at 37 °C: $kT = 4.3 \cdot 10^{-21}$ J. Thermal energy can be neglected when describing the behavior of magnetic particles within macrophages. Nemoto (1989) showed, that with lavaged hamster macrophages the cell energy could be influenced by the ATP-concentration of the cell.

The Newton-viscous properties of the cytoskeleton could be determined as (mean ± standard deviation): $\eta = 87.1 \pm 34.5$ Pa•s. The intracellular viscosity of macrophages reported in literature are much higher: Nemoto, 1982 (in vivo): η = 320 - 480 Pa•s, Valberg et al., 1987 (in vitro) η = 250 - 2700 Pa•s. Valberg's analysis on cultured macrophages show a high shear-rate dependence, that can not be analyzed in our experimental set up.

When describing the hydrodynamic behavior of the particles in the cells, elastic properties of the cytoplasm must be taken into account. Therefore the influence of magnetization duration on the relaxation phenomenon during primary magnetization was investigated (Stahlhofen et al., 1990a). Pulse magnetization and long term magnetization yield a slower decay. Only with a magnetizing duration of some seconds was a faster decay of the rmf recorded. The analysis with a double exponential relaxation model showed that only 10 % of the relaxation were affected by elastic properties.

It was shown that magnetopneumography is a noninvasive method to characterize two important properties of the alveolar macrophages, the cell-energy and the viscosity. Both properties should contribute to the defense-reaction of the macrophages. It should be interesting to investigate these phenomenon in healthy (smokers and non-smokers) and disease.

Beside this phenomenons the clearance behavior of the alveolar region of the human lungs is demonstrated in fig. 2. This retention curve was measured after localized field magnetization over one region of the right lung during a period of 300 days. During the initial phase an increase of the field of the lung was recorded that might reflect a redistribution of the magnetic material to peripheral lung structures (pleura). This redistribution near to the sensor results in a higher signal in the sensor. For non-smokers the retention process for the magnetite particles could be described with a half time of 110 days. This value is comparable to the results of Freedman et al., 1988. Freedmans results for smokers show a reduced clearance potency for smoking persons.

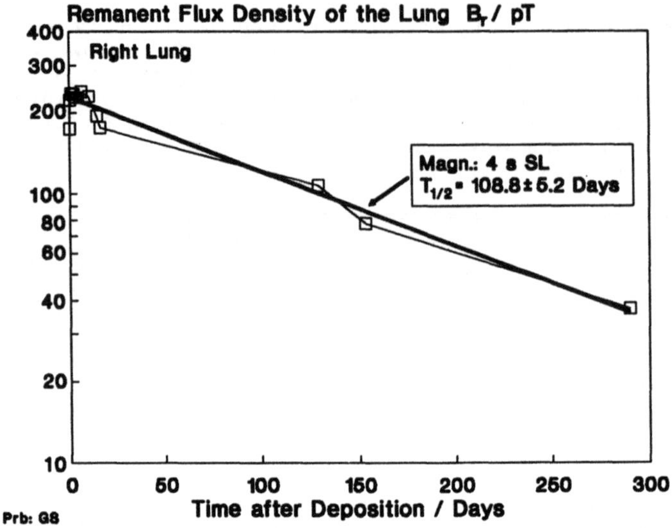

Fig. 2: Measured retention after voluntary inhalation of about 0.5 mg of magnetite particles.

Acknowledgement - This work was partially supported by the CEC under Contract No. PL 890082.

References

Cohen, D., I. Nemoto (1984), Ferrimagnetic Particles in the Lung-Part I: The Magnetizing Process, IEEE Transactions on Biomedical Engineering, Vol. BME-31, No. 3, 261-273.

Freedman, A.P., Robinson, S.E., Street, M.R. (1988), Magnetopneumographic Study of Human Alveolar Clearance in Health and Disease, Ann. Occup. Hygiene, Vol. 32, Suppl. I, 809-820.

Möller, W., W. Stahlhofen, C. Roth (1990), Improved Spinning Top Aerosol Generator for the Production of High Concentrated Ferrimagnetic Aerosols, Journal of Aerosol Sci, Vol. 21, S657-S660.

Nemoto, I. (1982), A Model for Magnetization and Relaxation of Ferrimagnetic Particles in the Lung, IEEE Transactions on Biomedical Engineering, Vol. BME-29, 745-752.

Nemoto, I., Kazuhito, O., Hiroyuki, W. (1989), Estimation of Energy of Intracellular Movement and Apparent Viscosity by Magnetometry, Advances in Biomagnetism (Ed.: S.J. Williamson, M. Hoke, G. Stroink, M. Kotani), Plenum Press, 469-472.

Stahlhofen, W., Möller, W. (1988), Description of a Biomagnetic Method for Detection of the Behavior of Magnetic Aerosols in the Human Lungs, Journal of Aerosol Science, Vol. 19, No. 7, 1087-1091.

Stahlhofen,W., Möller, W. (1990a), Magnetic Measurement of Macrophage Activity in Human Lungs, Environmental Hygiene II (Eds.: N.H. Seemayer et al.), Springer Berlin, 213-216.

Stahlhofen,W., Möller, W., Godleski, J. (1990b), Relaxation Measurements with Spherical Magnetic Particles in Human Lungs, Journal of Aerosol Sci, Vol. 21, No. 3, 355-362.

Stahlhofen, W., Möller, W. (1990c), Using Spherical Magnetic Particles for Testing the Intracellular Viscosity, Journal of Aerosol Sci, Vol. 21, Suppl., S435-S438.

Valberg, P.A., J.P. Butler (1987a), Magnetic Particle Motions within Living Cells, Part I: Physical Theory and Technique, Biophysical Journal, Vol. 52, 537-550.

Valberg, P.A., Feldman, H.A. (1987b), Magnetic Particle Motions within Living Cells, Part II: Measurement of Cytoplasmic Viscosity and Motile Activity, Biophysical Journal, Vol. 52, 551-561.

HUMAN EXPOSURE MONITORING

Cross Sectional Epidemiological Study on Cobalt Excretion in Urine of Children in Areas with Different Air Pollution

Eikmann, Th. and A. Krönung
Hygiene Institute of Ruhr-Area/Gelsenkirchen, Rotthauserstr. 19, D-4650 Gelsenkirchen, FRG

Introduction

Cobalt is a relatively rare element, which takes the 32nd place on earth in distribution of elements. It is mainly found in copper ores of Katanga in Zaire as well as in Pyrites of Ontario in Canada. The evident glitter of some parts of these ores pretended to contain silver, copper or tin, but didn't forward any precious metal when dressed and washed. Therefore the ancient miners believed in a specter called "imp" - in German called "Kobold".
Cobalt is mainly used in the carbide industries. Typical signs of carbides are most of all stability of temperature and corrosion-resistance. Cobalt is also a component of hard alloys, called stellit, which contain 50 to 60 % cobalt. In the USA 75 % of the cobalt reservoires were used to produce steel and alloys.
Cobalt belongs to the essential trace elements and shows its biological effect in the form of vitamin B_{12} which is cyanocobalamin.

Absorption, Metabolism and Toxicity

The intake of cobalt by man is about 140 to 580 μg/day, 20 to 95 % of this quantity can be absorbed. The main quantity of the absorbed cobalt is not available as vitamine B_{12}, but is component of food linked as inorganic compound. It is absorbed in the small intestine und excreted mainly in urine and partly in sweat or feces.
The intake of inorganic cobalt compounds is coupled with those of iron. Anemic patients given iron preparations in therapeutic doses showed an increase of the cobalt absorption. It is estimated that iron transport proteins can link cobalt with higher affinity than iron.
Pharmacological doses of cobalt stimulate the hematopoesis; there are indications that the element probably increases the production of erythropoetin. Thirty years ago cobalt-salts were therefore used for the treatment of refractory anemia; nowadays the use of this kind of treatment is discussed controversially. There are some indications that cobalt has an influence on the hormonal synthesis capacity of the thyroid gland.
The toxic effects of cobalt are extraordinary diverse; they extend from contact dermatitis, sensibilization, allergic reactions, epileptic attacks, testicular degeneration, lung fibrosis, cardiomyopathia with kidney failure up to cancer. In the frame of this paper it doesn't seem useful to expand the description of the different effects of cobalt.
Only one occurence should be presented shortly because of its special interest for the general population. In the sixties in Canada an increase of cardiomyopathia of beerdrinkers had been discovered. It was found that the reason was cobalt-chloride which had been added to the beer as stabilizer of froth (Schrauzer 1991).

Materials and Methods

The aim of the presented study was to examine the exposure of children in areas with different air pollution with the aid of cobalt excretion in 24 hour urine and to investigate simultaneously the influence of confounding factors.
The polluted area in Germany, chosen for this study, was Herne in the central Ruhr-area; this city is characterized by numerous industrial plants and a high flow of traffic. Borken in the western part of the Münsterland, a rural area with less traffic, was chosen as reference area.

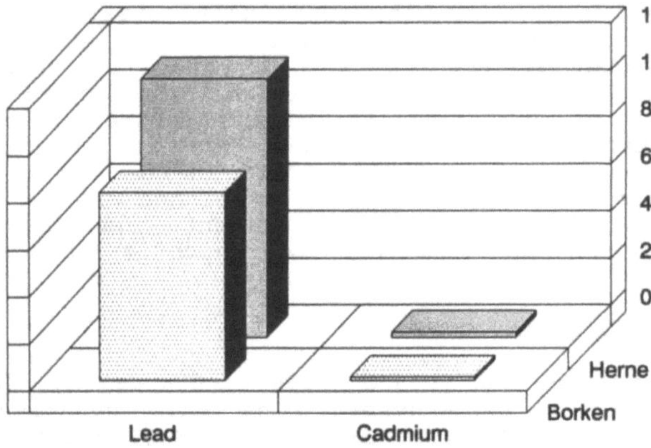

Fig. 1: Lead and Cadmium Deposition ($\mu g/m^2 d$) in the Areas Herne and Borken

There are only few data about the air pollution situation in both areas. The sulfur dioxide and nitrogen oxide concentration in the air show a tendency to higher concentrations in Herne, but the difference is not very distinctive. In contrast to this fact a clear difference in the deposition of lead and cadmium exists between these areas. In the industrialized region of Herne the deposition of heavy metals is statistically prooved higher than in the rural area of Borken.
In this study children between the age of 5 and 6, before first school year, were selected as volunteers to be investigated. In both areas 573 children took part in the study, the response rate was about 70 %.
The analysis of cobalt was carried out with atomic absorption spectrophotometry. As a model for the analytical statistics the main-effects-model SAS-GLM was used. This model represents a variance analysis for unequally distributed data.

Results and Discussion

As a first step in the evaluation of data the influence of confounding parameters on the mean cobalt excretion of the children in the different polluted areas is examined. None of the investigated confounding factors shows a statistically relevant influence on the cobalt excretion. Therefore it was not necessary to exclude any parameter from further evaluations.

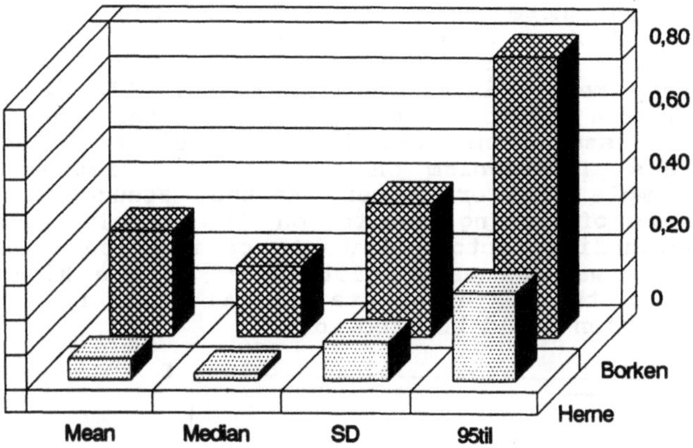

Fig. 2: Mean Cobalt-Excretion in Urine (μg/24h) of Children in Herne and Borken

In comparing the mean cobalt excretion of the children the lowest values are always seen in the polluted area of Herne. The average cobalt excretion of children from Borken is about ten times higher than those from Herne. The mean value of cobalt in urine in Borken amounts to 0.30 μg per 24 hours, in Herne to 0.06 μg. The median values show a similar difference between both collectives.

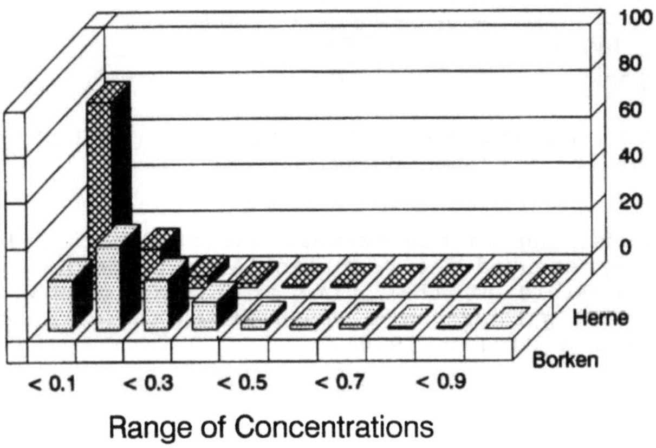

Fig. 3: Frequency Distribution (%) of Cobalt Concentration (μg/24h) in Urine Samples

The frequency distribution of the individual values of both collectives show this difference as well. The tendency to higher cobalt-concentrations is more distinct in Borken than in Herne. The values of our study correspond to the results of investigations of professionally not exposed persons to a sufficient extent.

The results of the demonstrated investigation show that the place of residence has an overwhelming influence on the cobalt excretion of children. The mean excretion of the children in Borken is about ten times higher than those of the children in Herne. An influence of any confounding factors could not be prooved. The data of the biological monitoring investigation are in contrast to air borne deposition of heavy metals in the areas. At this time it appears not possible to give a sufficient explanation for this phenomena. There are several assumptions: one is the possible influence of other heavy metals in organism on the cobalt-excretion of the children, for example zinc or copper. Another assumption is the possible influence of eating habits on the cobalt excretion; although the confounding factor vegetable consumption - asked in the questionnaire - was unequally distributed with a higher frequency in Borken in the different statistical tests no influence on the cobalt excretion could be prooved.

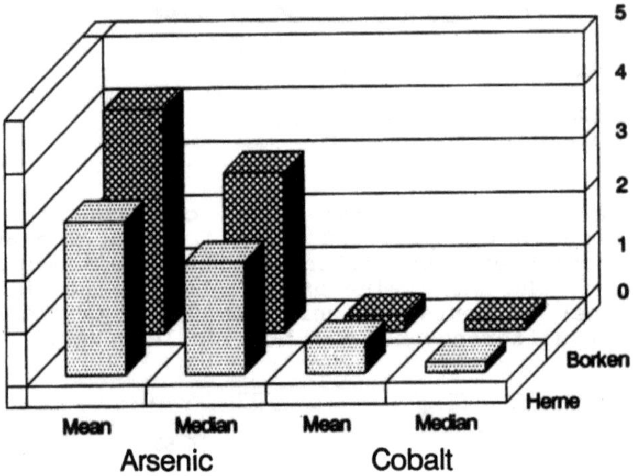

Fig. 4: Mean Cobalt- & Arsenic-Excretion in Urine (μg/24h) of Children in Herne and Borken

Similar results referring to the arsenic excretion of children in these areas were represented one year ago. An explanation of these results was not possible at that time and isn't possible until now. Further investigations are necessary to get adequate explanations (Eikmann & Michels 1990).

References

Eikmann, Th., and S. Michels (1990) Cross Sectional Epidemiogical Studies on Arsenic Excretion in Urine of Population in Areas with Different Air Pollution. In: Seemayer, N.H., Hadnagy, W. (Eds.) Environmental Hygiene II, pp. 149-152, Springer-Verlag, Berlin-Heidelberg

Schrauzer, G.N. (1991) Cobalt. In: Merian, E. (Ed.): Metals and their Compounds in the Environment. Occurence, Analysis, and the Biological Relevance, pp. 879-892. VCH Weinheim, New York, Basel, Cambridge

Human Exposure to Mercury - Effect of Amalgam Fillings

U. Ewers, Hygiene-Institut des Ruhrgebiets - Institut für Umwelthygiene und Umweltmedizin, Rotthauserstr. 19, D-4650 Gelsenkirchen

Human Mercury exposure from amalgam fillings and other sources

According to a WHO task group on environmental health criteria for methylmercury the average daily intake of mercury in the general population not occupationally exposed to mercury can be described as shown in Table 1. Taking into account the low mercury concentration in air the fraction of mercury absorbed by inhalation is very small. Mercury in non-fish foodstuffs normally occurs in the form of inorganic mercury compounds. The average daily intake is estimated to be 3,6 μg per day in industrialized countries. Mercury in fish mainly consists of methylmercury (mean intake: 2,4 μg per day). Therefore, fish consumption represents the main source of exposure to methylmercury.

In subjects with amalgam fillings the major part of mercury intake results from the release of mercury vapour from dental amalgam fillings. It has been clearly demonstrated by numerous studies that amalgam restorations emit trace amounts of mercury into the oral cavity. A substantial part of this amount is inhaled and efficiently absorbed into the blood-stream via the lungs. According to the WHO Task Group the daily intake of mercury released from amalgam fillings may vary between 3,8 and 21 μg. Estimates of similar magnitude were published by other authors. As shown in Table 1 the average daily intake of mercury from amalgam fillings clearly exceeds the intake of mercury from all other sources.

Table 1. Estimated average intake of mercury (μg/day) in the general population not occupationally exposed to mercury.

Exposure	Elemental mercury vapour	Inorganic mercury compounds	Methyl-mercury
Air	0.03	0.002	0.008
Food			
- Fish	0	0.6	2.4
- Non-fish	0	3.6	0
Drinking water	0	0.05	0
Dental amalgam	3.8 - 21.0	0	0
Total	3.8 - 21.0	4.3	2.4

Data from ICPS (1990)

Retention and distribution of mercury in the human body

Estimating the dose of mercury absorbed and retained in the human body the following assumptions (which are based on a number of experimental data) can be made:

- absorbtion of inhaled mercury vapour: 80 %
- absorbtion of inhaled inorganic mercury compounds: 50 %
- absorbtion of inhaled organic mercury compounds: 80 %
- absorbtion of inorganic mercury compounds in the gastro-intestinal tract: 5 %
- absorbtion of organic mercury compounds in the gastro-intestinal tract: 80 %.

Based on these assumptions, the average daily retention of mercury in the general population not occupationally exposed to mercury can be described as shown in Table 2. The data show that mercury retention from dental amalgam fillings represents the main source of internal mercury exposure in man.

Elemental mercury taken up into the bloodstream is physically dissolved in the blood. In this form it diffuses into various tissues and also through the blood-brain-barrier. Within a few minutes it is oxidized to divalent mercury ions. The largest amount of mercury is accumulated in the kidneys. Other organs, where significant accumulation occurs, are the lungs, spleen, heart, brain, pancreas and some other tissues.

Considering the relative importance of mercury exposure from amalgam fillings, it is not surprising that subjects with amalgam filling have significantly higher mercury concentrations in blood, urine and many tissues (including kidneys and brain) than subjects without amalgam fillings. Moreover, there is a significant correlation between the number of amalgam fillings and the levels of mercury in blood, urine and various tissues (for references see Weiner et al., 1990; Zander et al., 1990;

Table 2. Estimated average retention of mercury (μg/day) in the general population not occupationally exposed to mercury.

Exposure	Elemental mercury vapour	Inorganic mercury compounds	Methyl-mercury
Air	0.024	0.001	0.006
Food			
- Fish	0	0.042	2.3
- Non-fish	0	0.25	0
Drinking water	0	0.004	0
Dental amalgam	3 - 17	0	0
Total	3.1 - 17	0.3	2.31

Data from IPCS (1990)

Hanson and Pleva, 1991; Schiele, 1991). In a recent study performed in the present author's laboratory (Zander et al., 1990) the mean urinary mercury excretion of subjects with amalgam fillings (n = 72) was found to be 0,6 μg/l (range: < 0,1 - 7,7 μg/l). The mean urinary excretion of subjects without amalgam fillings (n = 21) was 0,2 μg/l (range: < 0,1 - 1,1 μg/l). Standardizing urinary mercury for creatinine the following values were obtained: Subjects without amalgam fillings: mean: 0,2 μg/g creatinine (range: < 0,1 - 0,7 μg/g creatinine); subjects with amalgam fillings: mean; 0,8 μg/g creatinine (range: < 0,1 - 14,0 μg/g creatinine). The upper normal limits were found to be 1 μg/g creatinine for subjects without amalgam filliungs and 4 μg/g creatinine for subjects with amalgam fillings.

Following application of certain chelating agents such as dimercaptopropione sulfonic acid (Dimaval^{R}) subjects with amalgam fillings excrete a significantly higher dose of mercury than subjects without amalgam fillings (Schiele et al., 1989; Zander et al., 1992).

The present data, which are reviewed very briefly here, clearly show that dental amalgam fillings represent the main source of mercury exposure in man, exceeding substantially the amount of mercury absorbed from food and other sources.

Health effects of mercury released from amalgam fillings

The present debate focuses on the question, whether exposure from amalgam fillings represents a health risk for man, at least for those, whose amalgam fillings are in bad condition und, therefore, exhibit a high degree of internal exposure to mercury.

The problem of adverse reactions related to dental amalgam fillings is that most cases are anecdotal and poorly investigated and documented. One reason for this situation is that the symptoms are very unspecific and that it is nearly impossible to relate them to a specific cause.

The symptoms commonly reported are related, in part, to the central nervous system including disabling fatigue, headaches, vertigo, visual disturbances, impaired short-term memory, insomnia, anxiety and anguish, difficulty in concentrating, and depression. Other symptoms, commonly reported are muscular pains, gastro-intestinal and renal disturbances, and increased susceptibility to infections. These symptoms are similar to those reported in cases of occupationally related chronic mercury poisoning. Based on dose-response relationships derived from studies of occupationally exposed populations it has been concluded, however, that the level of mercury exposure related to amalgam fillings is insufficient to cause such health effects. In a large-scale recent study no relationship between the number of amalgam tooth fillings and subjectively experienced symptoms could be demonstrated (Ahlquist et al., 1988). Similar results were reported by Mülller-Fahlbusch (IDZ, 1988).

Regarding the toxic effects of elemental mercury and inorganic mercury compounds there are numerous experimental data and case reports showing that individual susceptibility can vary over a wide range. Autoimmunity induced by inorganic mercury compounds in experimental animals appears to be a question of susceptibility rather than a question of dose. In the case of acrodynia, only a small fraction of children exposed to mercury from calomel powders developed serious and sometimes fatal symptoms of poisoning, whereas the majority of children appear to have tolerated the treatment without any ill effect . This also underscores the importance of individual susceptibility (for details see Weiner et al., 1990). Finally, it should be pointed out that certain mercury compounds can induce allergic contact dermatitis. Amalgam fillings, however, do not appear to play a role in this kind of allergic reaction. On the other hand, it can not be excluded that an increased internal exposure to mercury may exacerbate allergic reactions in sensitized individuals, a phenomen generally called systemically induced or hematogenous contact ekzema (Klaschka and Ring, 1990).

Conclusions

In conclusion, it can be stated that the release of mercury from dental amalgam fillings represents a significant source of mercury exposure in man, which may substantially exceed mercury exposure from food and other sources. At present, there are no data indicating that mercury from amalgam fillings constitutes a serious health risk at the population level. However, there are a number of experimental data and case reports showing that individuals react with unusual great variability, i.e. some appear to be very susceptible whereas other are not. Animal data suggest that host factors affecting susceptibility to mercury are related to the immune system. Thus, mercury could, via the immune system, have effects on many organs at doses which are not directly toxic.

References

Ahlquist, M. et al. (1988), Commun. Dent. Oral Epidem. 16:227-231.

Hanson, M., J.Pleva (1991), Experientia 47:9-22.

IDZ (= Institut der Deutschen Zahnärzte) (1988). Amalgam - Pro und Contra. Materialienreihe, Band 1. Deutscher Ärzte-Verlag, Köln.

ICPS (= International Programme on Chemical Safety) (1990), Environmental Health Criteria 101, Methylmercury. World Health Organization, Geneva.

Klaschka, F., J. Ring (1990), Seminars in Dermatology 9:210-215.

Koch, W., M. Weitz (1991) Amalgam - Wissenschaft und Wirklichkeit. Öko-Institut - Institut für Angewandte Ökologie, Freiburg. Werkstattreihe Band 70

Schiele, R., und A. Kröncke (1989), Zahnärztl. Mitt. 79:1866-1868.

Schiele, R. (1991), Dtsch. Zahnärztl. Z. 46:515-518.

Weiner, J.A., et al. (1990), Sci. Total Environ. 99:1-22.

Zander, D. et al. (1990), Zbl. Hyg. 190: 325-334.

Zander, D., et al. (1992), Zbl. Hyg. (in press).

Cytogenetic Studies of Chemically Exposed Workers in Sachsen-Anhalt

Hüttner, E.
Institute of Genetics and Crop Plant Research, Corrensstr. 3-5, 0-4325 Gatersleben, FRG

Introduction

The purpose of human genetic monitoring is to detect exposures to chemical mutagens and to evaluate genotoxic effects from different exposure situations. Cytogenetic changes in somatic cells of in vivo exposed workers are an acceptable approach to occupational risk assessment. 5 different groups of chemical workers employed in synthesis and processing of several industrial chemicals (parathion-methyl, dimethylsulphate, ß-naphthol, hair dyes, chromate) from one of the formerly most important factories of the Bitterfeld region in Sachsen-Anhalt and 2 non-exposed control groups were analyzed for chromosomal aberrations (CA) and sister chromatid exchanges (SCE) in lymphocytes. Altogether 103 exposed and 117 non-exposed individuals were investigated. The investigations were performed from 1977 - 1984. Till 1990 there was no possibility to publish the results in East Germany. Both genetic endpoints are widely used to monitor human populations to environmental mutagens. A lot of epidemiological studies have shown that workers occupationally exposed to mutagens have higher frequencies of CA and SCE than non-exposed controls (Nordic Study, 1990; Hüttner, 1990).

Material and Methods

Subjects

Chromosome analyses were performed on different workers involved in industrial synthesis and processing of parathion-methyl (n=36), dimethylsulphate (n=6), hair dyes of the type of aromatic amines (n= 19), ß-naphthol (n=23) and chromate (n=19). Two control groups, a population of non-exposed people employed in the same chemical plant (n=62) and a population of non-exposed people from a distant and nonindustrialized area (n=55), respectively, were investigated.

Lymphocyte cultures

Blood samples were sent coded to Gatersleben. Lymphocyte cultures were prepared according the method of Moorhead et al (1960) with slight modifications. Lymphocytes were cultured in medium after Eagle supplemented with 10 % calf serum and phytohemagglutinin.

Cytogenetic analysis

For analyses of CA the incubation time was 48 h and for analysis of SCE 72 h in the presence of bromodesoxyuridine, at 37°C. 4 h before harvesting colchicine was added. Hypotonic solution was 0.075 M KCl, and the cells were fixed in methanol: acetic acid 3:1. Air dried slides were stained with aceto orcein acid for analysis of CA and for SCE the FPG-technique was used (Perry and Wolff, 1974). Aberrations were recorded according to Savage (1975). In most cases, 100 well spread metaphases of each individual were scored for CA and 50 cells for SCE. Only cells containing 46 centromeres were scored. The genetic endpoint variables discussed in this paper are CA = number of cells with chromosome aberrations (gaps excluded) expressed in percent, and SCE = mean number of sister chromatid exchanges per cell. For statistical detection of differences between non-exposed and exposed groups Fisher's exact test was used for CA and the Mann Whitney u-test for SCE.

Results and Discussion

Table 1 shows the pooled frequency of chromosome aberrations and sister chromatid exchanges of the chemically exposed and non-exposed groups. In all groups of people occupationally exposed to the 5 studied chemical mutagens (Mayer, 1982; Leonard et al,1980; Clemmesen, 1981) increased frequencies for CA and SCE were found which were in each case statistically significant as compared to data obtained from appropriate non-exposed control populations. The highest frequency of chromosome damage (CA as well as SCE) occured in the population exposed to dimethylsulphate (5,8 % - 19,34 SCE/cell). The frequencies of chromosomal aberrations for exposed groups were found to be 40 - 400 % higher than in the control groups, while there was only a marginal SCE increase of 25 - 50 %. Reproducibility of data was proven for the exposure situation nitroaromatic hair dyes. Although in this case the increased frequencies of CA and SCE should be confirmed for the whole population; individual and group variability was observed during different investigation times (table 2). Unfortunately, no data as to the actual concentration of chemicals in the workers' environment were available.
Two control groups were investigated: (1) persons employed in the same chemical plant, but without direct exposure (office, medical service), (2) persons employed in a scientific research institute far distant from the chemical plant (scientists, technicians, service staff). In the 2 control groups we could find a clear correlation between the frequency of CA and SCE in peripheral lymphocytes and the smoking habit. Especially in the second control group there was a positive correlation between the dail cigarette consumption and the duration of smoking and the incidence of CA and SCE. A significant handicap in the present study was the smoking behaviour of the group members. Since many people included in this genetic monitoring programme of occupationally exposed groups were smokers it was sometimes difficult to differentiate between professional and lifestyle effects (in our case smoking). The data found might be a reflection of both, occupational and life style exposure. But in the case of ß-naphthol and chromate exposure we could find clear statistical significant differences between smokers and non-smoking individuals, as well as between chemically exposed nonsmokers and chemically non-exposed non-smoking individuals. Our data from cytogenetic monitoring of chemically exposed workers demonstrate a higher sensitivity for structural aberrations as compared to SCE in individuals potentially at risk in the chemical industry. Sister chromatid exchanges of in vivo exposed persons do not reflect genotoxic effects so dramatically. Genotoxic effects on somatic cells are useful and sensitive for identification of exposure to chemical mutagens and can also contribute to improve savety in technology (Sorsa et al, 1988). Exposure to the 5 chemical mutagens studied might involve a higher risk for genotoxic and probably carcinogenic effects for the employees under the conditions we found. At the end of 1990 these 5 investigated factories were closed.

Exposition	Parathion-methyl		Dimethyl-sulphate		Hair Dyes		ß-Naphtol		Chromate	
	E	C2	E	C2	E	C1	E	C1	E	C1
No. of individuals	36	55	6	55	19	62	23	62	19	62
Mean percentage of cells with CA	1,91	1,12	5,80	1,12	2,83	1,10	2,49	1,10	2,09	1,10
Stat. significance	*		*		*		*		*	
Frequency of SCE's	15,57	13,80	19,34	13,80	15,86	13,70	16,10	13,70	16,00	13,70
Stat. significance	*		*		*		*		*	

* $p \leq 0,05$; E: exposed population; C: control group

Tab. 1: Frequency of chromosome aberrations and sister chromatid exchanges of chemically exposed populations in Bitterfeld (Sachsen-Anhalt)

	1.	2.	3.
		24 months later	40 months later
No. of individuals	12	13	10
Mean percentage of cells with CA	2,60	4,30	1,60
Frequency of SCEs	16,60	15,80	15,20

Tab. 2: Frequencies of CA and SCE at different investigation times for exposure situation to hair dyes

References

Clemmesen J (1981) Epidemiological studies into the possible carcinogenicity of hair dyes. Mut. Res. 87:65-79

Hüttner E (1990) Die gegenwärtige Situation des Population Monitoring im Gebiet der DDR und zukünftige Entwicklungen. GUM-Zeitschrift 3-4:3-6

Leonard A, Lauwerys RR (1980) Carcinogenicity and mutagenicity of Chromium. Mut. Res. 76:227-239

Mayer VW (1982) Genetic effects of naphthylamines. Mut. Res. 99:349-369

Moorhead PS, Nowell PC, Mellman WJ, Battips DM, Hungerford DA (1960) Chromosome preparations of leucocytes cultured from human peripheral blood. Exp. Cell Res. 20:613

Nordic Study Group on the Health Risk of Chromosome Damage (1990) A Nordic data base on somatic chromosome damage in humans. Mut. Res. 241:325-337

Perry P, Wolff S (1974) New Giemsa method for differential staining of sister chromatids. Nature 251:156-158

Savage JRK (1975) Classification and relationships of induces chromosomal structural changes. J. of Med. Genetics 12:103-122

Sorsa M, Pyy L,Salomaa S, Nylund L, Yager JW (1988) Biological and Environmental monitoring of occupationally exposure to cyclophosphamide in industry and hospitals. Mut. Res. 204:465-479

Heavy Metals in the Blood, Urine and Hair of a Representative Population Sample in the Federal Republic of Germany 1985/86

Krause, C., Chutsch, M., Henke, M., Leiske, M., Schulz, C., Schwarz, E. and B. Seifert
Federal Health Office, Institute for Water, Soil and Air Hygiene, Corrensplatz 1, D-1000 Berlin 33, FRG

Introduction

In connection with the national Cardiovascular Prevention Study in the Federal Republic of Germany (FRG), which includes adults between 25 and 69 years of age, it was decided to take advantage of the availability of a representative population to know more about the relationship between health and the environment. Thus, in 1985/86, the nationwide representative survey "Environment and Health" was conducted. While in one part of the study questionnaires were used to acquire a better knowledge of the personal opinion of German citizens on their health status and the environment, the study was mainly directed at getting objective information by passing the participants through a medical examination and assessing their exposure to environmental pollutants. The final population sample included some 2700 persons and was representative for the West German population with regard to age, sex and community size.

The Population Sample

The population sample was selected from the reservoir of the national health survey for the German Cardiovascular Prevention Study using a stratified two-stage random procedure. Out of the selected 4000 persons, 259 persons could not be included in the study (deaths, wrong addresses, etc.), leaving a set of 3741 persons. With 608 persons not willing to cooperate (16.3 %) and 402 cases not being included for reasons of illness, absence, incomplete data or other, the final set included data from 2731 persons (73.0 %). The participating persons were distributed over 100 sampling points and were representative of the population of the Federal Republic of Germany with regard to age (25-69 years), sex and community size. Weighting factors were applied to compensate for any bias introduced by reduced participation of special population groups (e.g. elderly women in smaller communities).

Questionnaires, Medical Examination, Collection of Environmental and Biological Samples

Questionnaires including some 200 questions were administered to the sample by trained interviewers. Besides questions on health and on the personal opinion, knowledge and behaviour of the participants with regard to the environment, information was collected on a number of parameters linked to exposure, including potential sources of pollutants in the home, smoking and nutrition habits, location of the home, noise levels, etc. The participants underwent a general medical check-up including medical anamnesis in a mobile unit which stayed for several days at each of the sampling points. Samples of urine and blood were collected for clinical analyses (42 parameters). Table 1 gives an overview of the samples taken from the subjects, households and communities.

Table 1: Overview of collected samples.

Matrix	Subject	Household	Community
Air		- VOC/Passive sampler (indoor air; 2 weeks) - HCHO/Passive sampler (indoor air; 48 h) - Dust deposit - (indoor air; 1 year) - Content of vacuum-cleaner bag	- Dust deposit (outdoor air; 2 weeks)
Water		- Tap water a. Normal use b. After overnight stagnation	- Waterworks sample
Biological sample	- Morning urine (80 ml) - Blood - Hair (4 cm, occipital)		

Since this paper places emphasis on the results for some trace elements in hair, blood and urine, especially lead, cadmium, copper and arsenic, other matrices that were analyzed will not be addressed in detail. However, it should be noted that water samples were analyzed for 17 elements/anions, dust samples for 19 elements, and indoor air samples for 58 volatile organic compounds (VOC) or classes of compounds (Krause et al., 1991a-c).

Analysis and Quality Control

Analysis of elements was carried out by atomic absorption spectrometry (AAS) or inductively coupled plasma atomic emission spectrometry (ICP-AES). Prior to analysis, hair samples were washed with acetone and water, and ashed in a plasma. More details of all procedures used are given elsewhere (Krause et al., 1989). For routine quality control, different standard reference materials were run together with the samples on a daily basis. External quality control took place by regular participation in round robin tests organized by the German Society for Occupational Medicine.

Results and Discussion

A listing of all elements that were determined in hair, blood and urine is given in table 2. It characterizes the frequency distributions by percentiles. Further statistical characteristics such as the arithmetic and geometric mean not only for the total population but also for subpopulations stratified by age, sex, community size, smoking habits, etc. can be found in Krause et al., 1989. This publication is still available from the Federal Health Office. According to the above mentioned criteria, the data are representative of the general population of the FRG and serve as the basis for obtaining normal and guideline values. However, they can also be compared with those obtained in other studies.
As expected, the concentration of lead in blood is log-normally distributed with 10th, 50th, 90th percentiles of 0.2, 6.9, and 11.20 μg/100 ml, respectively. One of the most important sources

of lead is automobile exhaust. However, this influence is decreasing, because since 1976 the content of lead in gasoline has been reduced by law. Thus, blood lead levels have decreased from about 15 µg/100 ml in 1975 to about 7 µg/100 ml in 1985.

Table 2: Percentiles of element distributions in hair, blood and urine

	Hair (µg/g) Percentile				Blood (µg/l) Percentile				Urine (µg/l) Percentile			
Element	10	50	90	98	10	50	90	98	10	50	90	98
Al	1.6	4.5	11	21								
As	0.006	0.024	0.065	0.16					0.3	4.5	18	48
B	0.39	1.4	3.3	7.6								
Ba&)	0.29	0.75	4.2	8.7								
Ca	128	304	736	920								
Cd	0.03	0.10	0.29	0.67	0.05	0.40	2.5	5.2	0.03	0.19	1.0	2.0
Cr									0.05	0.25	0.81	2.2
Cu	7.2	12	38	140	730	940	1330	1650	3.3	9.2	20	39
Fe	3.7	8.4	18	34								
Hg					0.1	0.5	1.6	3.3	0.25*)	0.25*)	2.1	4.9
Mg	13	42	205	381								
P	97	135	174	240								
Pb	0.44	1.3	5.1	12	4.2+)	6.9+)	11+)	17+)				
Sr	0.36	1.7	8.8	18								
Zn	100	164	226	308								

&) Data not weighed (50 % below detection limit)
*) 50 % of detection limit
+) Concentration given in µg/100 ml

Nevertheless there is still an influence today, because inhabitants of larger communities show higher blood lead levels than those from smaller communities. Inhabitants from households with much lead in tap water, for example more than 20 µg lead/l, show slightly higher blood lead levels than inhabitants of households with lead content under 10 µg/l. We also checked the influence of smoking and consumption of wine/champagne and beer. The blood lead levels are higher in smokers than in non-smokers. The difference is more pronounced in women. The blood lead levels in persons consuming wine/champagne or beer are higher than in persons not consuming these beverages. This suggests that there may be an effect of alcohol that merits consideration.
The concentration of cadmium in blood is also log-normally distributed with 10th, 50th, 90th and 98th percentiles of 0.05, 0.4, 2.5 and 5.2 µg/l, respectively. Smokers have more than five times higher cadmium level in blood than non-smokers. Cadmium levels of the blood of ex-smokers are at about the same level as those of non-smokers. Contrary to the respective findings for lead, the difference between smokers and non-smokers is more pronounced in men.
Multiple regression analyses should provide information about the effects of some 800 variables on cadmium concentrations in blood, urine and hair. In the case of blood-cadmium, it is possible to explain 46 % of the total variation. The most important factor is smoking with 40 % followed by economic growth and other sources. The explanation of variation of hair cadmium is not so succesful,

only 9 % overall. For urine-cadmium we can explain 34 % of the variance: 11 % by smoking, 10 % by age, 7 % by sex, urine creatinine and serum urea, 5 % by regional factors, and 1 % by wine/-champagne consumption. With respect to mean concentrations, there were 4 % higher blood cadmium levels for the consumption of 1 filter cigarette per day, plus 5 % for 1 cigarette without filter per day, minus 13 % for each one percent increase in the regional economic growth rate per year, plus 1 % for each year of one's life, and plus 23 % for "ex-smoking" compared with "never smoking". Low correlation was found for cadmium in tapwater, job situation, and passive smoking for nonsmokers in the wintertime. No correlation was found for: type of heating, education, income, or number of persons in households. Analogous results for cadmium levels in urine were plus 44 % for females in relation to males, minus 9 % for 1 mMol serum urea, plus 50 % for living in early industrialized cities, plus 1 % for each additional ton of NO_x-Emission per 1000 inhabitants per year, plus 27 % for urban residents, and minus 5 % for each additional construction worker per 1000 inhabitants.

Acknowledgment

The study is conducted by the Institute of Water, Soil and Air Hygiene of the Federal Health Office, a project of the Federal Environmental Agency, commissioned by the Federal Ministry for the Environment, Nature Conservation and Reactor Safety.

References

Krause, C., M. Chutsch, M. Henke, M. Huber, C. Kliem, C. Schulz and E. Schwarz (1989) Messung und Analyse von Umweltbelastungsfaktoren in der Bundesrepublik Deutschland - Umwelt und Gesundheit-, Band I: Studienbeschreibung und Humanbiologisches Monitoring. WaBoLu-Hefte 5/89, Institute for Water, Soil and Air Hygiene of the Federal Health Office. Berlin.

Krause, C., M. Chutsch, M. Henke, M. Leiske, C. Schulz and E. Schwarz (1991a) Messung und Analyse von Umweltbelastungsfaktoren in der Bundesrepublik Deutschland - Umwelt und Gesundheit, Band IIIa: Wohn-Innenraum: Spurenelementgehalte im Hausstaub. WaBoLu-Hefte 2/91, Institute for Water, Soil and Air Hy-giene of the Federal Health Office. Berlin.

Krause, C., M. Chutsch, M. Henke, M. Leiske, E. Meyer, C. Schulz, E.Schwarz and R. Wolter (1991b) Messung und Analyse von Umweltbelastungsfaktoren in der Bundesrepublik Deutschland - Umwelt und Gesundheit, Band IIIb: Wohn-Innenraum: Trinkwasser. WaBoLu-Hefte 3/91, Institute for Water, Soil and Air Hygiene of the Federal Health Office. Berlin.

Krause, C., M. Chutsch, M. Henke, M. Huber, C. Kliem, M. Leiske, W. Mailahn, C. Schulz, E. Schwarz, B. Seifert und D. Ullrich (1991c) Messung und Analyse von Umweltbelastungsfaktoren in der Bundesrepublik Deutschland - Umwelt und Gesundheit, Band IIIc: Wohn-Innenraum: Raumluft. WaBoLu-Hefte 4/91, Institute for Water, Soil and Air Hygiene of the Federal Health Office. Berlin.

Urinary Alkylpurines as Markers of Alkylating Agent Exposure in Humans

Prevost, V., Shuker, D.E.G. and Bartsch, H.

Unit of Environmental Carcinogens and Host Factors,
International Agency for Research on Cancer,
150 cours Albert Thomas, 69372 Lyon cédex 08, France

Introduction

The mechanism of action of many carcinogens is considered to involve the formation of electrophilic alkylating agents which react with nucleophilic sites in DNA giving rise to a range of adducts (Miller, 1978). All of the nitrogen and oxygen atoms in purines and pyrimidines are known to form adducts with various carcinogens, although the relative amounts of each adduct vary greatly depending on the reactivity of the alkylating species (Singer and Kusmierek,
1982). O^6-alkylguanine adducts are promutagenic lesions which can be repaired by alkyltransferases which remove the alkyl group. In contrast, N7- and N3-purine adducts are repaired by glycosylases which excise the alkylated base (Karran and Lindahl, 1985). Excised alkylpurines are known to be resistant to catabolism and are excreted in urine. These observations are the basis of methods for the non-invasive measurement of human exposure to alkylating carcinogens using excreted DNA adducts (Shuker,1989). This approach has been particularly successful in studies on human exposure to aflatoxin B_1 (Groopman et al., 1991).

Craddock and Magee (1967) were the first to show that urinary 7-methylguanine was, in part, derived from rat liver DNA following administration of methylating agents. Exploitation of this idea for human biomonitoring had to await the development of analytical methodology based on the use of stable isotopes and gas chromatography-mass spectrometry (GC-MS, Shuker et al., 1984, Farmer et al., 1986). Studies on the excretion of 3-methyladenine (3-MeAde), long thought to be an ideal urinary marker of methylation, showed that humans excreted a substantial background of this adduct which was unrelated to carcinogen exposure, and of dietary origin (Shuker et al., 1987, Prevost et al., 1990). Nonetheless, analytical methodology based on immunoaffinity clean-up followed by GC-MS (Friesen et al., 1991) which was developed during the course of this work proved to be a powerful tool for human applications and suggested that higher homologues of 3-alkyladenines (3-alkAde) would be less susceptible to confounding by natural backgrounds.

This paper describes recent progress on 3-alkAde which is based on the availability of a monoclonal antibody which cross-reacts with a range of 3-alkAde (Eberle et al., 1990).

Materials and Methods

Immunoaffinity columns which retain a range of 3-alkAde were prepared by coupling monoclonal antibody EM-6-47 (Eberle et al., 1990) to protein A-Sepharose in a similar manner to that recently published (Friesen et al., 1991). For quantification by GC-MS a number of deuterated internal standards of 3-alkAde were synthesised by standard methods, namely, d_3-3-MeAde, d_5-3-ethyladenine (d_5-3-EtAde), d_4-3-(2-hydroxyethyl)adenine (d_4-3-HOEtAde) and d_7-3-benzyladenine (d_7-3-BzAde). GC-MS conditions were essentially the same as in Friesen et al. (1991).

Results and Discussion

A recently developed monoclonal antibody, EM-6-47 (Eberle et al., 1990) exhibited the unusual property of recognising a wide range of 3-alkAde with little discrimination based on the size of the alkyl group. This monoclonal antibody was covalently bound to protein A-Sepharose to give stable immunoaffinity columns which retained 3-MeAde, 3-EtAde, 3-HOEtAde and 3-BzAde equally well (Prevost, Shuker, Eberle and Rajewsky, in preparation).

In a typical procedure, 50 pmoles of each deuterated internal standard is added to 5 ml of urine, the pH adjusted to 7.4 by addition of concentrated phosphate buffer, which is then passed through a precolumn of unmodified Sepharose gel (0.5 ml) followed by the elution through the immunoaffinity column (0.5 ml). After a number of washing steps the fraction containing 3-alkAde is eluted with dilute acetic acid, evaporated to dryness and derivatised for GC-MS analysis (cf. Friesen et al., 1991). Using this procedure the four 3-alkAde described above were quantified simultaneously by selected ion monitoring (SIM) of characteristic fragment ions of the <u>tert</u>-butyldimethylsilyl derivatives, with a detection limit of 1 pmole/ml of urine. This method could also be applied to any 3-alkAde which was retained by the immunoaffinity column.

In order to extend previous observations that 3-MeAde was excreted intact in urine the metabolism of the four deuterated 3-alkAde was studied in two human volunteers. A mixture of 10 nmoles of each deuterated 3-alkAde was ingested orally by each volunteer and two consecutive 24 hour urines were then collected. The results are summarised in Table 1 and show that 3-MeAde and 3-HOEtAde were extensively excreted within the first 24 hours with 3-EtAde less so. Interestingly less than a quarter of the administered dose of 3-BzAde was recovered within 24 hours sugesting that some metabolism was occurring, possibly of the benzyl group. In all cases less than 10% of the administered dose was recovered in the second 24 hour period.

Table 1. Metabolism of deuterated 3-alkAde in humans

Alkyladenine	% Recovery
d_3-3-MeAde	93
d_4-3-HOEtAde	92
d_5-3-EtAde	70
d_7-3-BzAde	23

results are expressed as % of the administered dose excreted in the first 24 hours

Based on previous experience with 3-MeAde, where background levels of preformed adduct confounded low level exposures, it was decided to study the background levels of 3-alkAde in human volunteers who consumed both free-choice and controlled diets. The protocol of the experiments is similar to that recently described (Prevost et al., 1990). The results are summarised in Table 2 and show that background levels of higher homologues of 3-alkAde are much less influenced by diet than 3-MeAde with 3-EtAde being present at very low levels and only slightly affected by the change in diet. 3-BzAde was not detected in human urine under either dietary regimes.

Table 2. Background levels of 3-alkyladenines in urine from two human volunteers

3-Alkyladenine	Free-Choice Diet	Controlled Diet
3-MeAde	64 $\pm$26	1.9$\pm$0.2
3-HOEtAde	8.6$\pm$13.8	0.5$\pm$0.5
3-EtAde	0.6$\pm$0.2	0.2$\pm$0.06

values are expressed as nmoles per day

In view of these encouraging results a pilot study was undertaken in which two cigarette smokers collected 24-hour urines during a 15 day period in which they consumed their normal diet. They noted their cigarette consumption during each 24-hour period and in addition agreed to stop smoking for five consecutive days in the middle of the study (days 6-10). As expected there was no correlation between urinary 3-MeAde excretion and cigarette consumption due to dietary confounding. Interestingly there was no apparent correlation between 3-HOEtAde excretion suggesting that this adduct is not a sensitive marker of ethylene oxide exposure at the levels that occur in cigarette smoke. A surprising finding was that 3-EtAde excretion correlated strongly with cigarette consumption and did not appear to be an artefact (for example, careful analysis showed that 3-EtAde was not present preformed in tobacco smoke). The origin of this adduct is currently under investigation.

In summary, recent developments in analytical methodology for 3-alkAde, based on group-specific immunoaffinity purification from urine, allow studies to be carried out where exposures to different alkylating agents are involved. The monoclonal antibody which was used to prepare the immunoaffinity columns, EM-6-47 (Eberle et al., 1990), could be used to purify a wider range of 3-alkAde than those described in this paper and studies are curently under way to explore the potential of this approach.

Acknowledgments

This work was funded in part by a grant from the US National Cancer Institute (CA48473). V. Prevost received a scholarship from the Ligue Nationale Francaise Contre le Cancer. We also acknowledge the generous collaboration of Professor M.F. Rajewsky and Dr. G. Eberle (University of Essen) in providing us with monoclonal antibody EM-6-47.

References

Craddock VM and Magee PN (1967) Effect of administration of the carcinogen dimethylnitrosamine on urinary 7-methylguanine. Biochem. J. 104:435-440

Eberle G, Glüsenkamp KH, Drosdziok W and Rajewsky MF (1990) Monoclonal antibodies for the specific detection of 3-alkyladenines in nucleic acids and body fluids. Carcinogenesis 11:1753-1759

Farmer PB, Shuker DEG and Bird I (1986) DNA and protein adducts as indicators of in vivo methylation by nitrosatable drugs. Carcinogenesis 7:49-62

Friesen MD, Garren L, Prevost V and Shuker DEG (1991) Isolation of urinary 3-methyladenine using immunoaffinity columns prior to determination by low-resolution gas chromatography-mass spectrometry. Chem. Res. Toxicol. 4:102-106

Groopman JD, Sabbioni G and Wild CP (1991) Molecular dosimetry of human aflatoxin exposures. In: Groopman JD and Skipper PL (eds) Molecular Dosimetry and Human Cancer. CRC Press, Boca Raton FL, pp303-324

Karran P and Lindahl T (1985) Cellular defence mechanisms against alkylating agents. Cancer Surveys 4:585-599

Miller EC (1978) Some current perspectives on chemical carcinogenesis in humans and experimental animals: Presidential address. Cancer Res. 38:1479-1496

Prevost V, Shuker DEG, Bartsch H, Pastorelli R, Stillwell WG, Trudel LJ and Tannenbaum SR (1990) The determination of urinary 3-methyladenine by immunoaffinity chromatography-monoclonal antibody based ELISA: use in human biomonitoring studies. Carcinogenesis 11:1747-1751

Shuker DEG (1989) Nucleic acid-carcinogen adducts in human dosimetry. Arch. Toxicol. Suppl. 13:55-65

Shuker DEG, Bailey E and Farmer PB (1984) Determination of N-7-[$^{2}H_{3}$]methylguanine in rat urine by gas chromatography-mass spectrometry following administration of trideuteromethylating agents or precursors. Anal. Biochem. 140:270-275

Shuker DEG, Bailey E, Parry A, Lamb J and Farmer PB (1987) The determination of urinary 3-methyladenine in humans as a potential monitor of exposure to methylating agents. Carcinogenesis 8:959-962

Singer B and Kusmierek B (1982) Chemical mutagenesis. Ann. Rev. Biochem. 52:655-693

Total Exposure to Lead and Cadmium of a Group of People from a Lead Contaminated Area

Šarić, M., M. Fugaš and M. Blanuša
Institute for Medical Research and Occupational Health, University of Zagreb, 41000 Zagreb, Croatia

Introduction

In an area with a lead smeltery, concentrations of lead and cadmium in airborne particles, depositions and household dust, and biological indicators of lead exposure have been followed over 20 years. The environmental lead levels, particularly concentrations of lead in the air, have shown a marked decrease after the introduction (in 1978) of efficient bag filters for the control of smeltery emissions (Fugaš et al., 1984), while biological indicators of lead exposure have improved gradually over the years (Prpić-Majić et al., 1984). The purpose of this study was to assess approximately the range of possible total personal exposures to lead and cadmium and to determine the share of their absorption from air, food and beverages.

Subjects and Methods

Fourteen female subjects living and working at various distances from the lead smeltery (up to 7.5 km) were selected for the study. The subjects kept activity records over seven days in winter and nine of them also kept them in summer (1989).

Integrated samples of suspended particulate matter and deposited dust were collected over seven days in the kitchens and living-rooms of the subjects and outdoors. Grab samples of household dust and soil were also taken. Duplicate samples of daily diets were collected from each subject twice over the same week. In addition, samples of vegetables grown in own gardens were also gathered. All sampling took place in winter and again in summer. At the end of each sampling period blood samples were taken from the subjects and analysed for lead and cadmium (Telišman, 1985). Concentrations of lead and cadmium in airborne and deposited household dust and in soil were determined after digestion of samples with concentrated HNO_3 by flame AAS (Hršak, 1987) and in diet samples by flame or flameless AAS (Blanuša et al., 1990).

Results

The lead and cadmium concentrations in the air to which subjects were exposed were calculated from the data on time spent in a given microenvironment (based on activity record, average data shown in Table 1) and from respective concentrations of airborne lead and cadmium.

Table 1. Average percentage of time spent by subjects in a microenvironment

Season	Indoors: At home	Indoors: At work and other	In transport	Outdoors
Winter	66.3	23.9	2.7	7.1
Summer	60.3	25.2	1.2	13.3

Daily absorption of lead and cadmium from the air was calculated on the assumption that about 13 m^3 of air was inhaled daily in winter (20 m^3 in summer owing to outdoor activities) and that 50 per cent of the lead and cadmium that had been deposited in the lung was absorbed.
Lead and cadmium absorption from diet was assumed to be 10 and 5 per cent of the measured average daily intake through meals and beverages, respectively. Estimated daily absorption of lead and blood lead levels are presented in Table 2. Table 3 shows the estimated daily absorption of cadmium and cadmium levels in blood. In these Tables data on subjects residence, work place and smoking habit are also included.

Discussion and Conclusions

The results show that there is a wide range of estimated lead absorption (ratio of lowest to highest value is 1:11.5 from air and 1:44 from diet) which depends on residence and workplace location in relation to the smeltery and on the provenience of foodstuffs consumed. Blood lead levels although influenced by the same factors varied within the ratio 1:5.2.

Compared to the results of a pilot study on total exposure to lead and cadmium in an urban area, performed within a WHO-UNEP HEALs project in March 1988 (WHO/UNEP, 1990), where the concentrations of lead in air to which subjects were exposed varied from 0.14 to 0.84 μg/m^3, in the lead smeltery area they varied between 0.38 and 2.76 μg/m^3 in winter and from 0.17 to 1.97 μg/m^3 in summer. Much larger differences were found in daily dietary lead intake which was in the average 5-16 μg/kg wet weight in the urban area and 9-405 μg/kg in the smeltery area, indicating the influence of low emission sources in the close vicinity of the smeltery as well as contamination by soil in which lead concentration remained high all over the years despite control measures (144-2236 μg/g dry weight against 20-38 μg/g in the control area). The content of lead in household dust varied from 0.64 to 12.4 mg/g in the smeltery area against 0.11-0.39 mg/g in the control area.

Concentrations of cadmium in environmental samples varied less and were less extreme: absorption from the air was within the ratio 1:6, from the diet 1:4.7, but cadmium in blood was 1:27.7 - which may be attributed to the additional intake by smoking.

The concentrations of cadmium in the air (ng/m^3) to which subjects were exposed ranged from 3.7 to 18.1 in winter and from 2.0 to 6.7 in summer against 1.7-9.9 in the urban area (WHO/UNEP, 1990). Cadmium dietary intake (μg/kg wet weight) was 1.6-14.8 against 2.5-13.7 in the urban area.
Contamination of soil (μg/g of dry weight) with cadmium was higher than in the control area (2-12.5 against 1.1-2.3) and of household dust (μg/g) 6.9-42.8 against 1.1-8.5 in the control area.
Concentrations of lead and cadmium in typical vegetables (potatoes, lettuce, carrots, onions, parsley), grown in own gardens and used for preparation of meals by the subjects included in the study, were also much higher in the smeltery than in the control area. The highest values were obtained for lead in lettuce (up to 13.4 mg/kg wet weight) and parsley (up to 24.2 mg/kg wet weight) grown in the vicinity of the smeltery.

As seen from the presented data, calculated total absorption of lead in subjects from the smeltery area ranged from 3.5 to 77 μg/day depending partly on season (winter-summer). Although within wide ranges, relative contribution of lead intake through food (and beverages) was considerable. Calculated absorption for cadmium ranged from 0.21 to 1.18 μg/day and diet contributed to total cadmium absorption from 92 to 97%. The average concentrations in blood were 227 μg/L (97-602 μg/L) for lead and 2.6 μg/L (0.61-7.49 μg/L) for cadmium. The results show that people living in this area, particularly those who prepared and consumed

Table 2. Estimated daily absorption of lead from air, meals and beverages, and blood lead concentrations (winter-summer), and data on subject'sresidence and work place

Subject	Location of: Residence	Location of: Workplace	Employed in the lead smeltery: Subject	Employed in the lead smeltery: Member of the household	Lead/air µg W	Lead/air µg S	Lead/meals & beverages µg W	Lead/meals & beverages µg S	Lead/blood µg/L W	Lead/blood µg/L S
1	III	III	-	-	2.6	2.3	1.9	3.3	106	112
2	III	IV	-	+	2.5	1.7	2.6	1.8	97	106
3	II	IV	+*	-	10.0	-	2.5	-	136	-
4	II	I	-	+	8.8	9.2	1.9	1.5	260	227
5	II	II	-	-	9.2	2.0	1.3	1.8	107	103
6	II	-	-	-	5.1	1.7	6.2	3.1	134	129
7	II	II	-	-	6.9	-	3.7	-	136	-
8	II	I	+*	-	10.7	-	4.4	-	198	-
9	I	II	-	+	4.3	3.0	2.9	8.1	119	330
10	I	III	-	+	6.8	3.8	6.0	4.8	134	127
11	I	IV	-	+	18.0	19.7	44.8**	57.2**	272	309
12	I	I	+*	+	9.0	6.7	46.0	8.4	504	415
13	II	III	-	+	5.4		28.9	-	201	-
14	I	I	-	+	7.4	-	7.4	-	124	-

I close to the lead smeltery (within 500 m)
II 1.5-2.5 km S/SW of the smeltery
III 4 km N of the smeltery
IV more than 4 km of the smeltery
* office workers
** unrealistic, probably due to intentional contamination

Table 3. Estimated daily absorption of cadmium from air, meals and beverages and blood cadmium concentration (winter-summer), and data on subject's smoking habit

Subject	Present smoker: Subject	Present smoker: Member of the household	Cadmium/air µg W	Cadmium/air µg S	Cadmium/diet µg W	Cadmium/diet µg S	Cadmium/blood µg/L W	Cadmium/blood µg/L S
1	-	+	0.05	0.04	0.42	0.29	0.78	0.94
2	-	-	0.02	0.02	0.49	0.68	0.99	0.72
3	+	-	0.04		0.43	-	0.64	
4	-	+	0.03	0.07	0.22	0.26	0.27	3.84
5	+	-	0.04	0.03	0.66	0.18	5.97	7.49
6	+	-	0.04	0.02	0.54	0.48	1.44	1.30
7	-	-	0.12		0.39	-	0.82	
8	-	+	0.06		0.48	-	0.97	
9	+	+	0.04	0.03	0.29	0.56	0.71	2.32
10	-	-	0.04	0.04	0.66	0.39	0.73	0.86
11	+	-	0.05	0.04	0.68	0.54	4.42	5.56
12	-	+	0.04	0.06	1.14	0.66	2.83	2.65
13	-	+	0.04		0.58	-	0.76	
14	-	-	0.05		0.85	-	0.61	

meals with vegetables grown in their gardens, may have had higher lead intake than the Provisional Tolerable Weekly Intake (PTWI) defined by WHO (3.0 mg/person). Cadmium intake which shows similar trend and to which cadmium in meals contributed most, was within the range of control values and below TWI value (0.5 mg/person).

References

Blanuša, M., J. Hršak and M. Fugaš (1990) Lead and cadmium in human diet in lead polluted area. Environmental Contamination. Proceedings of the 4th International Conference - Barcelona, ed. J. Barceló, p. 490-492.

Fugaš, M., J. Hršak and P. Souvent (1984) The effect of a new emission control system on lead, zinc and cadmium concentrations in the environment of a lead smeltery. Staub-Reinhalt Luft. 44: 273-275.

Hršak, J. Emission and transformation of lead, zinc and cadmium aerosols in the environment of a lead smeltery (1987) PhD Thesis (in Croatian), University of Zagreb.

Prpić-Majić, D., D. Meczner, S. Telišman and A. Keršanc (1984) Biological monitoring and lead effects in a smelter community before and after emission control. Sci Total Environ. 32: 277-288.

Telišman, S. Determination of lead and/or cadmium in blood by an indirect method (deproteinization of blood samples with HNO_3) using electrothermal atomic absorption spectrophotometry - ET-AAS. In: D. Prpić-Majić, ed. Selected toxicological and chemical analyses of biological samples for the application in occupational medicine, clinical toxicology and ecology (in Croatian) Beograd - Zagreb. Medicinska knjiga, 1985: 20-31.

WHO/UNEP Human Exposure Assessment Location (HEAL) Programme (1990).Exposure monitoring of lead and cadmium: an international pilot study. Eds. Vahter, M., S. Slorach. Technical Report Series, GEMS. Nairobi:WHO and UNEP, pp. 82.

The Significance of Asbestos Bodies and Occurrence of Macrophages in the Sputum of Asbestos-Exposed Workers

Trošić, I., Šarić, M., Štilinović, L.
Institute for Medical Research and Occupational Health, Ksaverska cesta 2, Zagreb, Croatia.

Introduction

Chronic, pulmonary fibrosis caused by the inhalation of asbestos has a very clear etiology, although its pathogenesis remains obscure. Macrophages are a key element in the response of the host to asbestos. Dust clearance from the alveoli is the function of the macrophages. Whereas these cells phagocytize short fibres and remove them from the respiratory pathways they cannot encompass and transport the longer fibres, which tend to be retained in the lumens of the bronchioles and the alveolar ducts (Craighead and Mossman, 1987). The appearance of asbestos bodies represents the endogenous reaction of the pulmonary defense mechanism to the inhaled needlelike asbestos. Asbestos bodies are inhaled asbestos fibres which have been coated with ferroprotein by macrophages (Morgan and Holmes, 1985). It is generally accepted that coating neutralises the pathogenic potential of the fibre. The coated fibres are less toxic to pulmonary alveolar macrophages (McLemaire et al., 1980). The fibres, the lengths of which exceed the diameter of the human alveolar macrophage (ca. 15 µm) are poorly cleared. As asbestos bodies are rarely formed on fibres less than 10 µm in length, the majority will be retained in the lungs indefinitely, unless broken into smaller fragments, which undergo phagocytosis and are cleared by macrophages (Timbrel, 1982). Furthermore, attention is centred on the cytotoxic effect of asbestos on phagocytes. Pernis and coworkers are of the opinion that macrophage death is an extreme result induced by asbestos, and "that same molecular events under different conditions of dust concentration or latency time may instead induce a stimulation of macrophage function and cell generation" (Pernis and Vigliani, 1982). The Pneumoconiosis Committee of the College of American Pathologists has specified "that the minimum criteria for histologic diagnosis of asbestosis includes the identification of peribronchial fibrosis and at least two or more asbestos bodies in tissue sections". The justification for invasive treatment in the diagnostics of asbestosis can be questioned, particularly when occupational exposure to asbestos exists. For this reason it is important to determine the role played by the occurrence of asbestos bodies and macrophages in the sputum of workers. In the current study we investigated the occurrence of asbestos bodies and macrophages in the sputum of shipyard workers in order to determine how much, if any, the appearance of occupational asbestosis is related to cleared asbestos bodies and the occurrence of macrophages in the sputum of examined subjects.

Subjects and Methods

We examined the sputum of 494 male shipyard workers with a mean age of 46.3 years and mean period of exposure to asbestos dust of 22.5 years. The exposed group was divided into two separate well matched groups. The GI group consisted of 152 exposed subjects without asbestosis, and the G2 group consisted of 342 exposed workers with asbestosis, verified according to ILO classification. Identification of asbestos bodies and macrophages were provided by using light and phase contract microscopy of Prussian blue stained and native smears. The Prussian blue method stains the iron-containing coating material of the asbestos bodies a dense blue, making them easily visible against the red (safranin stain) background. Statistical evaluation was made by using the Student's t-test.

Results

Iron positive granular cells that contain asbestos bodies and iron positive coating of fibres were seen in macrophages (Fig.1). The findings of asbestos bodies were significantly more frequent in the diseased group ($p < 0.05$). On the contrary, the occurrence of macrophages in the sputum of the non-diseased group was statistically more frequent than in the diseased group (Fig.2).

Discussion

Asbestos is a family of fibrous minerals, each of which has distinctive physiological and chemical characteristics. Lung clearance of industrial dust such as asbestos is slow, as such material has been found in the sputum of workers 20 years after cessation of serious occupational exposure (Morgan et al., 1978). Sputum cytology is a practical, non-invasive and inexpensive method for identification of asbestos bodies, cell population and fibres in asbestos-exposed individuals. Uncoated fibres found in the sputum of asbestos-exposed individuals are not in a "free" state but rather are within the macrophages. Dodson and coworkers suggested that the numerous changes found in the cytoplasm of macrophages, interacting with fibres, are due to the functional role of cells involving active mobility and/or orientation of the engulfed fibre (Dodson et al., 1983). The more frequent findings of macrophages in the sputum of the non-diseased group is confirmation of a more active and consequently more efficient lung defense mechanism, because the group of workers in the same or similar conditions of exposure remain healthy. In our opinion this appears to be due to the higher generation of macrophages and their ability to handle the fibres and/or asbestos bodies in the macrophages, depending on the clearance rate. The findings of asbestos bodies and macrophages in the sputum of asbestos exposed individuals is an easy way to detect in vivo exposure to asbestos. The presence of asbestos bodies in sputum is a good hallmark of exposure but not a good hallmark of disease, because positive findings of asbestos bodies also occurred in the non-diseased group.

References

Craighead, J.E. and B.T. Mossman (1987) The Pathogenesis of Asbestos Associated Disease. New Engl. J. Med. 306:1446-1455.

Craighead, J.E. J.L. Abraham, A. Churg et al (1982) The pathology of asbestos-associated disease of the lungs and pleural cavities: diagnostic criteria and proposed grading scheme (Report of the pneumoconiosis Committee of the College of American Pathologists and the National Institute for Occupational Safety and Health). Arch. Pathol. Lab. Med. 106:544.

Dodson, R.F., M.G. Williams, J.W. McLarty and G.A. Hurst (1983) Asbestos Bodies and Particulate Matter in Sputum from Former Asbestos Workers. Acta Cytologica. 27:635-640.

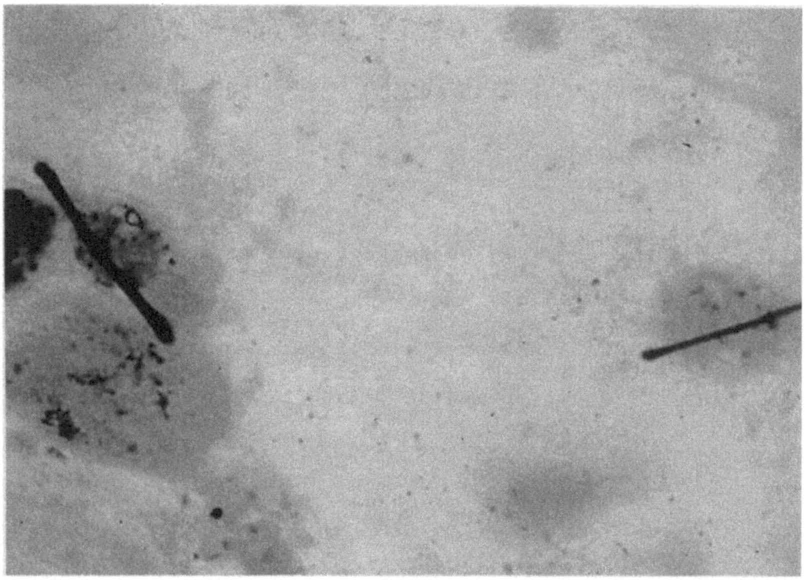

Fig. 1 Iron positive granular cells and asbestos bodies in the sputum of asbestos exposed workers

	G1 NON-DISEASED GROUP		G2 DISEASED GROUP	
	NUMBER	PERCENT	NUMBER	PERCENT
ASBESTOS BODIES (+)	43	28.3	206	60.2
ASBESTOS BODIES (−)	109	71.7	136	39.8
TOTAL	152	100	342	100

	G1 NON-DISEASED GROUP		G2 DISEASED GROUP	
	NUMBER	PERCENT	NUMBER	PERCENT
MACROPHAGES (+)	94	71.8	199	61.2
MACROPHAGES (−)	37	28.2	126	38.8
TOTAL	131	100	325	100

Fig. 2 Occurrence of asbestos bodies and macrophages in the sputum of non-diseased (G1) and diseased groups (G2) of asbestos-exposed workers

Morgan A. and A. Holmes (1985) The Enigmatic Asbestos Body: Its Formation and Significance in Asbestos-Related Disease. Environ. Res. 38:283-292.
Morgan A., R.J. Talbot and A. Holmes (1978) Significance of Fibre Length in the Clearance of Asbestos Fibres from the Lung. Brit. J. Ind. Med. 35:146-152.
McLemaire, T.L., V. Roggli, E.C. Marchael, S.D. Greenberg and P.M. Stevens (1980) Comparison of Phagocytosis of Uncoated Versus Coated Asbestos Fibers by Cultured Human Pulmonary Alveolar Macrophages. Chest. 399-429.
Pernis, B. and E.C. Vitliani (1982) The Role of Macrophages and Immunocytes in the Pathogenesis of Pulmonary Diseases Due to Mineral Dusts. Am. J. Ind. Med. 3:133-137.
Timbrel V (1982) Deposition and Retention of Fibers in the Human Lung. Ann. Occup. Hyg. 26:347-369.

ENVIRONMENTAL AND BIOLOGICAL MONITORING OF MERCURY FOLLOWING CHRONIC OCCUPATIONAL EXPOSURE

Widon, B.[1] Schweinsberg, F.[1], Holzapfel, G.[2] und Brückner, Ch.[2]
[1] Chemisches Labor, Abt. Allgemeine Hygiene und Umwelthygiene, Hygiene-Institut der Universität Tübingen, Eugenstr. 6, W 7400 Tübingen
[2] Inst. für Arbeitsmedizin, Friedrich-Schiller-Univ., Jahnstraße. 3, O 6900 Jena

INTRODUCTION

Mercury burden can be caused by intake of contaminated food (e.g. fish), by amalgam fillings and most importantly by occupational exposure, e.g. chlor-alkali plants, production of electrical equipment, production of thermometers. Hg vapor is the most common form of exposure from occupational sources and amalgam fillings. In food, Hg can be found in organic form (e.g. methylmercury). It has been shown that Hg intake from amalgam fillings has been reduced to levels that do not pose a health risk to dental patients (*Zander et al.*, 1990). Furthermore Hg in fish is regulated by threshold limit values that are very rarely surpassed in recent years.

The aim of the present study was to investigate occupational exposure to Hg vapor in a thermometer plant in Thuringia. To this end we conducted environmental and biological monitoring of Hg and compared the results with existing threshold limit values (TLV) for Germany (MAK and BAT).

MATERIALS AND METHODS

Air samples were drawn by static air samplers; Hg concentrations were taken as a time-weighted average. Hair, whole blood and urine (24h) samples were collected from about 100 male and female employees who provided informed consent to participate in the study. Two urine samples were taken: native urine, and urine after oral gavage with Dimaval®.

A special questionnaire was prepared for anamnesis.

Quantitative determination of Hg was performed sensitively and selectively by cold vapor atomic absorption spectrometry after amalgamation with gold in a gold trap. Results were obtained using standard addition and checked with Hg reference samples (lyophilized human whole blood from SERONORM™; lyophilized human urine from BIO-RAD). For the determination in air, Hg was collected by oxidative solution in an impinger flask with permanganate.

RESULTS AND DISCUSSION

The study is still under way; the results are therefore preliminary.

As reported elsewhere (*Holzapfel el al.*, 1991), the concentrations of Hg vapor in the air were in the range of 0.02 mg/m^3 in workplaces with minute contamination (e. g. calibration) and in the range of 0.1-0.2 mg/m^3 in workplaces with high contamination (e. g. adjustment section, processing of Hg). Here the TLV for Hg air (MAK) of 0.1 mg/m^3 was exceeded in several cases.
In 40 workers the Hg concentrations in native urine were in the range of 7-650 μg/l (mean: 108 μg/l; Figure 1), corresponding to 9-1110 μg/24h (mean: 135 μg/24h). In 3 cases, the TLV for Hg-U (200 μg/l) was exceeded.
The Hg concentrations in urine after oral gavage of Dimaval® (n = 40) were in the range of 100 - 7400 μg/24h (mean: 2200 μg/24h).
The Hg concentrations in whole blood (n = 20) were in the range of 8 - 114 μg/l (mean: 39.1 μg/l; Figure 2).
The results of Hg air, Hg-U and Hg-B indicate that the test persons were highly exposed to Hg. Therefore preventive measures to reduce Hg emissions are urgently recommended.
A correlation was found between Hg-U and Hg-B (Figure 3). Subjects with low Hg-U were low in Hg-B and vice versa.
The ratio Hg-U (native) and Hg-B of 2.76 was similar to that reported by *Roels et al.* (1987) in workers of a dry alkaline battery plant.
In 20 workers, the TLV for Hg-B of 50 μg/l (BAT) was surpassed in 7 cases. They were mainly occupied in the adjustment section where Hg is warmed. More workers exceed the established TLV of Hg-B (n=7) compared to Hg-U (n=3). For well weighed judgement of critical exposure using TLV's for different biological materials comparable results should be obtained.
No correlation was found between Hg-U and Hg hair (Figure 4). In workers with the highest Hg-U and Hg-B levels only medium levels of Hg hair was found. Hg hair is therefore not even useful as a qualitative indicator of high exposure.
In long-term exposure to Hg vapor, the central nervous system is the critical organ. Long-term exposure to 50 - 200 μg/m^3 Hg vapor in air - which was found at several workplaces - is believed to cause micromercuralism.
Symptoms reported on the questionnaire, which may have been associated with micromercuralism, were increased irritability and frequent headache in 40%. However, this high frequency might well be due to social insecurity of the employees at the time of questioning. Insomnia and hypomnesis were reported by 20-30%. More pronounced symptoms were tremor in 10 % and loss of appetite or cachexia in 2%. From these results no clear-cut conclusions with respect to adverse health effects can be drawn. To clarify this point clinical investigations are in progress and will be reported elsewhere.

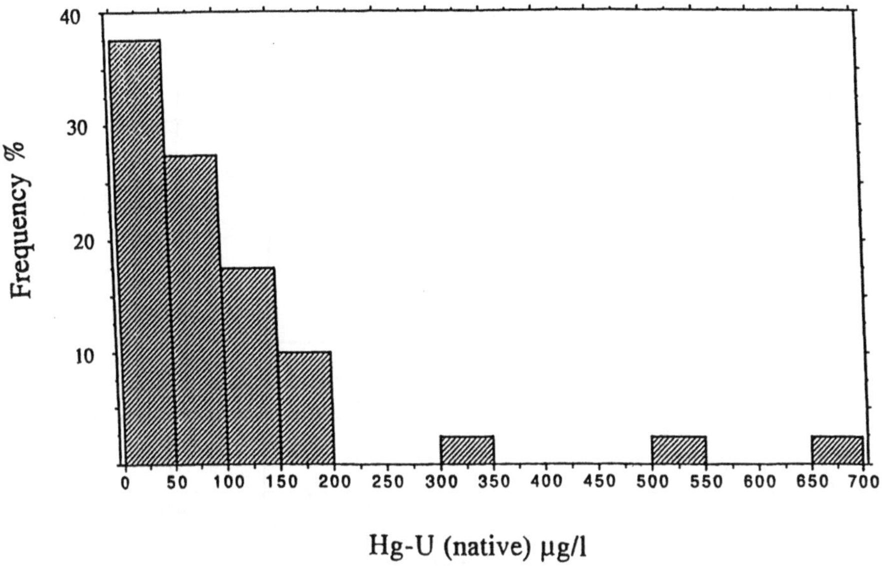

Figure 1
Frequency distribution of mercury levels (µg/l) in native urine (Hg-U) in 40 workers exposed to mercury vapor.

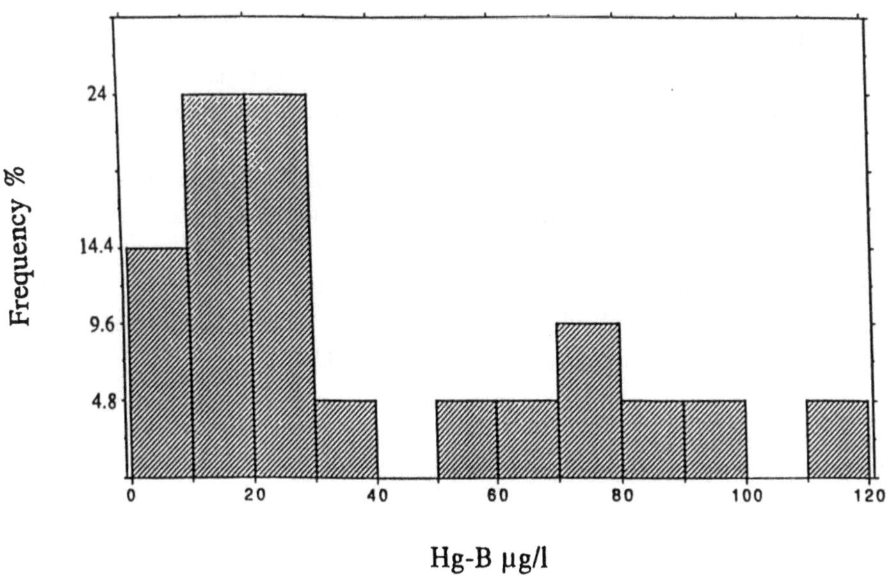

Figure 2
Frequency distribution of mercury levels (µg/l) in blood (Hg-B) in 21 workers exposed to mercury vapor.

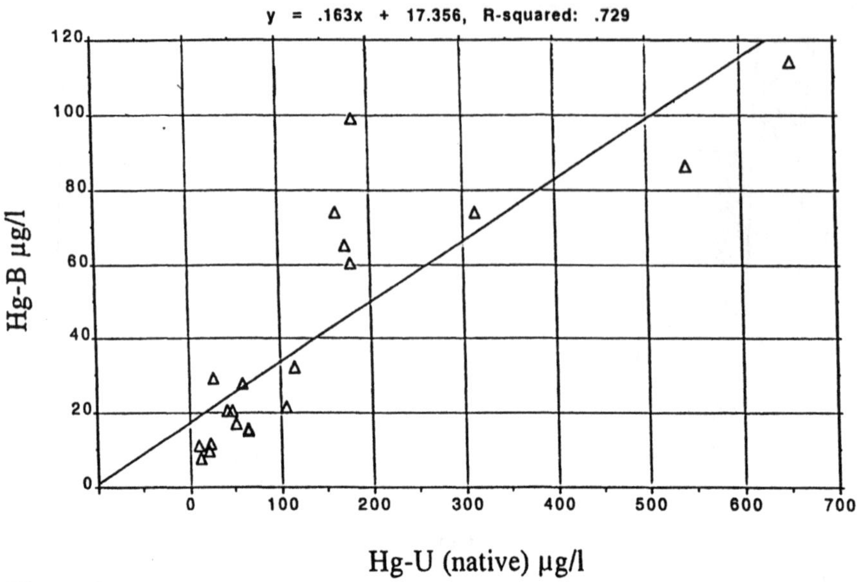

Figure 3
Correlation between mercury levels (µg/l) in native urine (Hg-U) and in blood (Hg-B) in 20 workers exposed to mercury vapor.

REFERENCES

Holzapfel G, Brückner C, Widon B, Schweinsberg F: Quecksilberbelastung an Arbeitsplätzen in einem Betrieb im Land Thüringen. 31. Jahrestagung der Deutschen Gesellschaft für Arbeitsmedizin 1991

Roels H, Abdeladim S, Ceulemans E, Lauwerys R (1987) Relationships between the concentrations of mercury in air and in blood or urine in workers exposed to mercury vapour. Ann occup Hyg 31: 135-145

Zander D, Ewers U, Freier I, Westerweller S, Jermann E, Brockhaus A (1990) Untersuchungen zur Quecksilberbelastung der Bevölkerung II. Quecksilberfreisetzung aus Amalgamfüllungen. Zbl Hyg 190: 325-334

ACKNOWLEDGEMENT

We thank Fa. Heyl, Berlin for generous supply of Dimaval®

ALUMINIUM EXPOSURE AND HEALTH RISKS

Wilhelm, M., Ohnesorge, F.K.
Institute of Toxicology, Heinrich-Heine-University Düsseldorf, Moorenstr. 5, 4000 Düsseldorf 1, FRG

INTRODUCTION (Gitelman, 1989; Lewis, 1989)

Aluminium (Al) is the most abundant metal. It is present in tap water, beverages, food, cosmetics and pharmaceutical preparations. Human exposure is unavoidable. An essential role has not been established but the element is now known to be toxic.
Strong acids from atmospheric deposition remobilize soil Al. There is strong evidence that increased Al concentrations in natural waters are responsible for some deleterious effects of acid rain on aquatic organisms and plants.
There is no doubt that Al accumulation in patients with chronic renal insufficiency causes encephalopathy, osteomalacia, and microcytic anemia.
Several pulmonary effects of Al in exposed workers such as bronchitis, alveolitis, and fibrosis have been reported. There is considerable dispute whether Al itself is pathogenic. The possibility of mixed dust effects are discussed.
Concern for general Al health risks has rapidly increased due to its possible association with Alzheimer's disease. There is no clear evidence that Al causes this disease, but it is more and more accepted that the element acts as a secondary factor in contributing to neuronal dysfunction.
Despite numerous reports with renally impaired patients little is known on the Al exposure and health risks for the general population.

DIETARY ALUMINIUM INTAKE (Table 1)

Dietary intake represents the major route of Al exposure. Al in food can derive from natural origin, from food additives and from contact with the metal used in containers and utensils. Recent estimates are generally below 20 mg Al/day (WHO, 1989). Our mean intake of 0.71 mg/day obtained with children (n=48, 5-9 years) accounts for only 3.4% of the provisional tolerable weekly intake (PTWI: 7.0 mg Al/kg b.w.) given by the WHO (1989). Higher intakes occur by using Al-containing drugs or under extreme conditions (frequent cooking of acidic foods in Al pots, excessive intake of foods with high Al content such as tea, coffee, herbes, spices).

Table 1. Dietary Al intake

Total diet studies (US FDA, WHO,1989)	2 - 14 mg Al/day
Duplicate study (own data)	0.71 mg Al/day
"Extreme conditions"	100 mg Al/day
Patients receiving antacids	100-5000 mg Al/day

ALUMINIUM IN TAP WATER (Wilhelm und Dieter, 1991)

The amount of Al ingested from the water supply is generally low. Since 1984 we have been monitoring Al levels in tap water of different suppliers around Düsseldorf (Table 2). The median was 12 µg/l. No value exceeded the EEC standard level. In our duplicate study Al in tap water was calculated to contribute to the total intake between 0.4 - 4%. In some European water authorities higher values due to the use of Al based coagulants in a technical insufficient manner for processing drinking water are reported. Very high levels have been found in different regions of the FRG in wells. It is apparent that in such extreme cases the resultant Al intake from water could exceed the dietary contribution for certain persons.

Table 2. Al in tap water

Al concentrations in tap water (n= 403; FRG; 1984-1989)	< 1.0 -157 µg Al/l
EEC standard level	200 µg Al/l
High levels of some European water authorities	2000 µg Al/l
Al levels in wells	2500-10090 µg Al/l

ALUMINIUM IN AIR

The intake from air is minor relative to that from food (Table 3). Al concentrations in ambient air are generally below 1 $\mu g/m^3$. In industrial areas Al levels as high as 20 $\mu g/m^3$ have been reported. In comparison, the maximum concentration at the work place in the FRG is 6000 $\mu g/m^3$ for different Al compounds and for aluminium oxide fume as fine dust. Al contribution from air for the general public is less than 5% of the total intake, but may reach 50% in polluted areas.

Table 3. Al in air

Unpolluted air	< 1 μg Al/m^3
Industrial area	up to 20 μg Al/m^3
MAK	6000 μg Al/m^3
Al contribution from air	< 5-50% of total intake

ALUMINIUM TOXICOKINETICS

There are great discrepancies on the basic toxicokinetic data (Wilhelm et al. 1990). Figure 1 shows our recent toxicokinetic model of Al which is based on experimental studies with isolated perfused organs like liver, kidney and intestine of the rat, an in vivo toxicokinetic study with rats, investigations with humans,

and some observations from other authors (Höhr et al. 1989, Jäger et al. 1991).
Al absorption in the gastrointestinal tract follows a 2-step mechanism with a rapid mucosal uptake followed by the release of a very small amount into the blood. No precise data exist for the absorption by the pulmonary system. Since lung Al levels are the highest found in any organs some of the inhaled Al must be trapped in the pulmonary tissue. A significant absorption of deposited Al into the blood is not assumed. Following exposition to Al fumes some of the element is absorbed as indicated by increases of Al in serum and urine in workers. There is no evidence that Al can be absorbed through the skin as a result of the application of Al containing deodorants, cosmetic creams and soaps.
Thus lung, gastrointestinal tract and skin are effective barriers to Al absorption. This is due to the insoluble nature of most Al compounds. On the other hand the inadvertent intravenous administration of Al contaminated solutions is of great clinical relevance (Wilhelm et al. 1987).
Once absorbed Al binds to albumin and transferrin. With respect to the renal elimination from the body it is important to note that the protein binding increases with the Al concentrations in plasma. This explains the limited capacity of the kidney to eliminate Al at high dose therapy. Nevertheless, for the normally exposed healthy population the absorbed amount of Al is effectively eliminated by the kidneys. Urinary excretion is normally about 10 µg/day but may increase substantially at high doses. In contrast to other suggestions we think that the kidney is the only effective pathway for the elimination of Al from the body. In our experiments the hepatic Al clearance was less than 1% of the renal Al clearance. When the renal capacity to eliminate Al is exceeded the element accumulates mainly in spleen, liver, soft tissues, and bone.

CONCLUSIONS

At present there are no Al health risks for the normal population. The risk of Al exposure has to be considered in following groups:
Persons obtaining tap water from highly Al contaminated wells at low pH.
All individuals with renal insufficiency including elderly persons with physiologically reduced excretory renal function.
Patients receiving contaminated intravenous solutions or who are on long-term treatment with Al containing medication.
Infants reveiving contaminated formulas may be also at risk since their gastrointestinal tract and kidneys are not fully developed and a greater proportion of alumimium is absorbed and retained.
Finally a risk of Al exposure for Alzheimer's patients can not be excluded.

REFERENCES

Gitelman HJ (ed) (1989) Aluminum and health. Marcel Dekker New York Basel

Höhr D, Abel J, Wilhelm M (1989) Renal clearance of aluminium: studies in the isolated perfused rat kidney. Toxicol Lett 45:165-174

Jäger DE, Wilhelm M, Witte G, Ohnesorge FK (1991) Intestinal absorption of aluminium: Studies in the isolated perfused rat intestinal preparation. J. Trace Elem. Electrolytes Health Dis. 5:81-85

Lewis TE (ed) (1989) Environmental chemistry and toxicology of aluminum. Lewis Publishers Chelsea Michigan

WHO (1989) WHO food additives series: 24. Cambridge University Press Cambridge New York

Wilhelm M, Sprenger KBG, Vossas U, Ohnesorge FK (1987) Aluminum load in chronic intermittent plasma exchange. Clin Toxicol 25:209-220

Wilhelm M, Jäger DE, Ohnesorge FK (1990) Aluminium toxicokinetics Pharmacol Toxicol 66:4-9

Wilhelm M, Dieter HH (1991) Vorkommen, Bedeutung und Nachweis von Aluminium. In: Aurand K et al. (eds) Die Trinkwasserverordnung. E Schmidt Verlag Berlin

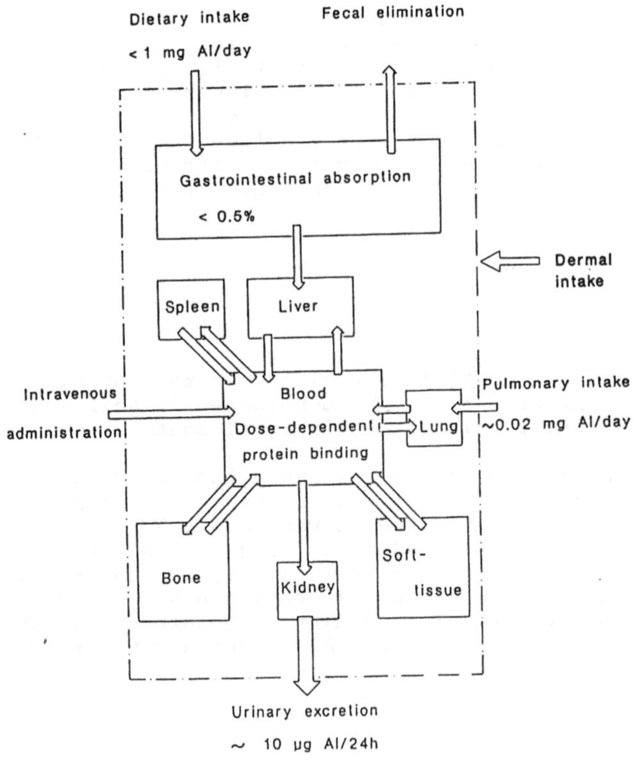

Figure 1. Toxicokinetic model of aluminium

EPIDEMIOLOGY

Air Pollution and Bronchial Reactivity in Schoolchildren with a Medical History of Respiratory Diseases

Englert, N. (1), W. Babisch (1), B. Niggemann (2), and U. Wahn (2)
(1) Institute for Water, Soil and Air Hygiene of the Federal Health Office, Corrensplatz 1, D-1000 Berlin 33, FRG
(2) University Children's Hospital of the Free University, Heubnerweg 6, D-1000 Berlin 19, FRG

Introduction

In Germany, possible associations between air quality and obstructive bronchitis as well as croup syndrome had been intensively discussed during the last decade. Meanwhile it is mostly accepted that air quality can only explain a relatively small part of the variance of these diseases (LAI, 1990), compared with the dominant role of viral infections and individual disposition.
Based on studies of croup syndrome and obstructive bronchitis in Berlin (Englert, 1988, 1989), we had the opportunity to conduct a follow-up study with respect to the following questions:
- Do schoolchildren with a history of croup syndrome develop bronchial hyperreactivity and show serum IgE values higher than other children ?
- Is bronchial hyperreactivity and elevated serum IgE concentration in schoolchildren influenced by indoor and outdoor air pollution ?

Subjects and Methods

Children included in this study were a subgroup of - in total - 794 children whose parents filled out a new questionnaire 3 to 5 years after having filled out a first one during studies of croup syndrome and obstructive bronchitis in Berlin between end 1984 and end 1986. These children were former patients the majority of whom underwent hospital treatment due to croup syndrome some years ago. They may not be understood as being a representative sample of schoolchildren.
Our final test group included 189 children (125 boys, 64 girls) aged 7 to 14 years. We classified them with respect to their medical history as shown in fig. 1. This classification is a hierarchical one. A child with bronchial asthma belongs to the first group even if suffering from hay fever, too, or in addition, from croup syndrome. So in total, 8 children had a medical history of bronchial asthma, 16 of hay fever, 27 of obstructive bronchitis, and 115 of recurrent croup syndrome, because some of them suffered from more than one of these diseases. 29 children never suffered from any of these respiratory diseases and were considered our "reference group".

189			
181	8	B.A.	Bronchial Asthma
170	11	H.F.	Hay Fever
148	22	O.B.	Obstructive Bronchitis
59	89	R.C.	Recurrent Croup Synd.
29	30	C.S.	Croup Syndrome (once)
0	29	R.G.	Reference Group

Fig.1 Test group classification

In order to test "nonspecific" bronchial reactivity, $PC_{20}FEV_1$ was determined by a bronchial histamine challenge. $PC_{20}FEV_1$ (histamine) is that provocation concentration of histamine which induces a 20 % reduction of the Forced Expiratory Volume in one second (FEV_1). In our procedure, histamine concentration was doubled from

step to step, starting with 0.125 mg/ml until a 20 % reduction of FEV_1 or a concentration of 4 mg/ml were reached. In addition, total Immunoglobulin E (IgE) in serum was measured provided that the children and their parents agreed to venous blood sampling.
By questionnaire we asked if the children or their siblings or parents had a medical history of bronchial asthma, hay fever, obstructive bronchitis or single or repeated attacks of croup syndrome. The diagnoses used in our study are based on questionnaire data, thus possibly being subject to some inaccuracy by incorrect classification performed by the parents. Questions concerning indoor air pollution in the homes of the children were included (Environmental tobacco smoke (ETS) and gas stove use as sources of pollutants, domestic animals as a source of allergens). Outdoor pollution was classified in three zones according to the sulfur dioxide concentrations in Berlin (fig. 2).

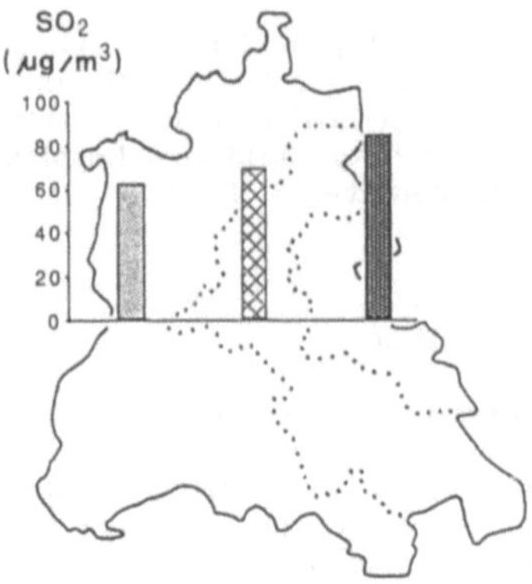

Fig.2 Sulfur dioxide concentrations in Berlin (West) 1982–1986 in 3 zones of pollution

Analyses were performed using univariate and multiple regression models as well as logistic regression models. In stepwise multiple regression models, we first included age, sex, and medical history, and in the next steps indoor pollution variables (ETS, gas stove use, domestic animals) and the classification of outdoor pollution by SO_2 according to the district where the children lived.

Results and Discussion

In accordance with other studies (Weiss et al., 1984), we found the histamine threshold concentration to be age-dependent, with age explaining 6.9 % of the variance in a multiple model. For IgE, such a dependency was not observed.
For boys the relative risk for asthma was 3.5 and for allergic rhinitis 2.2 (both n.s.), and they showed a tendency to higher IgE values.
Genetic disposition is an important factor for bronchial asthma, hay fever, obstructive bronchitis and croup syndrome (Peat et al., 1987; Zach et al., 1981). The relative risk for children to suffer from the same diseases as other family members do, was higher than 10 for asthma, and between 4 and 6 for hay fever and obstructive bronchitis. For croup syndrome, the relative risk was slightly, but significantly increased (1.4).
Associations between bronchial asthma and allergic rhinitis (Kaplan et al., 1987; Wüthrich et al., 1986) and obstructive bronchitis (Salome et al., 1987) are known, as well as those between allergic rhinitis and croup syndrome (Zach et al., 1981). Our study confirmed significant associations between bronchial asthma and allergic rhinitis as well as obstructive bronchitis without finding a relation between the latter ones.

Higher bronchial reagibility and increased total IgE in children with bronchial asthma and hay fever are to be anticipated (Backer et al., 1989; Cockroft et al., 1984). Higher bronchial reagibility is also described in children with recurrent croup syndrome (Zach et al., 1981) and even independently from the number of croup attacks in the past (Loughlin and Taussig, 1979). In our study, children with a medical history of asthma or hay fever showed lower histamine threshold concentrations and higher IgE values compared with the other children, but neither in univariate nor in multiple regression analysis, the subgroups of children with obstructive bronchitis, single or recurrent croup syndrome and the reference group showed any significant differences in histamine threshold concentrations or total IgE (fig. 3 and 4).

In total, socio-demographic and medical variables explained 10 % of the variance of histamine threshold concentrations and 5 % of IgE variance.

For indoor air, we considered three factors resulting in lower air quality: smoking, gas stove use (NO_x), and domestic animals (allergens).
In univariate analysis, the histamine threshold concentration tended to be higher in smoker's homes and in homes with domestic animals. Bronchial asthma and hay fever, too, were less frequent in those households. Gas stove use did not show significant effects. The lower reactivity found in children from smoker's homes did not necessarily show a "protective" effect of passive smoking, but may indicate that in families with members with hyperreactive airways, smoking was reduced for this reason. Likewise, this holds true for domestic animals.

Univariate analysis showed no significant associations between outdoor pollution and histamine threshold concentrations or IgE. In a multiple regression model, outdoor air quality alone resulted in no additional explanation of variance. However, introducing interaction effect terms resulted in a decrease of the histamine threshold concentrations attributable to combined outdoor and indoor pollution (smoker's home plus outdoor pollution and gas stove use plus outdoor pollution), comparable in magnitude with the influence of (slight) bronchial asthma (fig.5), and in an increase of explained variance (from 10.9 to 15.1 %, adjusted). This also increased the significance of the other factors included in the model.

Higher IgE levels in the first years of life are described in children from smoker's households (Kjellman, 1981). We found a higher IgE concentration in children from smoker's homes, being more marked when the interaction of passive smoking plus outdoor pollution was included. But, for comparison, hay fever was associated with a distinctly greater increase in IgE concentration (fig.6). The end model explains (adjusted) 10.0 % of

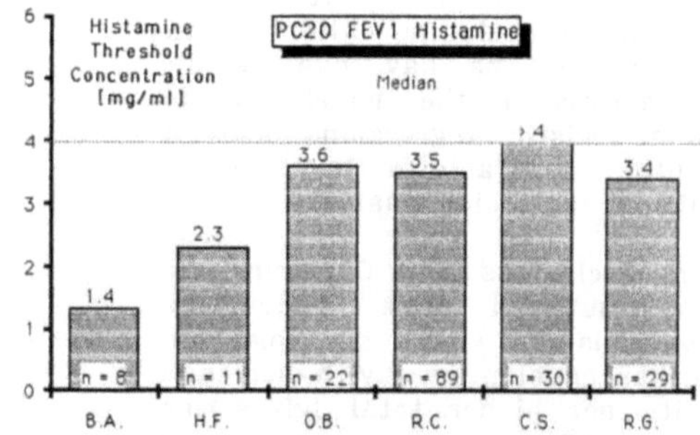

Fig.3 Histamine threshold concentrations, median values. Subgroups cf. fig.1

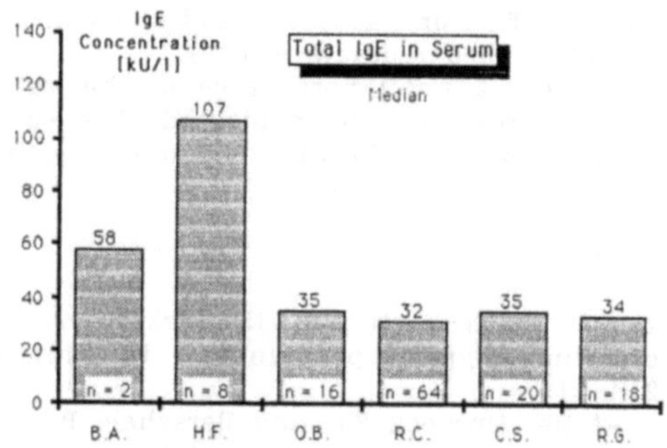

Fig.4 Total serum IgE, median values. Subgroups cf. fig.1

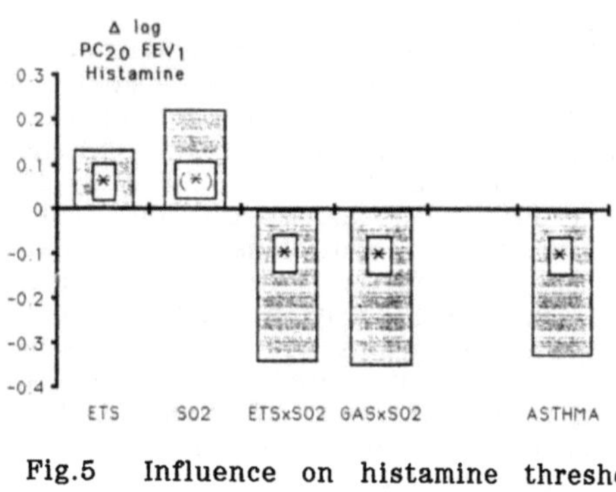

Fig.5 Influence on histamine threshold concentration; multiple regression model with interaction terms.

the variance of IgE logarithms.
In analyses excluding children with bronchial asthma, hay fever replaces asthma in the model. Models using logistic regressions showed results comparable with those of multiple regression analysis.

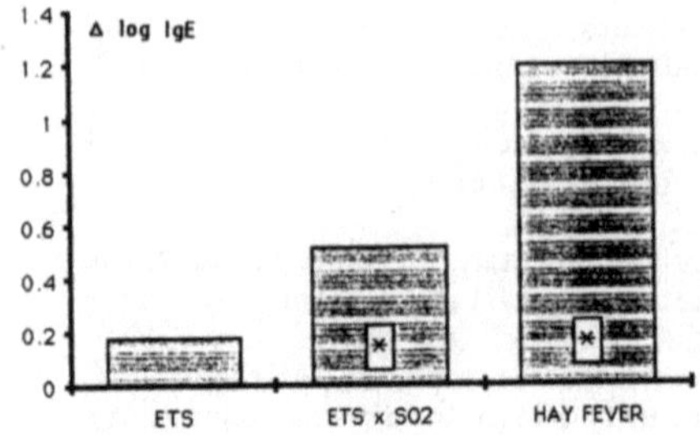

Fig.6 Influence on total serum IgE; multiple regression model with interaction terms.

Our conclusions are: Children who have suffered from obstructive bronchitis or croup syndrome do neither develop airway hyperreactivity nor higher total IgE serum concentration. However, a higher reactivity and higher total IgE is found in children with asthmatic or atopic symptoms, independently from former diseases.
Adverse effects of either indoor or outdoor pollution could not be shown with respect to bronchial reagibility and total IgE concentrations. However, combined exposure to ETS or gas stove emissions and, in addition, outdoor air polluted by SO_2, resulted in detectable effects.
In total, only a small percentage of the variance of bronchial reactivity or total IgE concentration can be explained. The contribution of environmental factors is detectable, but is explaining only a few percent of total variance.

References

Backer V, Bach-Mortensen N, Dirksen A (1989) Prevalence and predictors of bronchial hyperresponsiveness in children aged 7-16 years. Allergy 44: 214-219.

Cockroft DW, Murdock KY, and Berscheid BA (1984) Relationship between atopy and bronchial responsiveness to histamine in a random population. Ann Allergy 53: 26-29.

Englert N (1988) Pseudokrupp. Bundesgesundhbl. 31: 467-471.

Englert N (1989) Obstruktive Bronchitis in Berlin. Bundesgesundhbl. 32: 433-441.

Kaplan BA, Mascie-Taylor CGN (1987) Asthma and wheezy bronchitis in a British national sample. J Asthma 24: 289-296.

Kjellman NIM (1981) Effect of Parental Smoking on IgE Levels in Children. Lancet I: 993-994.

LAI Länderausschuß für Immissionsschutz (1990) Pseudokrupp und Luftverunreinigungen. Ed.: Minister für Umwelt, Raumordnung und Landwirtschaft des Landes Nordrhein-Westfalen, Düsseldorf.

Loughlin GM, and LM Taussig (1979) Pulmonary function in children with a history of laryngotracheobronchitis. J Pediatr 94: 365-369.

Peat JK, Britton WJ, Salome CM, and Woolcock AJ (1987) Bronchial hyperresponsiveness in two populations of Australian schoolchildren. II. Relative importance of associated factors. Clin Allergy 17: 283-290.

Salome CM, Peat JK, Britton WJ, Woolcock AJ (1987) Bronchial hyperresponsiveness in two populations of Australian schoolchildren. I. Relation to respiratory symptoms and diagnosed asthma. Clin Allergy 17: 271-281.

Weiss ST, Tager IB, Weiss JW, Munoz A, Speizer FE, and Ingram RH (1984) Airways Responsiveness in a Population Sample of Adults and Children. Am Rev Respir Dis 129: 898-902.

Wüthrich B, Schnyder UW, Henauer SA, Heller A (1986) Häufigkeit der Pollinosis in der Schweiz. Schweiz. med. Wschr. 116: 909-917.

Zach M, Erben A, and Olinsky A (1981) Croup, recurrent croup, allergy, and airways hyper-reactivity. Arch Dis Child 56: 336-341.

This study was supported by a grant of the Federal Environm. Agency (UBA).

Effects of Air Pollution on the Health of Children in East- and West-Germany: Aspects of Study-Design and Descriptive Results for some Questionnaire Items

Krämer[1)], U., R. Dolgner[1)], H. J. Willer[2)], H.-W. Schlipköter[1)]
[1)] Medical Institute of Environmental Hygiene Düsseldorf. [2)] Landeshygieneinstitut Magdeburg

Introduction

Since 1976, the Medical Institute of Environmental Hygiene conducts epidemiological studies the results of which form a part of the Clean-air plans (Luftreinhaltepläne) of North-Rhine-Westphalia. These studies are intended to monitor the effects of air pollution on human health. Since 1985, six year old children who take part in the compulsory school entrance examination are investigated by our group on a voluntary basis. Response rates of above 90 % could be achieved.

This design was adopted for a study in East-Germany with the special aims of
- monitoring the effects of air pollution on the health of children as a basis for control,
- corroborating dose-response relationships by studies in areas with highly different intensities of air pollution,
- monitoring the effects of air pollution levels changing with time.

The following areas in East-Germany were selected for study:
Halle, Leipzig - high pollution,
Magdeburg - moderate pollution,
Salzwedel, Gardelegén, and Osterburg (Altmark) - low pollution.

Characteristic parameters of outdoor pollution are given in table 1.

A comparison of the results with those of the Ruhr-area is planned. Every second to third year a comparable study will take place enabling us to monitor the changes expected as a result of improving environmental conditions.

Table 1: Air pollution in East- and West-German towns Mean values 1989 (1988)

	Leipzig Zentr. [2]	Leipzig SW	Halle Burgstr. [1]	Magdeburg (Raster) [3]	Duisburg-Walsum [4]	Düsseldorf-Reisholz [4]
SO_2 [µg/m^3]	210	310	220	70	36	26
dust-fall [mg/m^{2}d]	---	1070	780	350 (1988)	270 (1988)	110
particles [µg/m^3]	80		123	---	73	68
NO_x [µg/m^3]	41	---	73	36	87	154

Table 2: Target variables
Examination of 6 year old children 1991

body burden:	in urine: arsenic, nickel, mercury, chrome, cadmium, tri-chloro-acetic acid, fluorine, phenole in teeth: lead, cadmium in serum: selenium in blood: lead, cadmium
immunological parameters:	C3c, C4, IgA, IgG, IgM, IgE
lung function:	bodyplethysmographic measurements: airway resistance, vital capacity ...
respiratory diseases and symptoms:	questionnaire: bronchial asthma, pseudocroup, frequent cough ...
allergological parameters:	spec. IgE antibodies against house-dust /-mite, pollen, food skin-test
neuropsychological and neurophysiological parameters:	examination of memory and attention, reaction-time

Target variables and course of investigation

In East-Germany, the parents of 1876 children just starting school (about 6 years old) in the selected areas were asked by the local public health departments to have their children investigated and to fill in a questionnaire. Additionally, the parents of 2500 school-beginners were only asked to answer the questionnaire. The study in East-Germany took place from 4.2. to 22.3.91. The overall response was 93 %.

The parallel study in West-Germany was run from 8.4. to 7.6.91. The areas in West-Germany as fixed by the Clean-airplans were Borken, Duisburg, Essen, Gelsenkirchen, and Dortmund, with Borken as a reference area.
About 6350 children had been asked to participate, 3390 of them to deliver the questionnaire only. The mean response rate in West-Germany was 79 %.

Table 2 gives a summary of the outcome measures. Up to 250 informations were gathered per child. These relate to the questionnaire, the analyses of blood, tooth, and 24 h urine samples, a lung-function investigation with a mobile equipment, a test for allergies, and neuro-psychological and -physiological tests. The last two investigations were not done in all study-regions.
Nationality, age, gender, weight, social class, sources of indoor pollution, and acute infections were recorded as potential confounding factors.

Description of some results

Table 3 presents a description of informations extracted from the questionnaires. No correction for confounding has been done for this purpose. The results of the East-German towns are shown together with those of a study on six year old children in Cologne / Düsseldorf (Wirkungskataster 1990), because the data-set of the simultaneous study in West-Germany is not yet available.

Bronchial asthma and hay-fever were stated fairly seldom. The rates in East-Germany are below those in Cologne / Düsseldorf. Remarkably the rate for eczema is higher in all East-German towns than in the West-German areas. For these diseases with an atopic component no obvious association with the level of pollution can be seen. In contrast, the rates for frequent cough, bronchitis, and infectious airway diseases are higher in Leipzig or Halle than in the West-German towns, whereas the rates in the East-German control region are in three cases similar to those in Borken, the West-German control area.

The results given are descriptive in nature. A detailed analysis, which includes confounding variables, small area aspects and a comparison with objective parameters measured, is just being done.

Table 3 **Symptoms and diagnoses**
6 year old German children living more than 2 years in their present residential areas (% positive)

	Cologne/ Düsseld.	Borken	Leipzig	Halle	Magde-burg[1]	Osterb./Gardel./ Salzwedel
n	1700	364	513	1120	306	877
bronchial asthma [3]	3.5	1.5	2.2	2.1	1.0	1.1
hay fever [3]	2.6	1.5	1.0	1.3	3.3	1.2
eczema [3]	9.6	7.0	13.0	15.2	18.6	14.3
frequent cough	9.7	5.5	18.0	11.6	9.9	6.0
bronchitis [3]	36.3	41.1	51.9	60.3	56.6	51.1
cold with fever [2] [4]	28.6	22.7	75.3	67.0	66.6	54.0
tonsillitis [4]	16.0	15.5	37.9	23.5	25.1	17.8
> 4 colds [4]	19.3	10.5	24.9	18.5	17.2	8.9

[1] results from those only answering the questionnaire are not yet included
[2] for this question, the East- and West-German questionnaires differ
[3] ever diagnosed by a physician
[4] in the last year

References

Wirkungskataster zu den Luftreinhalteplänen Rheinschiene Süd und Rheinschiene Mitte 1990 Editor: Ministerium für Umwelt, Raumordnung und Landwirtschaft des Landes Nordrhein-Westfalen 1990/1991

AIR POLLUTION AND ALLERGIC REACTIONS

E.Marth[1], C.Hülße[2], J.Pelech[3]
[1]Institute of Hygiene of the University of Graz, Austria, [2]Institute of Hygiene of Rostock, FRG, [3]Institute of Hygiene and Epidemiology of the University of Prague, CSFR.

Introduction

Epidemiological studies provide evidence that air pollution influences incidence, prevalence, course of the disease and lethality of asthma (McDONNEL,1983; MENZEL,1984; MOSBECH, 1987). With an increasing degree of pollution, the results become more distinct, especially during episodes of extremely severe air pollution. A study in Donora (USA) showed that 87.6% of asthmatics developed acute symptoms.
The question which components of air pollution in polluted areas may trigger or cause asthma has been investigated in several studies. Correlations between medical data and exposure to air pollutants were demonstrated for SO_2, sulfates, NO_2 and O_3, and especially dust (ARANYI,1983; GILMOUR,1991). However, only some asthmatics reacted to certain air pollutants. High SO_2 concentrations may trigger an almost immediate respiratory obstruction that is rapidly reversible following exposure. O_3 exposure, on the other hand, causes an obstruction that develops more slowly and includes the lower respiratory tract.

Material and Methods

With an increase in air pollution over the past 20 years, epidemiological identification of health risks caused by air pollutants has become increasingly difficult. Studies on the impact on the immune system of children between 9 and 11 years of age were carried out in Halle, a city of the former GDR known for its high degree of pollution (see table 1), in Graz, an Austrian city, and on the waste disposal sites of Cairo.
The concentration of salivary IgA and the concentration of the immunoglobulins (IgG, IgA and IgM) in the serum of the children were the parameters used to study the state of the immune system. Total IgE concentration was determined in 148 children from Halle, 48 of which were from a non-polluted area near Halle. In addition, the reaction to 8 different allergens (see table 2) was studied.
For this determination of total serum IgE, a new testing system, Pharmacia Cap-System, was applied. This system, consisting of immunoassay reagents, instrumentation and information managment software was developed for the measurement of total and specific IgE in undiluted serum or plasma. The system is built around a new type of solid phase consisting of a flexible hydrophilic carrier polymer encased in capsule, the ImmunoCAP. The carrier consisit of a CNBr-activated cellulose derivative which can bind at least three time more protein than a corresponding paper disc and about 50 times more than can be absorbed to the inner surface of a traditional coated tube. The system utilizies IgE(Fc) specific combinations of polyclonal and monoclonal anti-IgE antibodies which are labelled with B-galactosidase generating fluorescence (LUNDKVIST, 1975). Compared to traditional methods, this system has the advantage of a larger linear regression and a higher sensitivity.

Results

Lung function as tested by spirometry was clearly impaired among children in the various polluted regions although the data were not pathological (BIENENSTOCK,1980; BUTLER,1982)(see table 1).
The proportion of children with obstructive bronchial changes could be further differentiated using a ß-mimetic. The direct toxic effect of air pollutants on the mucous membranes of the respiratory tract leads to mucosal edema. Constriction of the airway lumen narrowing is reflected by the increase in airway resistance. Mucosal edema membranes caused by inflammation without involvement of the smooth muscles is not influenced by the ß-mimetic. In addition to bronchial constriction, pollutants lead to a suppression of sIgA formation in the respiratory tract. This suppression then leads to an increased inflammatory response due to infection, equally resulting in mucosal edema. If bronchodilation occurs following inhalation of terbutaline sulfate, this can be explained by the action of the ß-mimetic on smooth muscle which can be interpreted as clear sign of asthma (see fig.1).
Again, there is a significant difference between the two groups with regard to the effect of the bronchodilator. 25% of the children in the control group with a restricted lung function reacted to the stimulation of ß-receptors with an increase of the original value. Almost 40% of the children of the affected group reacted positively to stimulation of the ß-mimetic.

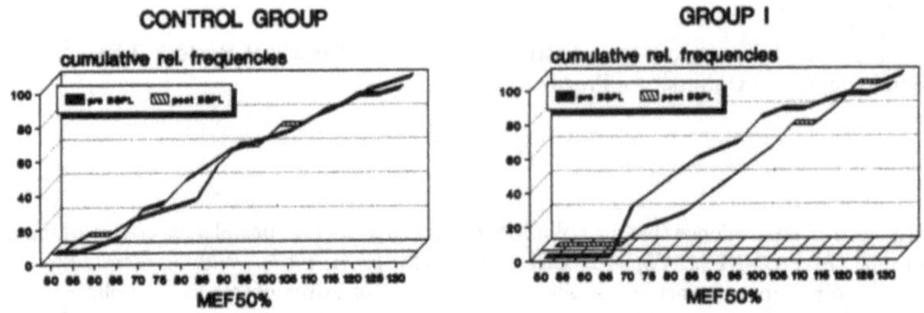

Figure 1: A) Cumulative relative frequency before and after inhalation of terbutaline sulfate. In the control children a barely measurable difference was documented. **B)** In the affected children the inhalation resulted in significant improvement in lung function.

CITY	SO2	CO	NOX	HCL	STAT.	FEV1%	IgA(IU/m	IgE(KU/l)
HALLE (CITY)	1.48	0.93	0.78	n.d.	MEAN	81.3	n.d	530.2
					STDEV.	6.2		232.5
					n	81		100
HALLE (COUNTRY)	0.36	0.66	0.25	n.d.	MEAN	95.12	n.d	147.7
					STDEV.	3.6		136.7
					n	48		48
GRAZ (CITY)	0.19	0.62	1.1	n.d.	MEAN	83.3	8.2	n.d.
					STDEV.	20.4	1.3	
					n	325	325	
GRAZ (COUNTRY)	0.01	0.08	0.06	n.d.	MEAN	93.8	9.8	n.d.
					STDEV.	11.7	2.2	
					n	125	125	
CAIRO (WASTE-PLACE	1.35	0.83	n.d.	0.96	MEAN	88.6	8.4	n.d.
					STDEV.	26.3	3.4	
					n	100	100	
CAIRO (CONTROL)	0.37	0.83	n.d.	0.11	MEAN	98.3	9.5	n.d.
					STDEV.	22.9	4.1	
					n	100	100	

Table 1: Concentration of the various pollutants in the different areas studied and immunologic data.

A significant difference in total IgE concentration between children in polluted areas and those in non-polluted areas was demonstrated (mean value 147.7 KU/l for the control group and 530.2 KU/l for the group I; see also table 1). The next question to be investigated was which additional information concerning air pollution could be gained by

ALLERGENS:

1. Timothy	**5. Cladosporium**
2. Birch	**6. Cat**
3. Rye	**7. Dermatophag. pteronyssimus**
4. Artemisia	**8. Dog**

Table 2: List of tested allergens.

testing for specific IgE's (again using the new Pharmacia Cap-System) against 8 different allergens (see table 2). Incidence and prevalence of positivity against an allergen was significantly higher in children from polluted areas than from non-polluted areas ($p < 0.05$).

Determination of specific IgE's is a very complex and, especially, very expensive method. We now need to investigate whether determination of total IgE is sufficient for epidemiological studies or if determination of specific IgE's is so much more advantageous that the method is indispensable. The figure 2 shows the Box-and-Whisker-plots clearly demonstrating the high correation between specific IgE and the concentration of total

IgE's. Therefore, we can postulate that determination of total IgE's is sufficient for epidimiological studies. There is a highly significant correlation between the concentration of unspecific total IgE and specific IgE ($r = 0.4733$ $p < 0.001$).

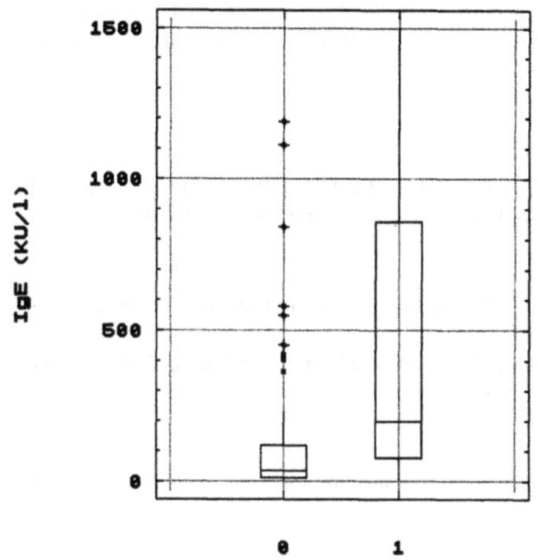

Figure 2: Comparison of the concentration of IgE of allergen-positive (1) and allergen-negative (0) children by Box and Whisker Plot.

DISCUSSION

The epidemiological studies suggest that air pollutants may cause hyperreactivity of the respiratory tract. The possible pathogenetic mechanism is as yet largely unknown. Various mechanisms are being discussed. Certain is merely that there is not one single cause but a combination of several factors leading to hyperreactivity and possibly also asthma. Moreover, the significant differences between individuals in developing hyperreactivity point to the important role of endogenous factors.

The question of the significance of damage to the epithelium of the respiratory tract for the development of hyperreactivity has received increased attention recently. Viruses, cigarette smoke and air pollutants such as NO_2 and O_3 lead to loosening of the epithelial cells and opening of the tight junctions. In the search for possible pathogenetic interaction between air pollutants and hyperreactivity of the respiratory tract, indirect factors must be considered as well. Hyperreactivity caused by viruses and bacteria is well documented.

In addition to the various changes of the alveolar macrophages (Van der HEIJDEN,1989), immunoglobulin A has a special significance in determining the impact of environmental pollutants on the immune system. In several studies we have repeatedly shown that IgA concentration in saliva is significantly reduced in polluted areas (MARTH, 1990).

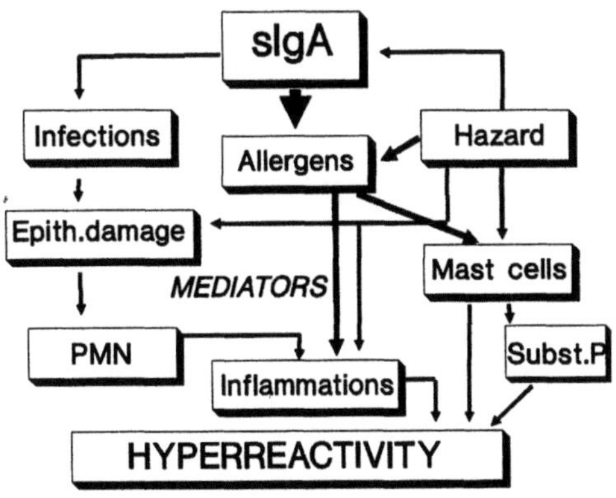

Figure 3: Possible interactions of air pollutants.

On the one hand, pollutants have a direct toxic effect on the various mucous membranes leading to production and release of various mediators of inflammation via the defect in and cumulation of PMN.
The result is increased hyperreactivity However, pollutants also suppress the concentration of sIgA in saliva leading to more frequent inflammations. The decreased possibility for neutralisation of allergens by IgA and the increased permeability of the mucous membranes lead to stimulation of specific IgE production. Hyperreactivity is the most important prerequisite for the development of bronchial asthma (see fig.4).

References

Aranyi, C., Vana, S.C., Thomas, P.T., Bradof, J.N., Fenters, J.D., Graham, J.A., and Miller, F.J.:1983. Effects of subchronic exposure to a mixture of O_3, SO_2, and $(NH_4)_2SO_4$ on host defenses in mice. J. Toxicol. Environ. Health 12:55-71.

Bienenstock, J., and Befus, A.D. 1980. Mucosal immunity. Immunology 41:249-270.

Butler, J.E., Swanson, P.A., Richerson, H.B., Ratajezak, H.V., Richards, D.W., and Suelzer, M. 1982. The local and systemic IgA and IgG antibody responses of rabbits to a soluble inhaled antigen. Am. Rev. Respir. Dis. 126:80-85.

Gilmour, M.I., Jakob, G.J. 1991. Modulation of immune function in mice exposed to 0.8 ppm ozone. Inhalation Toxicology 3:293-308.

Lundkvist,U. 1975. Research and development of the Rast technology. Advances in Diagnosis of Allergy: RAST, Evans,R.,(III Ed.), Miami, Symposia Specialists: 85, 1975

McDonnel, W.F., Horstman, D.H., Hazucha, M.J., Seal, E., Haak, E.D., Salaam, S.A., and House, D.E. 1983. Pulmonary effects of ozone exposure during exercise: dose response characteristics. J.Appl.Physiol. 54:1345-52.

Marth, E., Sixl, W., Schaffler, R., Zach, M. 1990. Die Wirkung von Luftschadstoffen auf den Organismus. Forum Städte-Hygiene 41:197-200.

Menzel, D.B. 1984. Ozone: An overview of its toxicity in man and animals. J. Toxicol. Environ. Health 13:183-204.

Mosbech, H., Dirksen, A., Madsen, F., Weeke, B. 1987. House dust mite asthma. Allergy 42:456-463.

Van der Heijden, P.J., Bianchi, A.T.J., Bokhout, B.A., Dol,M., Scholten, J.W., and Stok, W. 1989. Quantification of antigen-specific antibody-secreting cells in the small intestine and other lymphoid organs of mice after oral booster immunization. Immunology 66:404-409.

Effects of the extreme environmental pollution on the general health state of the population in Bitterfeld (former G.D.R.)+

Thielebeule,U., I.Winkler, R.Ebert und Chr.Wegner
c/o University of Rostock, Institut for Hygiene, Schillingallee 70, D-O2500 Rostock 1

Introduction

In order to integrate the environmental medicine into the necessary redevelopment programmes for the Bitterfeld region we tried to compile during the last half year every still available data for a first assessment of the complex effect. This study contains apart from a renewed analysis of the own field studies additional pilot examinations.

The county of Bitterfeld has an area of 454 km^2 with 130.000 inhabitants. The cities Bitterfeld and Wolfen are centers of the chemical industry. The most part of the population is living in areas where the annual mean concentrations of sulphur dioxide and dust exceed the "MIK"-limits 2-3 times. Additionally many specific burdens exist from products of the chemical industry and their old deposit sites. The cities Berlin-Pankow (1968, 1978), Rostock (1980-1988) and Wismar (1990) were selected as less polluted control areas.

Results and Discussion

The Study 1968 and the Rostock studies 1978 and 1980 -1988 already showed for pupils from the lower school in Bitterfeld clear differences of the physiological parameters in comparison to control groups from less polluted areas. The retarded bone age (Pelech,L.,1985) was particularly noticeable and correlated with the dust pollution of the area (Fig.1).

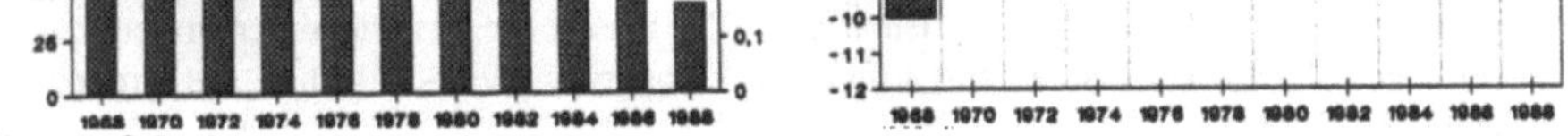

Fig. 1: SO-2, dust and bone age retardation

+ sponsored by UBA , Berlin

It`s true all hematological, immunological and other anthropometric findings were in the range of the norm for the age group, but these differences must be interpreted as a forced reaction on the permanent influence of the environmental stress. This is clearly shown by the approximation to findings from the less polluted control area under the conditions of a 4 week stay in clean air. About that we reported several times.
Especially important is the enlargement of the difference in simple lung function parameters as FVC and FEV 1,00 with increasing age, which was reported in the longitudinal study 1981 - 1983.
This objective finding is in good agreement with the higher frequency of respiratory tract diseases in Bitterfeld (Fig.2) resulting from a questionnaire as well as with the clearly higher prevalence of respiratory tract diseases from the occupational mass screenings (Ebert et al., 1990).

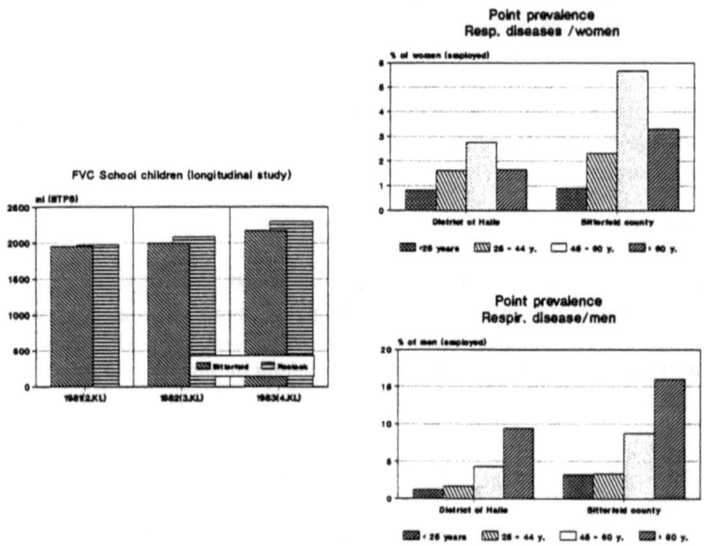

Fig.2:FVC in Bitterfeld children and Prevalence of resp.dis.(occup. medicine)

The evaluation of the routine statistic "Acute respiratory diseases" from Bitterfeld and the Wismar district as control area showed only for the age group 0 - 1 year differences to the unfavor of the Bitterfeld group. But this picture is moved with increasing age to a higher incidence of acute respiratory diseases in Wismar city. The reason for this phenomenon may be the different medical care for babies and infants in comparison with all other age groups (Hülße,Chr. et al.,1990).
The doctors of occupational medicine showed by mass screenings of the Bitterfeld workers that the need of treatment is increased from age step to age step and lies significantly above the mean of the Halle district. This commitment is very important for the evaluation of the general state of health as more as the half of the population is employed. Apart from respiratory diseases this concerns diseases of the digestive system (also in children born in 1982 and adults from the pilot study), the circulation system and the rheumatic diseases.
The risk of dermatological and respiratory diseases for workers in the Bitterfeld district is twice as high as in the total district of Halle. Just as for the occupational diseases the reason for these findings is probably the high degree of employment in the 2 large-scale works of the chemical

industry. Therefore the workers have several contacts with harmful chemical substances. Moreover it is important for the evaluation that multiple overloads especially by manual work too existed at many places of work in the technicaly obsolete chemical industry of the former G.D.R..

The occurence of allergic reactions in children showed no differences but a higher allergy rate in men from the pilot study seems to be a reflection of the environmental influence. The evaluation of the complete data pool of the year of birth 1982 seems to indicate that environment and place of work cause complications in pregnancy. Disturbances during the pregnancy were found in 20.2 % of the mothers from Bitterfeld, but only in 13.1 % of the mothers from Wismar.

Determinations of pesticides in human milk showed only for ß-HCH and total DDTs clearly higher concentrations compared with those from the Schleswig-Holstein county (Heinzow,B. et al.,1990). But all results of this investigation were below the limits recommended by the DFG (Acker ,L.et al.,1984).

Disturbances of the mental development were recorded for children by the "Children and Youth Health Service" using a questionnaire. Probably it may be that especially the psychic alterations will have a negative effect on the performance during the further life. Indications for that are also the unfavorable results of the children from Bitterfeld during the concentration-performance-test by KURTH,E.,1978, and the higher rate of psychic alterations diagnosticized by a questionnaire by HÖCK,K., HESS,H. and SCHWARZ,E.,1981.

If we take the life expectancy as an expression of the genetic potential for performance and a long life, then we find the above described disturbances of the health state adequately reflected by women. Compared with the less polluted areas of Wismar city, Wismar county and the district of Rostock the life expectancy of men in Bitterfeld is significantly higher, but for woman it is lower than in Wismar and in the districts of Halle and Rostock. For the interpretation of these results it should be taken, thet women exceeded the limit of livelong adaptability because therefold the load on 1. running the household, 2. theyr employment and 3 rd environmental pollution(Fig.3).

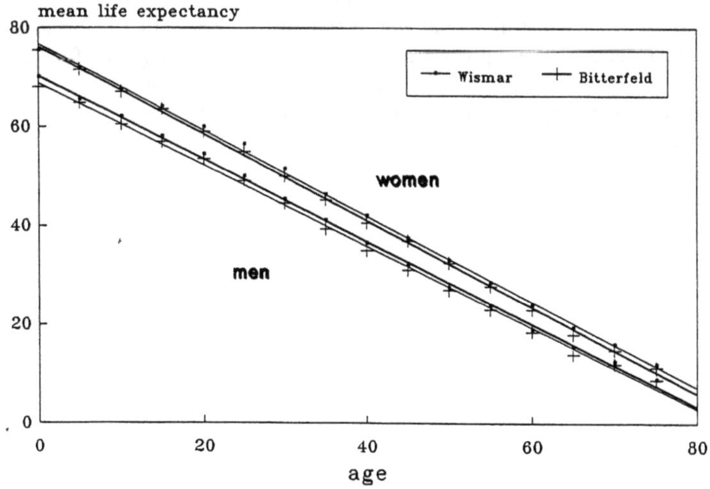

Fig. 3: Mean life expectancy

Certainly one should mention, that deaths were always assigned to that area where the person was living in the end. During this study it was not possible to get information about the duration of the stay in the death place, about the profession and how long the decided person was exposed to the environmental pollution. Moreover the environmental conditions got worse especially during the last 10 years of the G.D.R.. The resulting effects from this are expected in the future. The same statement is so for the cancer statistics. The fact that wether in malign tumors nor in lung cancer differences were found being unfavorable for the Bitterfeld area may be interpreted by migrations in the higher age.
Investigations in the National Cancer Register of the former G.D.R. showed that only in prospective studies all interferences in data sets can be eliminated. This enables recording and therefore the prevention of special cancers in relatively small areas.
This immediate programme Bitterfeld which used many records, references, statistics and mass screenings showed a broad spectrum of harmful effects of the environment and the working place on the health state of the population.
Apart from this there are many other factors overlapping the environmental influences as indicated by the mean life expectancy. At this time the complex of psychosocial reasons is not evaluable, which may have a higher importance than the measured environmental influences.

References

Acker,L. et al. (Eds.)(1984) Mitteilung XII der Kommission der Deutschen Forschungsgemeinschaft zur Prüfung von Rückständen in Lebensmitteln :Rückstände und Verunreinigungen in Frauenmilch. Verlag Chemie, 1984

Ebert, R. et al. (1990) Arbeitsmedizinischer Bericht für den Bezirk Halle, Halle

Heinzow, B. et al. (1990) Bericht der Untersuchungsstelle für Umwelttoxikologie des Landes Schleswig-Holstein, Jahresbericht 1988/89 ,Kiel

Höck, K.,H. Hess und E. Schwarz (1981) Der Beschwerdefragebogen für Kinder (BFB-K) und Kleinkinder (BFB-KK) - ein neurosenpsychologisches Siebtestverfahren für Psychologen, Kinder- und Schulärzte. Herausgegeben von der Psychodiagnostischen Kommission im Auftrage des Vorstandes der Gesellschaft für Psychologie der DDR; VEB Deutscher Verlag d. Wissenschaften. Berlin

Hüllße, Chr. et al. (1991) in: Sofortbericht Bitterfeld, UBA 91-116 09 003, Rostock

Kurth, E. (1978) Testreihe zur Prüfung der Konzentrationsfähigkeit (TPK) - Verfahren zur tätigkeitsgebundenen Erfassung von Leistungsmengen und Leistungsgüte für Schüler der 2. - 6. Klasse, Handanweisung - Psychodagnostisches Zentrum, Berlin

Pelech, L. (1985) Entwicklung von Kindern in luftverunreinigten Gebieten. Soz. pädiatr. Prax. Klin. 7 , 128-133

HAZARDOUS ENVIRONMENTAL AGENTS

1. Radon
2. Dioxin
3. Mineral Dust and Fibres

EFFECTS OF MOBILITY UPON THE DISTRIBUTION OF LIFETIME RADON EXPOSURES

Liu, K-S., Chang, Y-L., and Hayward, S.B.

Indoor Air Quality Program, Air & Industrial Hygiene Laboratory, California Department of Health Services, Berkeley, CA 94704, USA

Gadgil, A.J., and Nero, A.V.

Indoor Environment Program, Energy and Environment Division, Lawrence Berkeley Laboratory, Berkeley, CA 94720, USA

INTRODUCTION

Although various attempts had been made to assess the health risk from exposure to radon, most approaches have not differentiated concentration data from exposure data. The average indoor radon concentration, not the average radon exposure, was used to calculate the number of estimated lung cancer cases among the U.S. population (1). The justification for the use of concentration data for risk assessment was based on simple assumptions such as that people spend a fixed proportion of their time inside their dwellings and live in the same residence for their entire life. These assumptions are invalid since the time spent in various environments is age dependant (2) and the U.S. population is quite mobile (3). This study was designed mainly to study the effect of mobility upon the distribution of lifetime radon exposures by applying computer simulation techniques. The lifetime exposure considered here can be characterized by the following equations:

$$E_{lifetime} = E^{0}_{annual} + E^{1}_{annual} + E^{2}_{annual} + \ldots + E^{k}_{annual} + \ldots, \text{ where}$$

$$E^{k}_{annual} = C^{k}_{1} * T^{k}_{1} + C^{k}_{2} * T^{k}_{2} + C^{k}_{3} * T^{k}_{3}.$$

For each age k, E_{annual} is the annual exposure, C_1 is the indoor residential concentration, C_2 is the indoor non-residential concentration, and C_3 is the outdoor concentration. Likewise, T_1, T_2, and T_3 denote the proportions of time spent in indoor residential, indoor non-residential and outdoor environments respectively. Two cases of simulation were carried out: 'move' and 'no move', and a comparison of the results was made. For the 'move' case, simulated individuals were given a chance to move at all ages. The moving probabilities were estimated from a mobility survey, which was conducted specifically for this study. For the 'no move' case, simulated individuals were assumed to live in the same residence their whole life.

METHODS

Lifetime radon exposures for ten thousand individuals were simulated for each of the two cases, 'move' and 'no move', according to the formulas mentioned above. By definition, the lifetime exposure is the sum of annual exposures over the years exposed to radon. The number of years exposed was defined as the life-span minus five, which is gender dependent. From the 1988 vital statistics of California this number was estimated as 73 for females and 68 for males. In this study the State of California was divided into six regions according to the statewide radon survey (4), and region seven was defined as areas outside California. Each annual exposure was computed from the corresponding annual average concentrations in the three major microenvironments and the associated time/activity parameters for that age. These concentrations were determined as follows. The residential concentration was generated by the regional lognormal distribution obtained from the statewide radon survey. The indoor non-residential concentration was estimated by the average of the regional geometric mean of the residential concentrations and the outdoor concentration, where the outdoor concentration was assumed to be 7.4 Bq/m^3. Note that the residential radon concentration was region dependent. Parameters that were used in the simulation are given in the following tables:

Table 1 Parameters by Region

Region		Birth Prop. (%)	Cal. Birth Prob. (%)	Male Birth Prob. (%)	Lognormal Model	
					G.M. (Bq/m^3)	G. S.D.
CAL	1	49.5	17.55	51.42	24.42	1.71
	2		9.60	51.06	48.47	1.58
	3		58.57	51.07	27.01	1.82
	4		1.07	51.48	81.40	2.41
	5		9.69	51.26	43.29	1.62
	6		3.52	51.39	29.23	1.79
	7	50.5		51.00	33.30	2.80

Table 2 Time/Activity Parameters by Age Group

Age Group	T_1:Indoor Residential Prop. (%)	T_2:Indoor Non-Residential Prop. (%)	T_3: Outdoor Prop. (%)
0 - 4	73	21	6
5 - 19	62	25	13
20 - 64	61	25	14
>= 65	74	20	6

For each individual the following steps were carried out:
1. The region of the birthplace was chosen according to the birth probabilities gathered from the mobility survey and the California vital statistics. For the 'no move' case, only the birth probabilities for regions within California were used.

2. The gender was determined by the probability of having a male birth in the region.
3. The residential radon concentration that he/she was exposed to in the first year (age zero) was generated by the corresponding regional lognormal distribution.
4. The first year exposure was then computed directly from the formula with various concentrations estimated by the methods described above and the corresponding time/activity parameters for age zero.
5. i) For the case of 'no move', the annual exposure of each following year was computed with the birthplace residential concentration and the time/activity parameters for the corresponding age.
ii) For the case of 'move', each annual exposure for age greater than zero was computed as follows: At the beginning of every year, it was determined whether he/she would move. This was done based on the moving probability for the corresponding age. If the individual was going to move, then a destination was selected according to the matrix of transition probabilities within and among regions for that age. Both moving probability and matrix of transition probabilities were obtained from the mobility survey. The residential concentration needed in the computation for the annual exposure would have to be selected from the regional lognormal distribution of the destination should he/she move; otherwise it would be the same as that of the previous year. The rest of the calculation for the exposure was the same as before, with the proper time/activity parameters for the age.
6. Step 5 was executed repeatedly until the age reached 67(=68-1) for males or 72(=73-1) for females.
7. Lifetime exposure was then computed as the sum of the annual exposures obtained above.

RESULTS AND DISCUSSION

Under the assumption of fixed residence for the entire life, the 'no move' case, the arithmetic mean of the lifetime radon exposures was 2052 Bq/m^3 * year, and the standard deviation was 1378 Bq/m^3 * year. Allowing simulated individuals to move, the 'move' case, the arithmetic mean became 2448 Bq/m^3 * year, and the standard deviation became 1130 Bq/m^3 * year. Compared with the 'move' case, the average lifetime radon exposure of the 'no move' case was underestimated by approximately 20%. This implies that the health risk would also be underestimated by 20%, if a non-threshold linear model is assumed. The distributions of lifetime radon exposures under non-moving and moving assumptions are shown in Figure 1. As can be seen from this figure, moving has the effect of shifting the mode of the distribution to a higher value. However, the right hand tails of the two distributions are not very different from each other. This is consistent with the fact that the standard deviation of the 'no move' case is approximately 20% higher than that of the 'move' case. The results of this study clearly demonstrated that the distribution of lifetime radon exposures could be drastically affected by assumptions on mobility. Any exposure assessment without incorporating the moving patterns would result in biased estimates of the distribution. This work was based on California data, although similar findings had been observed in another study that was based on Minnesota data (5). California is a state with lower-than-average radon concentrations and Minnesota is a state with relatively higher concentrations. The similarity of the findings of these two studies indicates that it is important to distinguish concentration data from exposure data. Ignoring moving patterns and using concentration data alone for risk assessment is a simple but unrealistic approach.

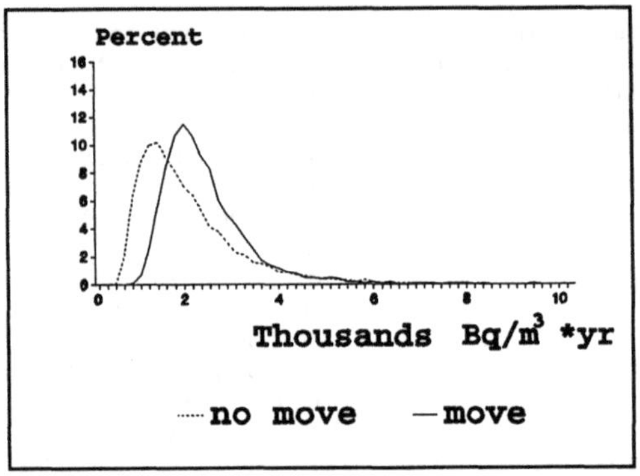

Figure 1. Simulated lifetime radon exposure distributions

ACKNOWLEDGMENTS

The authors wish to thank the U.S. Department of Energy for financial support of this research. They also wish to thank the California Air Resources Board for providing the time/activity data. Specific thanks go to David Quinton for assisting with the mobility survey.

REFERENCES

1. Mills, W.A., and Egan, D.J. Risk assessment and policy, in: Environmental Radon (Cothern, C.R. and Smith, J.E. eds) pp. 273-295, Plenum, New York, 1987.
2. Liu, K-S., Hayward, S.B., Kulasingam, G., Chang, B. and Sexton, K. Estimation of formaldehyde exposures for mobile home residences. Proceedings of the 79th Annual Meeting of the Air Pollution Control Association, Minneapolis, MN, June 1986.
3. Long, L. Migration and residential mobility in the United States. Russell Sage Foundation, New York, 1988.
4. Liu, K-S., Hayward, S.B., Girman, J.R., Moed, B.A., and Huang, F.Y. (1991) Annual average radon concentrations in California residences. J. Air & Waste Manage. Assoc. 41:1207-1212.
5. Gadgil, A.J., Rein, S., Nero, A.V., and Wollenberg, H. The distribution of lifetime radon exposures for Minnesota residents. Draft Report, Lawrence Berkeley Laboratory, Berkeley, CA, 1991.

Radon in the Human Environment

Müller, W.-U. and C. Streffer
Institut für Medizinische Strahlenbiologie
Uni-Klinikum Essen, Hufelandstr. 55, D4300 Essen 1, Germany

Introduction

Extensive screening of radon concentrations during recent years has revealed that these concentrations are markedly higher inside of buildings than outside (e.g. ICRP 50, 1987). The exposure to radon and to radon daughters results in an, on the average, tenfold higher radiation dose to the bronchial epithelium (i.e. about 10 mSv) than to most other tissues of the body due to background exposure. For children the models of calculation even predict double the dose estimated for adults (ICRP 50, 1987, page 17). This comparatively high burden for the population requires thorough studies of a possible risk exerted by radon and its progeny.

Physical Background

Radon-222 originates from uranium-238 via radium-226. The gaseous radon disintegrates by emission of alpha-rays to solid, but radioactive polonium-, lead- and bismuth-nuclides. In particular the polonium nuclides (polonium-218 and -214) make the major contribution to the radiation dose to the respiratory tract. This is mainly due to the fact that these solid alpha-emitters are bound to the aerosolic fraction of the air and are deposited in the respiratory ducts after inhalation, whereas radon itself (physical half-life = 3.82 days) is predominantly exhaled shortly after entering the respiratory tract.

Uranium and consequently also radium are distributed almost ubiquitously, though in varying concentrations, in soil and rock. Thus, human beings are exposed to radon everywhere, however, to grossly different doses depending on the environmental conditions.

Radon Sources and Concentrations

As outlined above, the outdoor concentration of radon is caused by the radium-226 content of rock and soil. The concentration is about 9 Bq/m^3 worldwide (Gesell, 1983). There are considerable differences when the concentrations are compared either with regard to time of the day (2 to 5 times higher values during night than around noon) and to time of the year (2 to 4 times higher values during late summer and autumn than during spring time). Also concentration diminishes with increasing distance from the ground, so that children are exposed to slightly higher concentrations than adults.

Most of the indoor radon derives from the radium located in the soil below the buildings; variable amounts are contributed by the building material, air from outside, tap water, and natural gas. This sums up to a mean value of approximately 50 Bq/m^3 in dwellings (for reviews see UNSCEAR, 1986; ICRP 50, 1987; BEIR IV, 1988). Variability, however, is pronounced: the indoor range is between a few Bq/m^3 to several thousand Bq/m^3. With the exception of a specific building material used in Sweden, it is almost exclusively the soil underneath the buildings that is responsible for unusually high radon concentrations indoors.

When estimating radiation risk, one has to take into account that indoors no equilibrium between radon and its daughters is reached. Therefore, one has to correct for this fact by multiplying the radon concentration with an equilibrium factor (usually in the range between 0.3 and 0.5); radon concentration is then expressed as "equilibrium-equivalent radon concentration" (or abbreviated: Rn-eq).

Health Risks that May be Expected

In general, four different types of risk can be expected after exposure to ionizing radiation: lethality, teratogenicity, mutagenicity, and carcinogenicity (detailed information by Streffer, 1989). For various reasons (Müller and Streffer, 1990), only the carcinogenic risk must be considered in the context of radon exposure.

The major target organ is, of course, the respiratory tract. As our current knowledge suggests that there is no threshold dose with regard to radiation-induced tumour risk, one should expect a certain contribution to the lung tumour frequency by radon and radon daughters. In the remaining part of the paper, the procedure will be described, how to achieve an estimate of this contribution.

Epidemiologic Studies and Risk Coefficient

Already in the 16th century radiation-induced lung tumours caused a high frequency of deaths in miners ("Schneeberger Krankheit"), though nobody, of course, was aware of the nature of the disease and the involvement of ionizing radiation in its genesis at that time. Only at the end of the 19th century, the disease was identified as a lung tumour and only some decades ago the causation by radon was realized.

Recent epidemiologic studies of miners (in particular uranium miners) clearly point to an increase in the frequency of lung tumours in dependence of the amount of radon exposure (for reviews see ICRP 50, 1987 and BEIR IV, 1988). The relative risk coefficient derived from these studies amounts to about 1.6% per 1 million Bq h/m^3 Rn-eq. That is: the radon concentration (in Bq/m^3) at the place of work is multiplied with the duration of stay (in hours); this time integrated concentration gives the exposure of the individual (in Bq h/m^3) and is, after consideration of the equilibrium factor, the basis for risk estimation.

Estimation of Risk due to Indoor Radon

Estimation of radiation risk of indoor radon exposure is difficult, despite the information available from the studies in miners. The difficulties are mainly due to the necessity of extrapolating to lower exposure levels in dwellings and to the hardly comparable conditions in mines and households (e.g. differencies in respiration rates, in particle sizes of the aerosol, in additional agents that may be involved in lung tumour genesis like heavy metals). Correcting for the major known differences in both surroundings leads to an estimate of the absolute risk value of between 1 and 3 lung cancer deaths per 1 million people inhaling 1 Bq/m^3 (Rn-eq) for 1 year (Jacobi, 1986). Expressed in terms of the relative risk this means about 1% increase of the spontaneous frequency per 1 million Bq h/m^3 (Rn-eq).

If a population is exposed to 15 Bq/m^3 (Rn-eq; derived from a radon concentration of 50 Bq/m^3 and multiplication with an equivalent factor of 0.3) for 80% of the time (probability of staying inside of buildings) for 70 years (life expectancy), an exposure of about 7 million Bq h/m^3 (Rn-eq) can be calculated. Thus, taken together with the relative risk coefficient mentioned above, between 5 and 10% of all lung cancer deaths may be attributed to radon inside of dwellings. This, of course, is true only if a number of prerequisites are adequate, e.g.: 1. the linear extrapolation from high to low doses; 2. the transfer of risk coefficients of miners to non-miners after adjusting these coefficients to non-mining conditions; 3. the models used for calculating dose from exposure data. In any case, the high risk estimate underlines the necessity of further studies of the impact of radon-222 on human health.

Steps to Reduce Radon Concentrations in Dwellings

In most houses regular ventilation through the windows will be sufficient to keep the radon concentrations within a reasonable range. Only in houses with extraordinarily high radon concentrations, additional, more sophisticated steps are necessary, like construction or reconstruction of the foundation, artificial ventilation underneath the foundation and/or of the basement etc. Recommendations for carrying out such actions are different in various countries; in most cases, radon concentrations exceeding 250 to 500 Bq/m^3 are thought to require measures aiming at a reduction of radon concentrations below these values.

References

BEIR IV (1988) Health risks of radon and other internally deposited alpha-emitters. Report of the BEIR IV Committee of the National Research Council (USA); National Academy Press, Washington D.C.

Gesell TF (1983) Background atmospheric 222-Rn concentrations outdoors and indoors: a review. Health Phys 45: 289-302

ICRP 50 (1987) Lung cancer risk from indoor exposures to radon daughters. Annals of the International Commission on Radiological Protection. Publication 50, Pergamon Press, Oxford, 1-60

Jacobi, W. (1986) Lungenkrebs nach Bestrahlung: Das Radon-Problem. Naturwissenschaften 73, 661-668

Müller W-U, Streffer C (1990) Strahlenrisiko durch Radon-Exposition in Häusern. Der Kassenarzt 30, Heft 22, p.40-47

Streffer C (1989) Ionisierende Strahlen: Biologisch-medizinische Wirkungen und Abschätzung des Risikos. Der Kassenarzt Heft 7, p.43-52; Heft 9, p.48-60; Heft 11, p.42-53

UNSCEAR (1986) Genetic and Somatic Effects of Ionizing Radiation. Annex B: Dose-response relationships for radiation-induced cancer. United Nations Scientific Committee on the Effects of Atomic Radiation. United Nations Publication E.86.IX.9, New York, 165-262

Dioxin Exposure and health effects on people at Seveso, Italy.

Mocarelli, P.[1], Needham, L.L.[2], Marocchi A.[1], Patterson, D.G. Jr.[2], Brambilla P. [1], Gerthoux P.M.[1], Meazza L.[3] and Carreri V. [3]

[1]University Department of Clinical Pathology, Hospital of Desio, Milan, Italy;
[2]Center for Environmental Health and Injury Control, Centers for Disease Control, Atlanta, GA, USA;
[3]Health Department of the Regione Lombardia, Milan, Italy.

Introduction

2,3,7,8-tetrachlorodibenzo-*p*-dioxin (TCDD) is a by-product of polychlorinated phenol manufacture, especially the germicide hexachlorophene and the herbicide 2,4,5-trichlorophenol (TCP) and is considered one of the most powerful man-made toxic molecules.
Even if TCDD is widespread in the environment, in humans the food chain is the main route of contamination; a lesser source derives from waste disposal incinerators.
Data on dioxin toxicity have been collected mostly from adult males professionally exposed to the compound (Webb et al., 1989; Kimbrough, 1990) Until recently, TCDD exposure levels were measurable only indirectly.
In July, 1976, an incident at the ICMESA factory, located near Seveso, Italy, caused a fallout of TCDD which contaminated the area. At that time, no techniques to measure TCDD in body fluids were available, and only an indirect estimate of exposure could be made on the basis of TCDD deposited in soil.
Different areas of contamination were defined, based on TCDD concentration:

Zone	Population	Area	mean TCDD ($\mu g/m^2$)	Highest TCDD ($\mu g/m^2$)
A	736	80	230	5,447
B	4,737	269	3.0	43.8
R	31,800	1,430	0.9	9.7

Because the half-life of TCDD in humans is about 7 years (Pirkle et al, 1989), the acute and chronic results of exposure tend to overlap. In Seveso, the residents of A Zone were evacuated within 15 days of the accident, while those of B Zone were allowed to remain in the area. This population assumed a dosage from the cloud and a low dose, if any, through dust inhalation and dermal absorption during the following years. We summarize here the 9-year follow up health project (1976-1982; high-exposure individuals monitored until 1985), and new data on the correlations between levels of TCDD in serum (collected in 1976) (Facchetti et al., 1981; Patterson et al, 1987) and clinical laboratory results both of individuals of the most contaminated A Zone and of B Zone, as well as TCDD levels in the blood of people who immigrated into B Zone after the accident and lived there for some years.

Results and discussion

The clinical follow-up demonstrated no significant increase in miscarriages, perinatal mortality and morbidity, low birth weight or congenital malformations (Mastroiacovo et al., 1988; De Carli et al., 1982); and no effects in internal medicine and pediatrics, with the exception of transient cases of chloracne (Assennato et al, 1989).

Laboratory tests did not show higher prevalence of pathologic results in the exposed population when compared with controls. Test results often ranged between the lower end of the normal range and slightly above it. Their comparison against controls demonstrates some biochemical "acute effect" in the first year (1976-77) after the accident (Mocarelli et al., 1986).

In A Zone, subjects with maximal body loads of TCDD (up to 56,000 ppt) did not demonstrate pathologic laboratory results or clinical effects, except chloracne. Among individuals with similar TCDD body burden, some developed chloracne while others did not. Both groups presented similar slight alterations in biochemical values, demonstrating that chloracne is not a constant and sensitive indicator of TCDD exposure (Mocarelli et al., 1991).

Preliminary data from adults (26-49 years) of B Zone did show serum levels of TCDD about 8-10 times higher than those of people in industrial areas. Between 1976 and 1982, laboratory results from these people were within normal limits.

Subjects of comparable age groups, who immigrated in B Zone after July 1976, were tested, after some years of residence, for their TCDD level in blood. They demonstrated extremely low levels of TCDD, that is, below 15 ppt. These values are perfectly comparable with those of the general population of urban and industrial areas.

These results, while supporting data on the surprisingly low effects (chloracne) of high doses of TCDD in humans at least at Seveso, demonstrate that, although TCDD remained within the soil, it did not seem to escape from it.

Cancer was a pathology not considered in this study.

References

Assennato, G., Cervino, D., Emmett E.A., et al. 1989. Follow-up of subjects who developed chloracne following TCDD exposure at Seveso. Amer. J. Ind. Med. 16: 119-124.

De Carli, L., Mottura, A., Nuzzo, F., et al. (1982). Cytogenetic investigations of the Seveso population exposed to TCDD. In: Dardanoni L, Miller RW, eds.

Plans for clinical and epidemiological follow-up after area-wide chemical contamination: proceedings of an international workshop. Washington, DC, National Academy of Sciences, 292-317.
Facchetti, S., Fornari, A., and Montagna, M. (1981). Distribution of 2,3,7,8-tetra-chlorodibenzo-p-dioxin in the tissues of a person exposed to the toxic cloud at Seveso. Forensic Environ. Appl. 1:1406-1444.
Kimbrough, R.D. (1990). How toxic is 2,3,7,8-tetrachlorodibenzodioxin to humans? J Toxicol Environ Health 30:261-271.
Mastroiacovo, P., Spagnolo, A., Marni, E., et al. (1988). Birth defects in the Seveso area after TCDD contamination. JAMA 259:1668-1672.
Mocarelli, P., Marocchi, A., Brambilla, P., et al. (1986). Clinical laboratory manifestations of exposure to dioxin in children. A six-year study of the effects of an environmental disaster near Seveso, Italy. JAMA 256: 2687-2695.
Mocarelli, P., Needham, L.L., Marocchi, A., Patterson, D.G., Brambilla, P., Gerthoux, P.M., Meazza, L., Carreri, V. (1991). Serum concentrations of 2,3,7,8-tetrachlorodibenzo-p-dioxin and test results from selected residents of Seveso, Italy. J. Toxicol. Env. Health. 32:357-366.
Patterson, D.G. Jr, Hampton, L., Lapeza, C.R. Jr, et al. (1987). High-resolution gas chromatographic/high-resolution mass spectrometric analysis of human serum on a whole-weight and lipid basis for 2,3,7,8-tetrachlorodibenzo-p-dioxin. Anal Chem 59:2000-2005.
Pirkle, J.L., Wolfe, W.M., Patterson D.G. Jr., et al. (1989). Estimate of the half-life of 2,3,7,8-tetrachlorodibenzo-p-dioxin in Vietnam veterans of Operation Rank Hand. J. Toxicol. Environ Health 27:165-171.
Webb, K.B., Evans, R.G., Knutsen, A.P., et al. (1989). Medical evaluation of subjects with known body levels of 2,3,7,8-tetrachlorodibenzo-p-dioxin. J Tox Environ Health 28:183-193.

Comparison of cytotoxic effects of asbestos and glass fibres on the cytoskeleton of epithelial lung cells in vitro

M. Aufderheide, M. Riebe-Imre, J. Hertes and U. Mohr
Institute of Experimental Pathology, Hannover Medical School, Konstanty-Gutschow-Str. 8, 3000 Hannover 61, FRG

Introduction

Asbestos and other mineral fibres including glass fibres are known to cause pleural mesotheliomas and pulmonary carcinomas in experimental animals and humans (Environmental Health Criteria, 1986). The physical characterization of fibres (size, shape) and their chemical properties have been implicated in determing fibrogenic and tumorigenic potential (Allison et al., 1975; Beck and Bignon, 1985; Hesterberg and Barrett, 1985; Jaurand et al., 1986; Mossman et al., 1983). Some in vitro studies have been carried out with different cell systems to elucidate further the mechanisms of interaction between the cells and fibres. In the present study the effects of crocidolite asbestos and two glass fibres on the cytoskeleton were tested. The cytoskeleton represents a complex structure which is of major importance for the structural integrity of the cell and for its physiologic processes. Therefore damage of the various filament systems of the cytoskeleton, the actin-, tubulin- and vimentin system may contribute to the pathologic potency of fibres in the respiratory tract.

Material and Methods

Cells of the Syrian hamster epithelial cell line (M3E3/C3) were grown in a medium consisting mainly of RPMI 1640 (80%, Flow, Meckenheim) and fetal bovine serum (BSA; 20%, Flow) as described previously (Emura et al., 1982). Cultures were maintained at 37° C in a humidified atmosphere with 10% CO_2 and 90% air. Confluent cultures were harvested by trypsination (0,25% Trypsin, 2% EDTA in PBS) and resuspended in the medium after low speed centrifugation. The epithelial character of the cell line and its application to pulmonary toxicology have been described elsewhere (Emura et al., 1989, 1990).
All fibers tested were provided by Prof. Pott from the Medical Institute for Environmental Hygiene (University of Düsseldorf). The characterization of the three fibres was performed by Dr. Bellmann, Fraunhofer Institute of Inhalation Toxicology and Aerosol Research and is given in table 1.

Table 1: Characterization of fiber dusts

Fiber	mass of critical fibers (%)	critical fibers/ng mean	fiber length (µm) 10%<	50%<	90%<	fiber diameter (µm) 10%<	50%<	90%<
Crocidolite	42	99	0.9	1.8	4.7	0.11	0.19	0.32
JM 104/475	88	158	0.7	2.3	8.2	0.06	0.14	0.38
B-1M	15	27	6.7	10.7	21.1	0.88	1.68	2.49

Cells were grown on 13-mm-diam glass coverslips in 8 cm² petri dishes for 86 hs. Mineral fibers sterilized by autoclaving were suspended in growth medium. The resulting suspensions were added to a final volume of 5 ml at a final concentration of 10 µg crocidolite asbestos/ml, 100 µg JM104/475 /ml and 100 µg B-1M/ml..

This study was supported by the Federal Minister of Health (Grant No. 417-6080-1/44))

Immunfluorescence microscopy was performed after 14, 32, 62 and 86 hs of exposure. Cells were fixed in 4% paraformaldehyde in PBS (20 min) for actin and tubulin detection. The whole preparation procedure was at 37°C. After washing the cells in the same buffer, they were permeabilized with 1% NP-40 in PBS for 10 min at room temperature, washed three times with PBS, incubated for 20 min with BSA to avoid unspecific reactions and washed again with PBS. For actin labeling probes were incubated for 60 min with monoclonal anti-actin IgM (Amersham) at a dilution of 1:200 in PBS. After intensive washing, cells were incubated with the secondary anti-mouse IgG conjugated with AMCA (Aminomethyl-coumarin acetic acid; Dianova) for 45 min. For tubulin labeling a primary monoclonal antibody directed against a-tubulin (Sigma) was used after dilution to 1:200 in PBS (60 min) followed by an incubation with a second anti-mouse IgG-AMCA for 45 min. The procedure was exactly the same as for the actin. For the detection of vimentin, cells were fixed in methanol for 10 min at -20°C. Permeabilisation was performed in acetone for 2 min at -20°C. Anti-vimentin IgG (Boehringer) in a dilution of 1:20 was used as the primary antibody (60 min) followed by incubation with an anti-mouse IgG-AMCA for 45 min.

Results and Discussion

Aktin

The application of crocidolite asbestos and the two glass fibres affected the actin filament system in a time-dependent manner. Different degrees of degeneration were observed in the tubular and vimentin filament organisation. The untreated cells demonstrated characteristic features of actin stress fibre arrangement which appeared as a well-represented network of fibres throughout the cytoplasm (figure 1a). After 38 hs of exposure to crocidolite and JM 104/475 glass fibres remarkable changes occurred in the actin filament organization. In particular, the filamentous structures were broken down into a granular configuration (figure 1b). An early stage of restoration of the filaments occured in cultured cells approaching 86 hs of exposure.

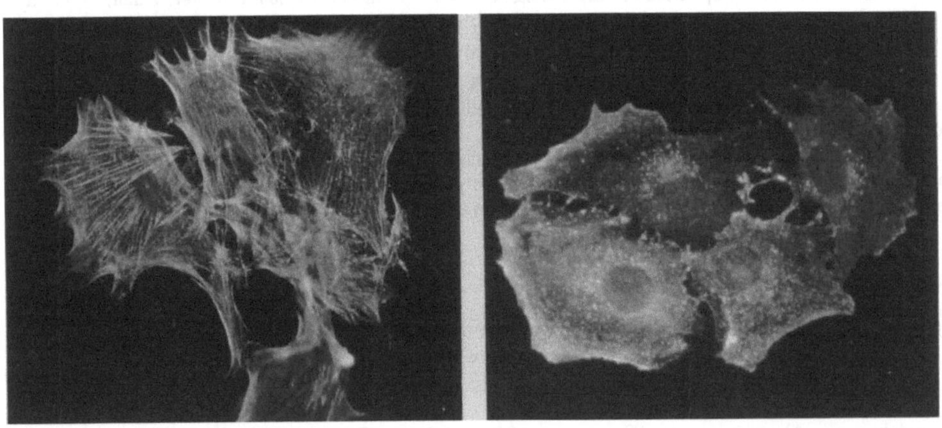

Figure 1: Fluorescence photomicrograph of actin in untreated M3E3/C3 cells showing a normal arrangement of actin fibres (1a) and in cells exposed to crocidolite asbestos for 38 hs (1b).

After treatment with B-1M glass fibres the epithelial cells revealed a delayed reaction in actin depolymerisation. Only at 86 hs did the cells show an actin organisation resembling that observed after crocidolite exposure. The changes in actin configuration correspond to a marked reduction in cell adhesion. This effect might be due to a damage in the organisation of adhesion plaques, a region where the cultured cells adhere most tightly to the underlying substratum (Burridge et al., 1987). Such specialized areas of the cytoplasmic membrane serve as attachment sites for stress fibres and may also provide regions where regulatory signals are activated and transmitted across membranes. The phosphorylation of adhesion plaque proteins is discussed as a major event leading to the disruption of adhesion plaques and stress fibers in transformed cells (Rohrschneider, 1980). As crocidolite asbestos is known for its transforming capacity, the disruption of the actin filament system may be a sensitive morphological marker for the susceptibility of cells to transformation.

Tubulin

The microtubule apparatus underwent no alterations in its organisation after crocidolite exposure. In contrast, both glass fibres induced a complete depolymerisation of the filamentous tubulin system after 62 hs of fibre treatment and persisted through 86 hs (figure 2a, 2b).

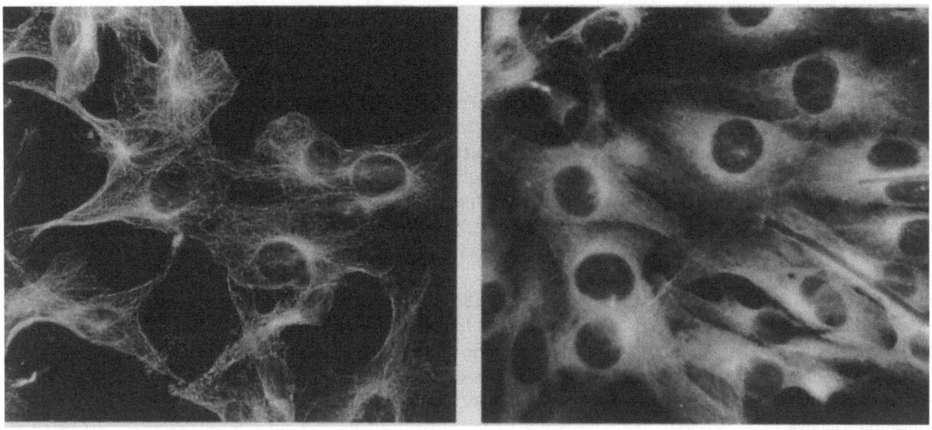

Figure 2: Fluorescence photomicrographs. M3E3/C3 control cells showing the normal tubulin network (2a). After 62 hs the exposure to B-1M glass fibres induced a depolymerisation of the microtubular apparatus.

Vimentin

The vimentin scaffold, which normally appeared in a filigree-like pattern throughout the cytoplasm (figure 3a), concentrated in thick bundles (figure 3b) after treatment with crocidolite asbestos, whereas no changes could be noted in cells exposed to the glass fibres.

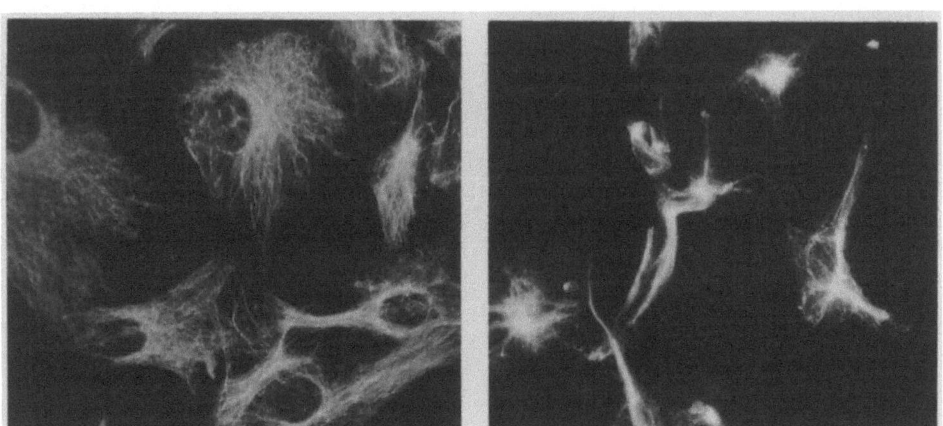

Figure 3: Fluorescence photomicrograph of vimentin in M3E3/C3 cells. Exposure to crocidolite asbestos induced a concentration of vimentin filaments within the cells at the expense of the normally anastomosing network.

Crocidolite asbestos and the glass fibres have different effects on various components of the cytoskeletal system suggesting that different mechanisms are involved in interactions between of special fibre types and epithelial cells. Furthermore it seems obvious that the cytoskeletal components responds sensitively to mineral fibres and can serve as morphological and mechanistic indicators of cytotoxic effects. The in vivo biological implications of the cytoskeletal changes remain to be determined.

Acknowledgements
The authors wish to thank Ms. B. Melching and Mr. M. Tauscher for excellent technical assistance.

References

Allison, A.C., Hatington, J.S., and Badami, D.V. (1975). Mineral fibers: Chemical physiochemical and biological properties. Adv. Pharmacol. Chemother. 12, 291-402.
Beck, E.G and Bignon,J. (1980). In vitro effects of mineral dusts. NATO ASI Ser.G No3
Burridge, K., Molony, L., and Kelly, T. (1987). Adhesion plaques: sites of membrane interaction between the extracellular matrix and the actin cytoskeleton. J. Cell Sci. Suppl. 8, 211-229.
Emura, M., Richter-Reichhelm, H.-B., Böning, W., Eichinger,R., Schoch, C. Althoff, J., and Mohr, U. (1982). A fetal respiratory epithelial cell line for studying some problems of transplacental carcinogenesis in Syrian golden hamster. J. Cancer Res. Clin. Oncol. 104, 133-144.
Emura, M., Riebe, M., Germann, P., Brockmeyer, C., Aufderheide, M., and Mohr, U. (1989). Functional culture of hamster and human airway epithelial cells and its application to pulmonary toxicology. Exp. Pathol. 37, 224-227.
Emura, M., Riebe, M., Ochiai, A., Aufderheide, M., Germann, P., and Mohr, U. (1990). New functional cell-culture approach to pulmonary carcinogenesis and toxicology. J. Cancer Res. Clin. Oncol. 116, 557-562.
Environmental Health Criteria 53 (1986). Asbestos and Other Natural Mineral Fibers. World Health Organization, Geneva.
Hesterberg, T.W., and Barrett, J.C. (1985). Induction by asbestos fibers of anaphase abnormalities: Mechanisms for aneuploidy induction and possible carcinogenesis. Carcinogenesis 6, 473-475.
Jaurand, M.C., Kheuang, L., Magne, L., and Bignon, J. (1986). Chromosomal changes induced by chrysotile fibers or benzo-3,4-pyrene in rat mesothelial cells. Mutat. Res. 169, 141-148.
Mossman, B.T., Eastman, A., Landesman, J.M., and Bresnick, E. (1983). Effects of crocidolite and chrysotile asbestos on cellular uptake, metabolism, and DNA after exposure of Hamster tracheal epithelial cells to benzo(a)pyrene. Environ. Health Perspect. 51, 331-336.
Rohrschneider, L.R. (1980). Adhesion plaques of Rous sarcoma virus-transformed cells contain the src gene product. Proc. Natn. Acad. Sci. U.S.A. 77, 3514-3518.

HL-60-Cell derived Macrophages are resistant to Cytotoxic Effects of Quartz

Brückner-Nieder, U., A. Schmidt, J. Schimmelpfeng and A. Seidel
Kernforschungszentrum Karlsruhe, Institut für Genetik und Toxikologie. Postfach 3640, 7500 Karlsruhe 1, Germany

Introduction

The cytotoxic effects of quartz particles towards alveolar macrophages (AM) are well known and have been studied extensively. According to the literature and to our own recent interspecies comparison (Seidel et al., 1990), there is no principle variation of the response between different animal species. For several reasons, especially for studies of a more basic character, proliferating macrophage cell lines instead of AM with their inevitable variation in properties and response have been used. In an attempt to develop a human cell model we have studied extensively the use of HL-60 cells, a human promyelocytic line. These cells can be differentiated by TPA (12-0-tetradodecanoylphorbol-13-acetat) into non-proliferating macrophage-like cells (HL-60-M). In this paper we will describe the stages of this differentiation in more detail as well as speculate about the complete lack of any cytotoxic effect of quartz on these cells.

Material and Methods

Undifferentiated HL-60 cells were differentiated for 20 h with 10 ng/ml or 100 ng/ml TPA in Polysterene dishes containing RPMI 1640 medium with foetal calf serum and non-essential amino acids. The specific activities of lactate dehydrogenase (LDH), N-Acetylglucosaminidase (NAG) and acid phosphatase (AP) were determined. After 48 h this medium was replaced by TPA-free medium and the cells were allowed to react further for up to four days. Cells at different stages of differentiation were incubated for 20 h with either the original Sikron F600 Quartz (Palas, Karlsruhe) or with a subfraction from the forementioned medium with or without FCS. The reaction of the HL-60-M cells has been compared with that of freshly isolated alveolar macrophages from different animal species. Details of these methods can be seen in our earlier publication (Schimmelpfeng et al., (1991)). These experiments were performed with DQ 12 Quartz, but the effects of the Sikron F600 used in this paper and this quartz are very similar (Drosselmeyer et al., (1988)). After incubation with quartz, the cells were rinsed with 0.9% NaCl and stained with Acridin Orange (10 µg/ml in 0.9% NaCl) for 15 min. For transmission electron microscopy the cells were differentiated on collagen-coated Nuclepore filters, incubated with quartz and embedded in Epon/Araldit using standard procedures. For the experiments concerning the Tumor Necrosis Factor (TNF), the cells were differentiated for 48 hours with 10 ng/ml TPA followed by a further 48 hours without TPA. Thereafter they were treated with Lipopolysaccharide (LPS, 10 ng/ml), and/or 500 µg/ml quartz in RPMI 1640 with 2% FCS for 4 hours. TNF release was determined by testing the cytotoxicity of the supernatants to L-929 cells.

Results and Discussion

After 20 h of TPA treatment the HL-60 cells had adhered to the plastic surface. They became strongly positive to alpha-Naphtylbutyrate staining and were also able to phagocytose Latex particles. Their actin-skeleton appeared after Rhodamin-Phalloidin staining just as that resembling typically adherent macrophages. The specific activities of LDH, NAG and AP increased about threefold. The staining of the cells with May-Grünwald-Giemsa (Hemacolor, Merck) became increasingly difficult when the duration of differentiation was increased.
HL-60-M cells also phagocytose quartz particles as was shown by vital staining with Acridin Orange: The particles were clearly taken up into the lysosomes. This effect can also be seen when electron microscopy is used, where the quartz is seen inside the phagolysosomes. In sharp contrast to the cytotoxic effects of Sikron F600 on the ultrastructure of alveolar macrophages (Seidel et al., 1989), the morphology of quartz containing HL-60-M remains normal after 20 h.

In Figures 1 and 2, the release of LDH and NAG into the culture medium after incubation of alveolar macrophages from various animal species with quartz is shown. As can be seen, in each case, except for the HL-60-M cells, all the alveolar macrophages from the different species had reacted to varying extents. In contrast to the results with alveolar macrophages (Seidel et al., 1990) the viability of HL-60-M is also not influenced by quartz. In Fig. 3 the influence of time upon differentiation and of FCS is shown. With increasing time of differentiation the relative LDH content in the culture medium also increases, but there is no obvious or systematic difference between controls and quartz treated cells. FCS has no influence on these results. When evaluating the data it should be noted that in the case of alveolar macrophages, 50 - 80% of the total LDH was released by quartz under the same conditions (Schimmelpfeng et al., 1991).

Fig. 4 shows that even the Sikron F600 fraction with very small grain size has no effect on the release of cytoplasmatic LDH by HL-60-M cells.

HL-60-M cells produce TMF after stimulation with LPS (Fig. 5) but not with quartz. This is in contrast to the results with bovine alveolar macrophages (Fig. 5b) where quartz was effective. Quartz itself does not influence the TNF release by HL-60-M cells. We have also tried to differentiate HL-60 cells to granulocyte like cells by culturing together with DMSO for five days (Collins et al., 1978). Once again, after using this method of differentiation, there was no reaction of the cells after incubating for 20 h with 250 µg/ml Sikron F600.

At present, these results cannot be understood. The explanation given by Behrendt et al., (1987) for the reduced quartz toxicity in human monocyte derived macrophages is probably not valid here, since we have not observed the "fingerprint structures" as described by these authors. The reason for this lack of influence is completely unknown but clearly further investigations would help us to improve our understanding of the mechanism of the action of quartz itself.

References

Behrendt B, Seemayer NH, Braumann A, Nissen M (1987) Elektronenmikroskopische Untersuchungen zur Wirkung von Quarzstaub DQ12 auf menschliche Monozyten/Makrophagen in vitro. Silikosebericht Nordrhein-Westfalen 16:213-222.

Collins S, Ruscetti F, Gallagher R, Gallo R (1978) Terminal differentiation of human promyelocytic leukemia cells induced by dimethyl sulfoxide and other polar compounds. Proc. Natl. Acad. Sci 75:2458-2462.

Drosselmeyer E, Hofheinz V, Pätzold S, G. Polzer G, J. Schimmelpfeng J, Seidel A (1988) Parameters influencing the effects of quartz on alveolar macrophages. J. Aerosol Sci. 19:1157-1160.

Schimmelpfeng J, Seidel A (1991) Cytotoxic effects of quartz and chrysotile asbestos: In vitro interspecies comparison with alveolar macrophages. J. Toxicol. Environm. Health 33:131-140.

Seidel A, Wilczek W, Pätzold S, Schimmelpfeng J (1989) Biochemische Untersuchungen zum Einfluß inhalierbarer Partikel auf Lysosomen von Alveolarmakrophagen. Bericht des Kernforschungszentrums Karlsruhe, KfK-PEF 50:413-422.

Seidel A, Nieder U, Pätzold S, Schimmelpfeng J, Schmidt A, Wilczek W (1990) Effects of quartz and asbestos on alveolar macrophages: Interspecies comparison and cell biological studies. Environmental Hygiene II, Springer, Berlin, 199-203.

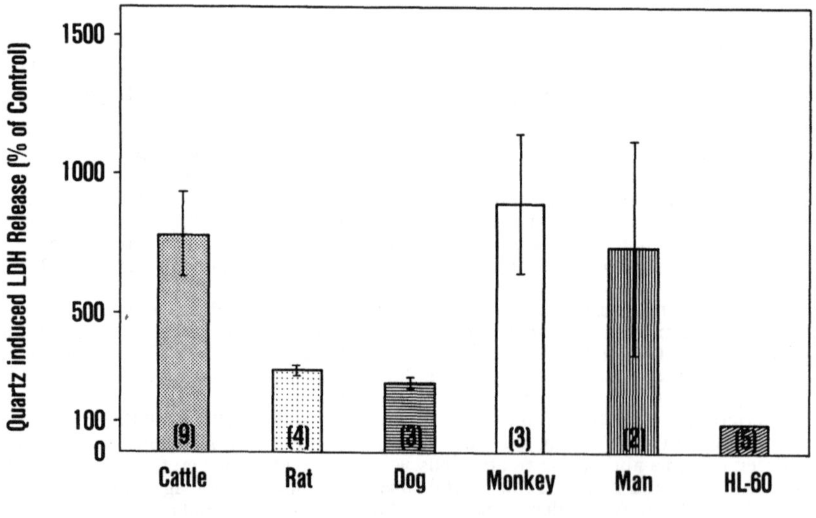

Fig. 1: Release of lactate dehydrogenase after 20 h of incubation of alveolar macrophages of different species and "HL-60-macrophages" with 250 µg/ml quartz. Mean values with standard error, number of individuals or experiments in brackets.

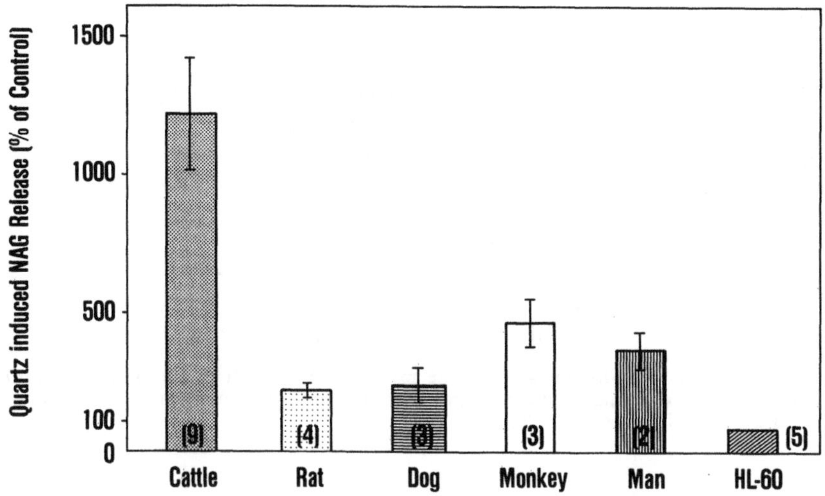

Fig. 2: Release of N-Acetylglucosaminidase after 20 h of incubation of alveolar macrophages of different species and "HL-60-macrophages " with 250 µg/ml quartz. Mean values with standard error, number of individuals or experiments in brackets.

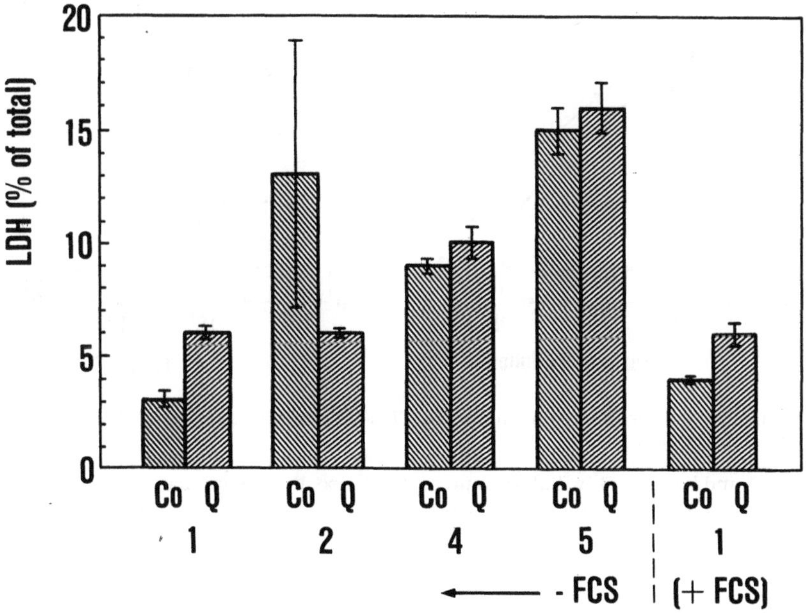

Fig. 3: Release of lactate dehydrogenase after 20 h incubation of "HL-60-macrophages" with 250 µg/ml quartz. Means with standard error from three values.
Co: control; Q: quartz. Number below the colums indicate days of differentiation.
± FCS: Presence or absence of fetal calf serum during incubation with quartz.
100% = total enzyme activity in cells plus medium.

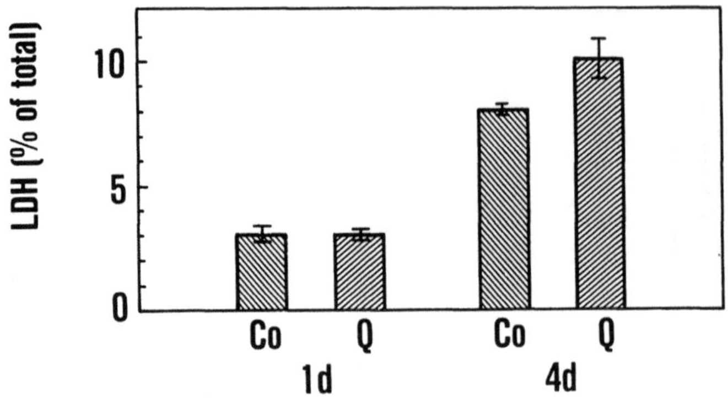

Fig. 4: Release of lactate dehydrogenase into the medium after 20 h incubation of "HL-60-macrophages" with 250 µg/ml quartz, grain size 0.3 µm (AMAD). Means with standard error from three values, see also Fig. 3.

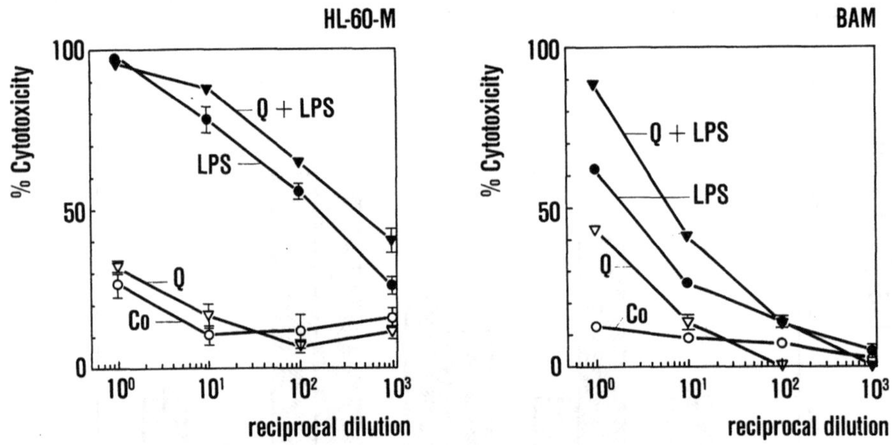

Fig. 5: Release of Tumor Necrosis Factor by "HL-60-macrophages" and bovine alveolar macrophages (BAM).
Co: unstimulated control; Q: Quartz, 500 µg/ml. LPS: Lipopolysaccharide. Incubation with Q and/or LPS: 4 h. Values represent cytotoxicity of supernatants to L-929 cells. Means with standard error from 4 values.

In vitro correlates of the pathogenicity of a long fibre sample of amosite asbestos compared to a short fibre sample

Kenneth Donaldson
Institute of Occupational Medicine Ltd, 8 Roxburgh Place, Edinburgh, Scotland.

Introduction

At the Institute of Occupational Medicine we have utilised two samples of amosite asbestos to study the role of fibre length in pathobiological activity. As shown in Table 1 these two samples, which differ only in length, are very different in their ability to cause tumours, fibrosis, inflammation and immunosuppression in animals exposed *in vivo*.

Table 1. Biological activity of the long and short fibre amosite asbestos samples.

	SHORT	LONG	Reference
ANIMALS WITH LUNG TUMOURS ON INHALATION	0%	32.5%	a
ANIMALS WITH MESOTHELIOMAS ON INTRA-PERITONEAL INSTILLATION	0%	88%	a
AVERAGE FIBROSIS SCORE ON INHALATION	0	15.6	a
INFLAMMATION SCORE ON INTRA-PERITONEAL INSTILLATION	-/+	++++	b
IMMUNOSUPPRESSION SCORE ON INTRA-PERITONEAL INSTILLATION	+	++++	c

References:- a) Davis et al 1986; b) Donaldson et al 1989; c) Donaldson et al in press.

We have attempted here to detect *in vitro* cellular response correlates of the *in vivo* activities shown by the two amosite samples. We utilised two cell systems that reflect important aspects of lung defence against fibres and that are important in pathological change:-

1) epithelial cells, which maintain the integrity of the barrier between the airspace and the interstitium;

2) alveolar macrophages, which phagocytose any depositing fibres and can then release, as a response, a number of substances (cytokines) capable of modulating the activities of other lung cells. We measured the *in vitro* effect of long and short fibre amosite asbestos on a) the integrity of an epithelial monolayer; b) alveolar macrophage secretion of the cytokine Tumour Necrosis Factor (TNF).

Materials and Methods

Long and short fibre amosite. The size distributions of the long and short fibre amosite samples are given in Figure 1 showing that these samples differ substantially in length.

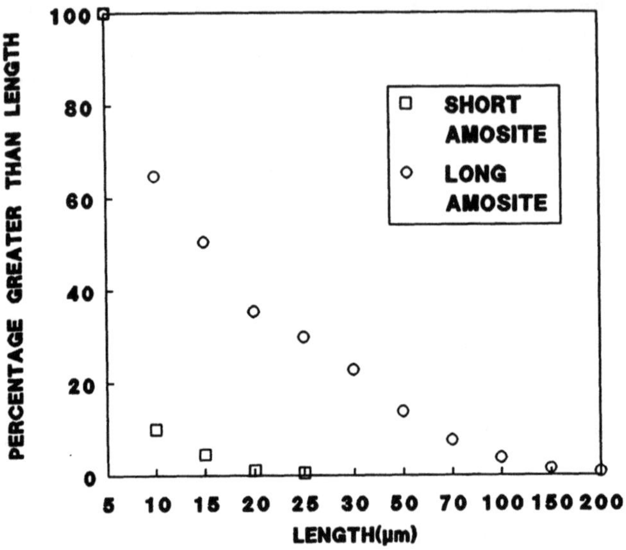

Figure 1. Phase contrast optical microscope counts of fibre lengths of the long and short amosite preparations. Fibres defined as > 5um long, <3 um in diameter and with an aspect ratio of greater than 3:1; over 600 fibres counted in total.

The assay for the ability to compromise a monolayer of alveolar epithelial cells was essentially that described by Donaldson et al (1988). Briefly, cells of the A549 alveolar epithelial cell line were labelled with ^{51}Cr and the ability of increasing doses of the two amosite samples to cause the cells to detach, live, from their substratum in 6 hours, was assessed.

For the measurement of TNF, control alveolar macrophages were obtained by bronchoalveolar lavage from HAN Wistar rats. These were allowed to attach for one hour and then cultured overnight in 50 ug of long or short amosite asbestos. The culture supernatants were collected and assayed for the presence of TNF using the L929 cell-line with recombinant TNF as the standard and checking the specificity of the assay with anti-TNF antibody.

Repeat experiments (3 for TNF, 7 for epithelial detachment) were tested by analysis of variance for effects of treatment with the two different fibre samples.

Results

As shown in Figure 2, the long fibre amosite asbestos sample caused a dose-dependent detachment injury to the cells of the alveolar epithelial cell line, whilst short amosite was inactive in this respect. Long amosite also stimulated more TNF release from alveolar macrophages than short amosite (Figure 3).

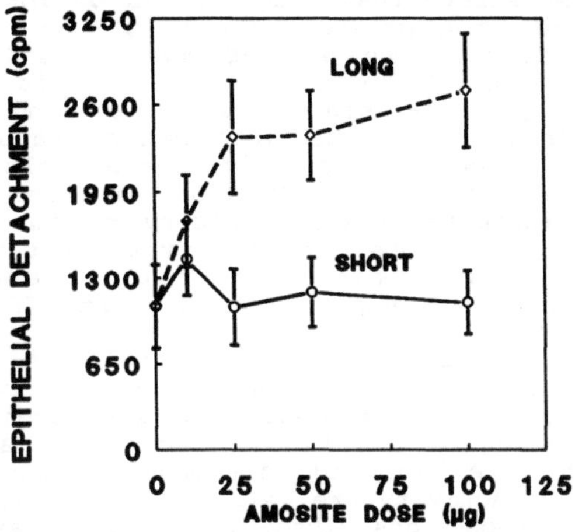

Figure 2. Detachment injury caused to epithelial cells by long or short fibre amosite (mean and SEM). Significant difference between the two types of fibre treatment, P<0.01.

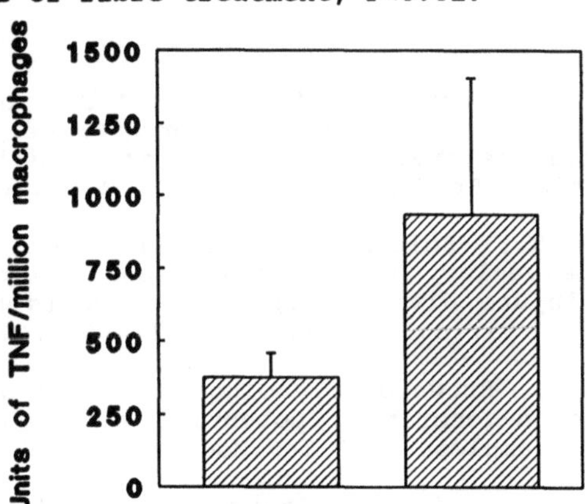

Figure 3. The production of TNF by alveolar macrophages treated in vitro with 50 ug/ml of long or short fibre amosite asbestos (mean and SEM). Significant difference between the two types of fibre treatment, P<0.05.

Discussion

Contact between epithelial cells and fibres occurs in the alveolar region following deposition and this interaction could result in compromise to the integrity of the epithelium if the fibres are long, as shown above.

The mechanism of this injury to the epithelial cells *in vitro* has yet to be elucidated. Impairment to the barrier function of the epithelium could result in a leak of fluid from the interstitium to the alveolar space, which could enhance inflammation. There could also be greater access of long fibres to the interstitium, from where they would be less likely to be cleared, and where they could exert their toxicity.
The above results demonstrate that phagocytosis of long fibres by macrophages could lead to a greater release of TNF than is the case with the short fibres. The pro-inflammatory and fibrogenic effects of TNF (Larrick and Kunkel 1988) could explain the greatly heightened inflammatory (Donaldson et al 1989) and fibrogenic (Davis et al 1986) properties of the long amosite sample compared to the short sample, although other activities are also likely to be important. Increased levels of TNF would seem unlikely to be associated with the increased carcinogenicity of the long fibre sample. However the gradients of TNF that obtain in the respiratory tract following inhalation of fibres are unknown and concentration differences may explain the diverse effects of TNF (Larrick and Kunkel 1988).
The present research points out the importance of correlating responses to fibres *in vitro* with the *in vivo* activity of the same fibres. Studies such as those described here suggest *in vitro* approaches to the testing of new fibrous materials, alternatives to asbestos, which are finding increased use. If their relationship to *in vivo* activity are well enough characterised, *in vitro* tests such as those described here, along with careful physico-chemical analysis of fibre size, structure and bio-persistence, could identify potentially hazardous fibres.

References

Davis,J.M.G., Bolton,R.E., Donaldson,K., Jones,A.D., and Smith,T. (1986)The pathogenicity of long versus short fibre samples of amosite asbestos administered to rats by inhalation and intraperitoneal injection. Br J Exp. Pathol. 67: 415-430.

Donaldson,K., Brown,G.M., Bolton,R.E. (1988) The ability of inflammatory bronchoalveolar cell populations elicited with microbes or dust to injure alveolar epithelial cells and degrade extra-cellular matrix in vitro. Br.J.Exp.Pathol. 69: 327-388.

Donaldson,K., Brown,G.M., Brown,D.M., Bolton,R.E. and Davis,J.M.G. (1989). The inflammation-generating potential of long and short fibre amosite asbestos samples. Br.J.Ind.Med. 46: 271-276.

Donaldson,K., Szymaniec,S., Li,X.Y., Brown,D.M. and Brown,G.M. (in press) Inflammation and immunomodulation caused by short and long amosite asbestos samples. Proceedings of the International Workshop on Mechanisms of Fibre Carcinogenesis. Albuquerque, New Mexico, October 1990.

Larrick,J.W. and Kunkel,S. (1988). The role of Tumour Necrosis Factor and Interleukin 1 in the immunoinflammatory response. Pharm. Res. 5: 129-139.

THIS RESEARCH WAS FUNDED BY THE COLT FIBRE RESEARCH PROGRAMME.

Quartz and Coal Mine Dust treated Human Monocyte-Derived Macrophages Release a Proliferation Factor for Human Lung Fibroblasts (WI-38)

Hübner, Karin and N.H. Seemayer*
Ruhrlandklinik Essen and *Medical Institute of Environmental Hygiene at the Heinrich-Heine-University Düsseldorf, Gurlittstr. 53, D-4000 Düsseldorf, F.R.G

Introduction

Silicosis is a chronic fibrotic lung disease caused by inhalation of quartz containing dusts in coal mine workers, stone masons and quarriers. It is characterized by an increase in collagen content of the lung, leading to parenchyma destruction and impairment of pulmonary functions. In the lung alveolar macrophages are the primary target of the noxious effect of quartz containing dusts. Macrophages form a large number of more than 100 "mediators" or "biofactors" which are involved in inflammation, tissue repair and immunological processes (Nathan, 1987). It is assumed today that mononuclear cells, especially monocytes and macrophages are involved in fibrosis by mediators or cytokines which induce enhanced growth and collagen formation by fibroblasts (Heppleston, 1991). In the present study we report about the kinetics of DNA-synthesis of human lung fibroblasts (Wistar 38) after addition of supernatants from quartz dust DQ 12 or coal mine dusts treated human macrophages. The percentage of DNA synthesizing human lung fibroblasts was determined by an immunochemical staining method after incorporation of the thymidine-analogous 5-bromo-desoxyuridine.

Material and Methods

Quartz dust and coal mine dusts: Dörentruper crystal quartz flour (grinding no. 12) named DQ 12, of particle size less than 5 µm was used as quartz dust. 3 samples of coal mine dusts from the Ruhr field, TF-1, TF-3 and TF-4 were kindly supplied to us by the Hauptstelle für Staubbekämpfung und Pneumokonioseverhütung des Steinkohlenbergbauvereins in Essen (F.R.G.). The composition and particle size distribution of the samples according to analyses by Reisner et al. (1981, 1985) has been described (Hübner and Seemayer, 1989). All dust samples were suspended in Iscove's medium without serum in a concentration of 500 µg/ml and treated with ultrasound for one minute (Sonifier B 12 from Branson Sonic Power Company, USA) as previously described (Hübner and Seemayer, 1989).
Cultivation of human macrophages and preparation of supernatants: The human monocytes were isolated by a Ficoll-Hypaque density gradient from enriched whole blood samples and cultivated in vitro as described previously in detail (Seemayer and Braumann, 1988; Hübner and Seemayer, 1989). After maturation of monocytes to cells with characteristics of macrophages, cell cultures were rinsed with serum-free complete Iscove's medium and freshly fed and the suspended dust samples were added at a concentration of 30 µg/ml. Supernatants were collected after an incubation period of 24 hrs. centrifuged at 2,000 rpm for 10 min. and, after heat treatment at 56°C for 60 min. subjected to sterile filtration through a Millipore filter Millex-HA (pore width 0,45 µm) and stored at -20°C until used. In this study, the culture supernatants of dust treated cultures are designed MO+DQ12 (or TF) supernatants and the culture supernatants of untreated cultures are named MO supernatants.
Fibroblast cultures: The Wistar 38 (WI-38) human lung fibroblasts were employed in the 24th to 35th passage. In Lab-Tek tissue culture chambers (No. 4803) cultures of WI-38 cells in quiescent state were used as has been elsewhere described in detail (Hübner and Seemayer, 1989). The culture supernatants were added to the fibroblast cultures in various dilutions and lengths of time, as described in results. An untreated cell control, a medium control (Iscove) and the addition of culture supernatants of untreated macrophages served as controls of the test system. In each case two to four replicative cultures were employed.
Determination of the rate of DNA-synthesizing WI-38 cells: The thymidine analogue 5-bromdeoxyuridine (BUDR, Calbiochem-Behring Corp.) in a final concentration of 15 µg/ml was added for 16 hours before the end of culture of WI-38 fibroblasts. Thereafter, following fixation, bonding of the primary monoclonal antibody "anti-BUDR-DNA" (Partec, Arlesheim) to the single stranded DNA was achieved. In order to render this bonding to BUDR-containing DNA visible, a gold-silver intensification reaction was subsequently carried out as described in detail by Hübner (1988). In each case 1,000 cells were evaluated microscopically in 2-4 stained

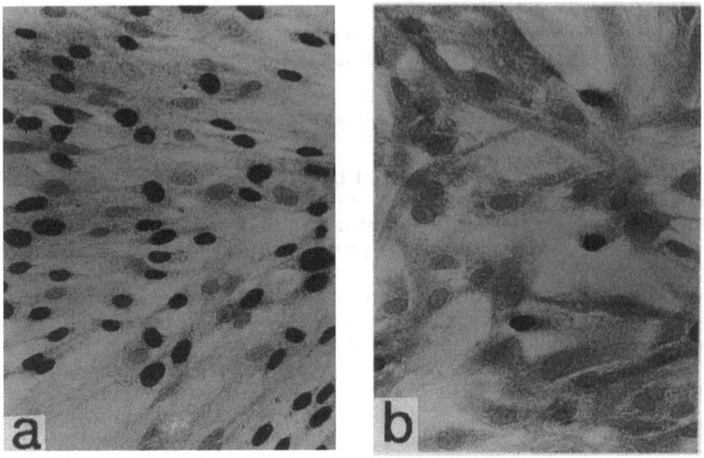

Fig.1: a) DNA-synthesizing lung fibroblasts WI-38 incubated for 48 hours with supernatant from macrophages exposed to quartz dust DQ12. Magnification about 240 x. **b)** DNA-synthesizing lung fibroblasts WI-38 of control culture after 48 hours. Magnification about 240 x.

parallel preparations. The data determined were subjected to the Student's t-test and simple variance analysis.

Results

Human diploid lung fibroblasts (WI-38) in quiescent state were incubated for 24 to 168 hours with supernatants of human macrophages incubated with quartz dust (MO+DQ12 supernatant) and corresponding control supernatants. Fig. 1a demonstrates a strong increase of DNA-synthesizing WI-38 cells after incubation for 48 hours with a supernatant of macrophages exposed to quartz DQ12. In contrast, control cultures of WI-38 cells reveal only few DNA-synthesizing cells after an incubation period of 48 hours (Fig. 1b). Kinetics of DNA-synthesizing WI-38 cells in presence of supernatant of quartz dust DQ 12 treated human macrophages at a dilution of 1:5 are presented in Fig. 2. The maximum of DNA-synthesizing human lung fibroblasts after an incubation period of 48 hours proved to be typical of the effect of supernantants of macrophages treated with quartz dust DQ12. A sequential decrease of DNA-synthesizing cells was recorded over the incubation period reaching nearly control values after 168 hours. In further experiments it was shown, that up to a dilution of 1:80 in presence of supernatants MO+DQ12 the number of DNA synthesizing fibroblasts increase significantly over control (Hübner and Seemayer, 1989). The culture supernatants of human macrophages treated with coal mine dusts TF-1, TF-3 and TF-4 increased the rate of DNA-synthesizing human

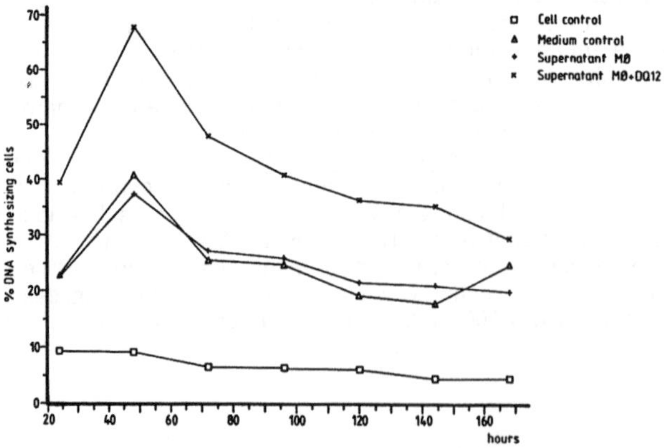

Fig.2: Effect of supernatant of quartz dust DQ12-treated and untreated human macrophages on the rate of DNA synthesizing human lung fibroblasts WI-38

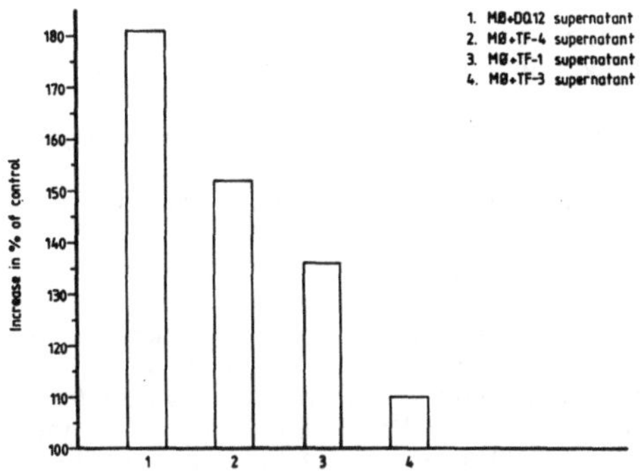

Fig.3: Comparison of activity of supernatants of human macrophages treated with quartz dust DQ12 and coal mine dust TF-1, TF-3 and TF-4 on the rate of DNA synthesizing human lung fibroblasts WI-38

fibroblasts of WI-38 cells .The kinetics of DNA synthesis were followed over an incubation period of 96 hours. The supernatants of mine dust TF-1 (mineral content : 95 wt%, quartz content: 10.6 wt%) increased the rate of DNA synthesizing fibroblasts to a maximum of 47.8% (1:10 dil.) after an incubation period of 48 hours. The supernatant of mine dust TF-4 (mineral content: 91.7 wt%, quartz content: 11.6 wt%) increased the rate of DNA-synthesizing fibroblasts to 57.1 % at 48 hours of incubation. The supernatant of mine dust TF-3 with a low mineral content of 10.6 wt% and quartz content of 0.6 wt% induced only a slight increase in the rate of DNA-synthesizing WI-38 cells of 42.3% after an incubation period of 48 hours. At a 1:10 dilution, the maximum of 47.0% synthesizing cells was reached after 72 hours. In comparison to supernatant MO+DQ12, the long lasting stimulation of the lung fibroblasts WI-38 to DNA synthesis by the supernatants from coal mine dust treated macrophages seems remarkable (Hübner and Seemayer, 1989). The activity of culture supernatants of human macrophages incubated with quartz dust DQ 12 and coal mine dusts TF-1, TF-3 and TF-4 are presented in Fig. 3 in comparison. Data are presented after 48 hours of incubation of WI-38 cells and expressed in per cent of control of supernatants of untreated macrophages. The strongest activity revealed the supernatant from quartz dust DQ12 treated macrophages, followed by supernatants of mice dusts TF-4, TF-1 and TF-3 in a declining sequence.

Discussion and Conclusion

Results presented demonstrate that human macrophages in culture, derived from blood monocytes by cultivation in vitro release a soluble mediator after incubation with quartz dust DQ12 and coal mine dusts, which significantly increases the rate of DNA-synthesis of "quiescent" human lung fibroblasts WI-38. As we earlier reported, this mediator which was designated as "Fibroblast proliferation factor" (FPF) leads to a significant increase of total cell count, mitotic index and cell density of human lung and dermal fibroblasts in culture (Seemayer et al., 1987; 1988; 1989; Seemayer and Maly, 1990). The mediator "FPF" proved to be relatively heat-stable (56° C, 1 hour) and preliminary data indicate a molecular weight of more than 30,000 Dalton (Seemayer, 1989). Previous results have shown, that the continous presence of the "fibroblast proliferation factor" is absolutely essential to guarantee both the high DNA synthesis rate and the increased cell replication of human fibroblasts (Hübner and Seemayer, 1989), indicating a growth factor of the "progression type" (Stiles et al., 1979). The increase in the action by growth factors of the "competence type", as our results in complementation experiments with "plateled derived growth factor" and "fibroblast growth factor" demonstrated, make it appear obvious to classify the "Fibroblast proliferation factor" as a growth factor of the "progression type". Beside FPF in supernatants of quartz DQ12 or coal mine dust treated human monocytes/macrophages a "Granulocyte activating Mediator" was detected causing a long lasting Lucigenin-dependent chemiluminescence of human granulocytes (Maly et al., 1988). The growth stimulation of fibroblasts by mediators from activated macrophages is of great importance for fibrotic lung processes and appears to be a generally applicable pathobiological principle (Benson et al., 1986; Bittermann et al., 1982; Heppleston and Styles, 1967; Lugano et al., 1984; Martinet, 1989; Schmidt et al., 1984).

References

Benson, S.C., J.C.Belton, L.G.Scheve (1986) Regulation of lung fibroblast proliferation and protein synthesis by bronchoalveolar lavage in experimental silicosis. Environ.Res. 41:61-78.

Bittermann, P.B., S.I. Rennardt and G.W. Hunninghake (1982) Human alveolar macrophage growth factor for fibroblasts. J. Clin. Invest. 70: 806-822.

Heppleston, A.G. (1991) Minerals, fibrosis, and the lung. Environmental Health Perspectives 94: 149-168.

Heppleston, A.G. and J.A. Styles (1967) Activity of macrophage factor in collagen formation by silica. Nature 214:521-522.

Hübner, K. (1988) Untersuchungen zur Bildung eines Fibroblasten-Proliferationsfaktors von quarz- und grubenstaubexponierten menschlichen Makrophagen: Stimulation der DNS-Synthese menschlicher diploider Lungenfibroblasten. Dissertation, Universität Düsseldorf.

Hübner, K. and N.H.Seemayer (1989) The role of a "Fibroblast Proliferation Factor" in the pathogenesis of anthracosilicosis. I. Stimulation of DNA synthesis of human lung fibroblasts (cell line WI 38) Silicosis Report North Rhine Westfalia 17:181-197.

Lugano, E.M. J.H. Dauber and J.A. Elias (1984) The regulation of lung fibroblast proliferation by alveolar macrophaghes in experimental silicosis. Am. Rev. Resp. Dis. 129: 767-771.

Maly, E., N.H. Seemayer, H. Behrendt, N. Manojlovic, A. Braumann and F.E. Maly (1988) Strong effects of cytokines from quartz dust exposed human mononuclear cells on human mesenchymal cells and polymorphonuclear leukocytes. In: N.H. Seemayer and W. Hadnagy (Eds) Environmental Hygiene, Springer Verlag Heidelberg, Berlin, pp. 83-86.

Martinet, Y.: Macrophage modulation of mesenchymal cell accumulation in lung fibrosis. 24th Annual Meeting "Societas Europeae Physiologiae Clinicae Respiratoriae (SEPCR`89) Lausanne, Switzerland, June 26-30. 1989, Summary Book, pp.47-50.

Nathan, C.F. (1987) Secretory production of macrophages. J. Clin- Invest. 79: 319-326.

Reisner, M.T.R., J. Frigge, M. Gade, R.-W. Schliephake, E. Wolff-Fischer (1981) I. Analysen von Feinstäuben und Kohlenproben. Silicosis Report North-Rhine Westphalia, Vol. 13: 172-185, Essen, Verlag Glückauf.

Reisner, M.T.R., L. Armbruster and L.Kühn (1985) Untersuchungen zum Einfluß der Korngröße und des Mineralgehaltes auf die spezfische Schädlichkeit von Feinstäuben des Steinkohlenbergbaus. I. Herstellung und Analyse der Staubfraktionen. Silicosis Report North-Rhine Westphalia, Vol. 15: 169-184, Essen, Steinkohlebergbauverein.

Schmidt. J.A., Oliver. C.N., Lepe-Zuniga. J.L., Green. I., Gery (1984) I.: Silica-stimulated monocytes release fibroblast proliferation factors identical to interleukin 1. A potential role for interleukin 1 in the pathogenesis of silicosis. J. Clin. Invest. 73:1462-1472.

Seemayer, N.H, Braumann, A., Maly, E. (1987) Development of an "in vitro" test system with human macrophages and fibroblasts for analysis of the effect of quartz dusts and coal mine dusts. I. Formation of a fibroblast proliferation factor. Silicosis Report North-Rhine Westfalia Vol. 16:143-156.

Seemayer, N.H. and A. Braumann (1988) Effects of particle size of coal mine dusts in experimental anthracosilicosis. In Vitro studies on human macrophages. Ann. occup. Hyg. 32: 1178-1180.

Seemayer, N.H., H. Behrendt, E. Maly, K. Hübner (1988) The role of quartz- and coal mine dust induced mediators from human macrophages in pathogenesis of silicosis. J.Aerosol. Sci. 19: 1129-1132.

Seemayer, N.H. (1989) The role of a "Fibroblast proliferation factor" in the pathogenesis of anthracosilicosis II. Biological characterization of a "Fibroblast proliferation factor" formed by human macrophage cultures following exposure to quartz dust and coal mine dust Silicosis Report North-Rhine Westphalia 17: 199-208.

Seemayer, N.H. and E. Maly (1990) Release of a "Fibroblast Proliferation Factor" from human macrophages in vitro treated with quartz dust DQ12 or coal mine dusts. In: Proceedings of the VIIth International Pneumoconioses Conference. DHHS (NIOSH) Publ. No. 90-108. Part II. Eds. US Department of Health and Human Services pp. 926-929.

Stiles, C.D., G.T. Capone, C.D. Scher, H.N., Antoniades, J.J. van Wyk, W.J. Pledger (1979) Dual control of cell growth by somatomedines and platelet-derived growth factor. Proc. Natl. Acad. Sci. 76:1279-1283.

DUST AND LPS INDUCED STIMULATION OF NITRITE AND NITRATE RELEASE IN BOVINE ALVEOLAR MACROPHAGES

SCHLÜTER, T., I. BERG, G. GERCKEN
Institut für Biochemie und Lebensmittelchemie, Abt. Biochemie, Universität Hamburg, Martin-Luther-King-Platz 6, D-2000 Hamburg 13, FRG

INTRODUCTION

Several types of cells have been demonstrated to generate nitric oxide using the guanidino group of L-arginine. In peritoneal macrophages and monocyte cell lines this pathway is non-constitutive (Stuehr and Marletta, 1987). The activation of the nitric oxide synthesis during macrophage development differs from that of the cytokine synthesis (Hauschildt et al.,1990) and is independent from the respiratory burst during phagocytosis (Iyengar et al.,1987). Authentic nitric oxide gas causes the same pattern of cytotoxicity in L10 hepatoma cells as is induced by cytotoxic activated macrophages in vitro (Hibbs et al.,1988). Thus nitric oxide and its stable oxidation products nitrite and nitrate are thought to be involved in the cytostatic and killing properties of macrophages.
The aim of the present study was to evaluate possible stimulating or inhibiting effects of airborne particulate contaminants on the L-arginine pathway.

MATERIALS AND METHODS

Cell isolation: Bovine alveolar macrophages (BAM) were isolated by post mortem lung lavage. Cells were cultured in Eagles' Minimal Essential Medium (pH 7.4, 25 mM HEPES, 100 U/ml penicillin, 100 µg/ml streptomycin, 2.5 µg/ml amphotericin B) without phenol red.
Dust preparation: Two city waste incinerator fly ashes (CWIA I,II), a power station fly ash (PSA), stone dust from a gravestone workshop (STD) and a dust from sevage fludge incineration (SFA) were separated into different grain size fractions by sieving and sedimentation in butanol. Dusts were analysed for element contents by AAS and AES/ICP. A quartz dust (QUSO G 32, Philad. Quartz Comp.) was used as a cytotoxic control dust. Macrophages were incubated with the fine fractions (diameter < 4µm).
Incubation procedure: Incubation began either immediately after isolation or after a resting period of 14 h. EMEM was supplemented with the indicated concentrations of LPS (E.coli lipopolysaccharide, serotype 0127:B8), zymosan A, concanavaline A (all purchased from Sigma Chemie, Deisenhofen), and/or dusts.
Assay for nitrite and nitrate: The production of nitric oxide (NO) was determined by analyzing the concentration of its stable oxidation products in the supernatants. Nitrite and nitrate, after enzymatic reduction (reagent kit Bioquant 12464,E.Merck,Darmstadt), were measured by a colorimetric assay based on the Griess reaction (Stuehr and Marletta,1985).

RESULTS

Supposing the NO-generating pathway to be non-constitutive, the first experiments were carried out to find the optimum inducing agent (Tab.1). A relative high basic production of nitrite in the controls can be seen. LPS was the only agent which significantly increased the production of nitrite. The optimum concentration was found to be 1 µg of LPS per $1x10^6$ cells.

Table 1: Accumulation of nitrite during BAM incubation for 24 h

		Control	LPS			Zymosan		Con.A	
Stimulant	[µg/10^6cells]	0	0,1	1	10	100	1000	10	50
Nitrite, means	[nmol/10^6cells]	5,31	5,95	6,50	6,13	4,27	4,21	4,01	6,02
± SD	[nmol/10^6cells]		0,21	0,49	0,38	1,17	1,21	1,36	1,27

The results of the time course experiments are shown in Fig.1. The curves for controls and for cells treated with LPS are parallel. A lag phase can not be detected (a). After a 14 h resting period control cells show a much smaller production of nitrite (b). The amount of nitrite generated by LPS stimulated cells was similar to the amounts occuring immediately after the isolation procedure.

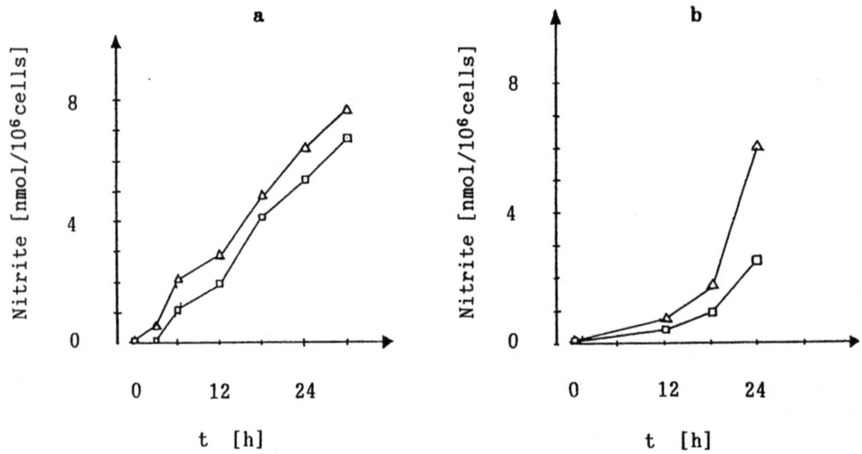

Figure 1: Nitrite accumulation by BAM after incubation with LPS. Beginning of incubation immediately after plating (a) or after a 14 h resting period (b). (□ = control, △ = 1 µg LPS/10^6cells)

The effect of heavy metal containing dusts on the NO-generating system was investigated with incubations beginning immediately after cell isolation (Fig.2). None of the dusts caused a stimulation of NO production (3-1000 µg/10^6cells). In comparison to the controls the accumulation of nitrite was inhibited in a concentration dependent manner. The dusts can be devided in two groups in respect of their inhibiting capacity. The first group consists of CWIA I and II, which caused a significant decrease in nitrite production at concentrations of 6 µg/$1x10^6$ cells. Nitrite production was inhibited almost completely by dust concentrations > 25 µg/$1x10^6$ cells. In the second group, at least four times higher concentrations of PSA, SFA and STD were necessary to cause the same inhibition.

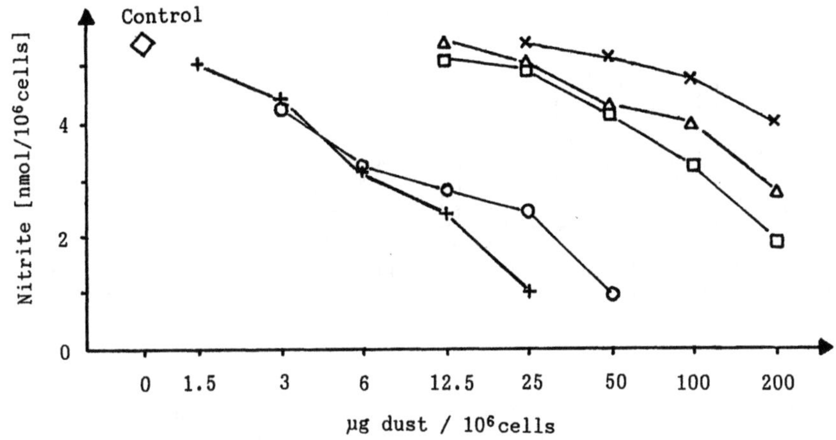

Figure 2: Nitrite accumulation by BAM after incubation with different dusts
+ = CWIA I, o = CWIA II, □ = STD, Δ = SFA, x = PSA

Tab.2 shows the number of cells still adherent after rinsing the dishes four times and the viability of the adherent cells, determined by trypan blue exclusion. Neither the cell number nor the viability were significantly effected by the dusts. The quartz dust QUSO G32 markedly decreased cell number and viability.

Table 2: Cell viability after incubation (24h) with dusts

		Control	CWIA I	PSA	Quartz
Dust	[µg/10^6cells]	0	25	200	50
Adherent cells	±SD [%]	100	103 ±8	81 ±9	58 ±11
Viability	±SD [%]	74 ±5	65 ±10	69 ±7	48 ±12

The sum of accumulated nitrite *and* nitrate is shown in Tab.3. There is no contradiction to the results obtained by determination of nitrite alone. The dusts did not change the relationship between the both metabolites.

Table 3: Sum of accumulated nitrite and nitrate

		Control	CWIA I		PSA		STD	
Dust	[µg/10^6cells]	0	3	25	25	200	25	200
Nitrite + nitrate	[nmol/10^6cells]	23,1	15,4	4,8	19,9	18,0	19,6	6,2

The observed inhibitory effect and the heavy metal contents of the dusts were correlated (Fig.3). The best correlation was found for zinc, antimony and cadmium.

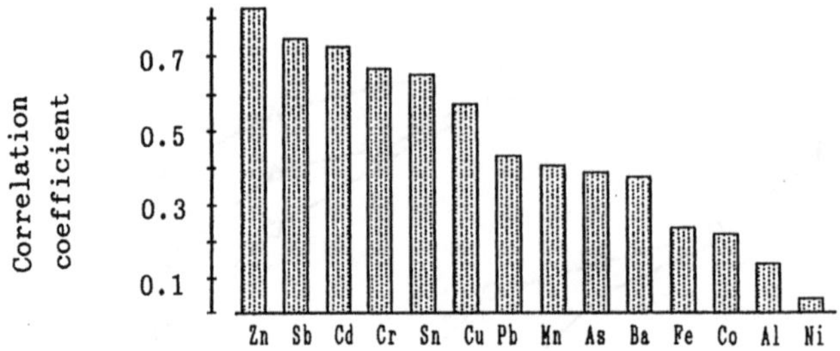

Figure 3: Correlation coefficients for linear regressions. y = a + b ln x (x = element mass, y = inhibition of nitrite release in % of control)

DISCUSSION

Allthough the NO-generating pathway is thought to be non-constitutive, a basic level of nitrite production was observed. Immediately after isolation cells could not be stimulated by any agent, whereas LPS was proved to be an activator after a resting period of 14 h. Heavy metal containing dusts inhibited the basic nitrite accumulation in a concentration dependent manner without impairing the cell viability. The same inhibitory effect was seen when the cells were incubated with dusts in the presence of LPS 14 h after isolation (Data not shown). Zinc, antimony, and cadmium show the best correlations between element contents in the dusts and the observed inhibitory effects. These results are in contrary to the observations made by Gulyas et al. (1990) who found the highest correlation coefficients for lead, antimony, and arsenic by investigating the inhibitory effects of dusts on the zymosan induced H_2O_2 and O_2^- release by alveolar macrophages. This is a further hint for the independence of the oxidative burst on the activation of the L-arginine pathway. The investigation of the NO formation could serve as another assay to evaluate possible toxic effects of metal containing air pollutants in vitro.

REFERENCES

Gulyas H, Labedzka M, Gercken G (1990) Depression of alveolar macrophage hydrogen peroxide and superoxide anion release by mineral dusts: correlation with antimony, lead, and arsenic contents. *Environmental Res.*, **51**:218-229

Hauschildt S, Bassenge E, Busse R, Mülsch A (1990) L-arginine-dependent nitric oxide formation and nitrite release in bone marrow-derived macrophages stimulated with bacterial lipopeptide and lipopolysaccharide. *Immunology*, **70**:332-337

Hibbs JB, Taintor RR, Vavrin Z, Rachlin EM (1988) Nitric oxide: a cytotoxic effector molecule. *Biochem.Biophys.Res.Commun.*, **157(1)**:87-94

Iyengar R, Stuehr DJ, Marletta MA (1987) Macrophage synthesis of nitrite, nitrate, and N-nitrosamines: Precursors and role of the respiratory burst. *Proc.Natl.Acad.Sci.USA*, **84**:6369-6373

Stuehr DJ, Marletta MA (1985) Mammalian nitrite biosynthesis: mouse macrophages produce nitrite and nitrate in response to E.coli lipopolysaccharide. *Proc.Natl.Acad.Sci.USA* **82**:7738-7742

Stuehr DJ, Marletta MA (1987) Induction of nitrite/nitrate biosynthesis in murine macrophages by BCG infection, lymphokines, or interferon-gamma. *J.Immunol.*, **139**:518-525

This work was supported by Projekt Europäisches Forschungszentrum Maßnahmen zur Reinhaltung der Luft, Grant No. 89/006/1C

INFLUENCE OF COAL MINE DUSTS AND SURFACTANT ON CHEMILUMINESCENCE-ACTIVITY OF ALVEOLAR MACROPHAGES IN VITRO

J.Wackes, F.Tilkes, E.G.Beck
Hygiene-Institut, Justus-Liebig-Universität, Friedrichstr.16, D-6300 Gießen, FRG

Introduction

Coal mine dusts are highly heterogeneous with regard to composition and particle sizes. In spite of numerous investigations it has to date not been possible to relate single dust components with the noxiousness of coal mine dusts to the lung.
Beyond its commonly known biophysical functions in the lung surfactant represents a complex defence system. Until now only few results were obtained using surfactant in in vitro studies. The special capability of macrophages to generate reactive oxygen species (ROS) which can be demonstrated by measuring their chemiluminescence (CL) activity offer a possibility for the assessment of the unspecific toxicity of xenobiotics in cell culture systems. The present study shows the effects of several coal mine dusts of the Ruhr and Saar area, also in combination with surfactant, on the primary CL activity of alveolar macrophages.

Material and Methods

Alveolar macrophages of guinea pigs were obtained by lung lavage and hundred thousend cells in each cuvette were preincubated for 90 minutes. Than the medium was changed to Dulbecco`s MEM for chemiluminescence or MEM plus 1 % surfactant and the cells were incubated at 38.1 °C and 5 % CO_2 for 20 hours. Measuring chemiluminescence, the samples were transferred to a 6-channel CL measuring instrument, the Biolumat LB 9505. By starting the measurement, 100 µg coal mine dust and 100 µl of the CL indicator DMNH were pipetted into the cuvettes. The measuring time was 30 minutes. After processing the measuring data by a personal computer the resulting curves were printed. The diagrams present the average CL peaks of the curves. Electro-corundum and Dörentruper quartz (DQ12) were used as control dusts.

Results and Discussion

All dusts were examined at different dose levels. With a rising dust concentration an increasing CL activity could be proved. At the concentration of 100 µg / 10^5 cells the dust groups D01 and D02 showed a slightly reduced CL in relation to the corundum control. The dust groups D03, D07 and D09 elicited a distinctly lower CL activity (Fig. 1 and 3). If surfactant is present the CL peaks are presented in relation to the not-surfactant incubated control. We see slightly different CL activities of the single dusts but an average CL reduction of the whole dust groups like the corundum control. The CL of the control dust DQ12 is obviously low (Fig. 2 and 4).
The demostrated assay is an unspecific method for the quantification of the toxicity of inhalative noxious factors. The ROS generated by macrophages through the contact or phagocytosis of these

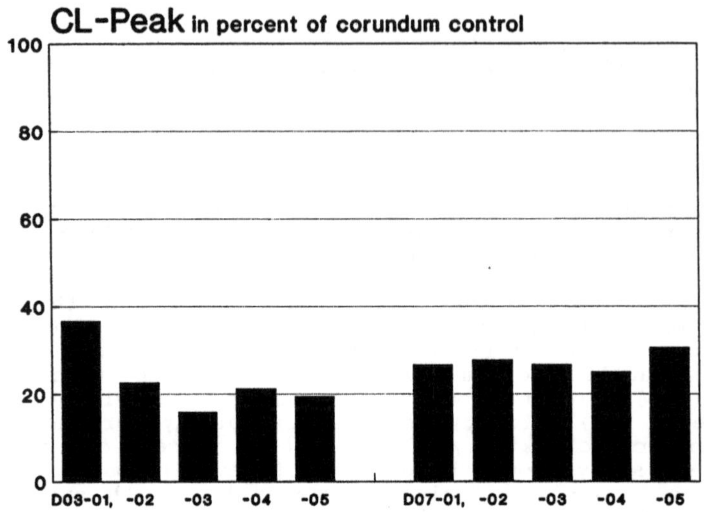

Fig. 1. shows the CL peaks of the dust groups D03 and D07 in relation to the corundum control. The CL activity of both groups is distinctly reduced. A relativly high CL of the D03-01 and D07-05. A relativly low CL of the D03-03 and D07-04.

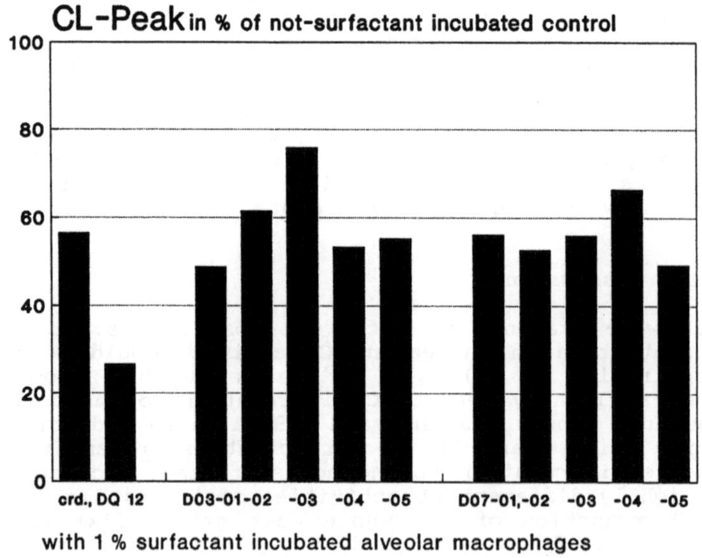

Fig. 2. shows the CL peaks of the dust groups D03 and D07 when surfactant is present. It represents a low DQ12 CL activity, an overall reduced CL to the corundum control niveau, a relativly low CL of the D03-01 and D07-05, and a relativly high CL of the D03-03 and D07-04.

noxious factors are correlated with the immunregulatory, cytotoxic and microbicidial activities of macrophages. Measuring the CL allows an assessment of the oxydative metabolism of these cells.
If a dust causes a high primarily CL activity we will also find a high contact and/or phagocytosis rate. This corresponds to a high ROS production and by that to a secundarily high toxcicity. In this case the cell damage can also be caused by the ROS themselves. But if the primarily toxicity is that high, that the primarily cell damage occurs quicker than a maximum of ROS can be built we will get CL activities which do not correspond to the real toxicity. These competitive proportions do complicate the interpretation of primarily CL activity caused by dusts. In the CL the dust groups D03 and D07 show half as much toxicity than the groups D01 and D02. These results correspond to LDH investigations of our institute.

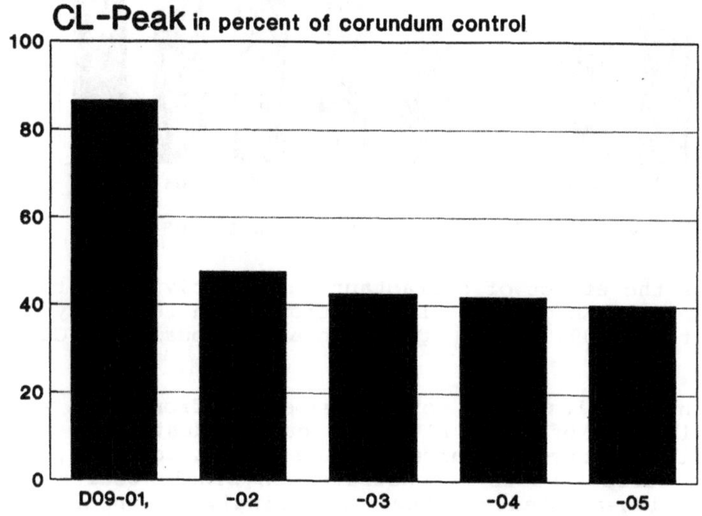

Fig. 3. shows the CL peaks of the dust group D09. The reduced CL activity is obvious exeptionally the D09-01.

The D09 group with its different ash, quartz and other components in the single dusts shows with exception of the D09-01 less differences than expected from dust analysis (Fig. 3). These results are in opposition to the LDH release which corresponds to the dust analysis.
The toxicological examination of a dust in vitro should aim at adapting the in vitro conditions to the in vivo conditions as far as possible. Therefore it is important to consider the influence of surfactant on biological effects.
These results evaluated statistically by the three factorial analysis of variance demonstrate a significant influence of surfactant on the CL activity which is evident in all examined dusts. Dusts with the lowest toxcicity like the D03-03, D07-04 (Fig. 1) and others do show least influence when surfactant is present (Fig. 2) which means a high production of ROS. This might be caused by the contact and/or phagocytosis situation petronized by

surfactant. The D09 group shows also a distinct influence of surfactant, however the influence on the single dusts is hardly differentiated (Fig. 4).

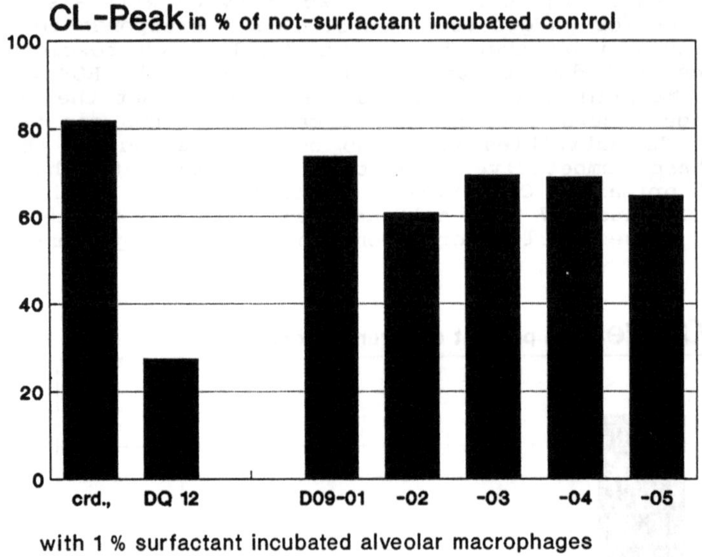

Fig. 4. shows the effect of surfactant on CL activity of the dust group D09. An overall CL reduction and a relativly higher CL of the D09-01. The DQ12 with an obviously low CL.

Even here the results were not as expected from dust analysis. With an inhibition of over 70 % the quartz dust DQ12 shows the most distinct influence of surfactant on the CL activity. A lesser influence of surfactant is demonstrated on single coal mine dusts in this "in vivo-adapted" system. Accordingly we should consider the possibility that the coal mine dusts may exhibit greater toxcicity in vivo than assumed until now using DQ12 as a reference.
Therefore the question must be asked in how far it is justified to use DQ12 as the sole reference dust in further investigations.
The results of this study confirm our opinion that the adaptation to in vivo conditions for example with surfactant, is an important objective.

References

Armbruster, L., Robock, K. (1989) Persönliche Mitteilung.
Habel, W. (1990) Der Einfluß verschiedener Kulturbedingungen auf die Zytotoxizität von Grubenstäuben in der Alveolarmakrophagenkultur. Dissertation Justus-Liebig-Universität Gießen, Fachbereich Humanmedizin.
Schmidt, J (1990) Einfluß von faserigen Mineralstäuben auf die DMNH- und Lucigenin-abhängige Chemilumineszenzaktivität von Ratten-Alveolarmakrophagen in vitro. Dissertation Justus-Liebig-Universität Gießen, Fachbereich Veterinärmedizin.

ENVIRONMENTAL CONTROL, PREVENTION AND LEGISLATION

Industrial Risk Assessment and Corresponding Management Scenarios-Case Study at the Aluminum Casting Plant

M.Boutoussov, M.H.A.Hamdy, S.Luken
United Nations Industrial Development Organization (UNIDO), A-1400 Vienna, Austria;
W.Gaubinger
Consultant in Environmental Engineering, Priesterhausgasse 1/14, 9020 Klagenfurt, Austria;
H.Friza
Institute of Environmental Hygiene, University of Vienna, Kinderspitalgasse 15, A-1095, Vienna, Austria;
A.Jernelov
Stockholm Environment Institute, Box 2142, 10314 Stockholm, Sweden;
H.Meisterhofer
Alpenconsult, Colloredogasse 24 , A-1180, Vienna, Austria;
I.Stankovich
Consulting Engineer Corp. P.O.Box 11182, Piedmont, CA 94611, USA;

Introduction

The proper estimation of industrial risk associated with the activities of industrial enterprise can be a complicated problem for the enterprise managerial staff. The reason is that the risk comprises several relevant components and there are no friendly methodologies at hand for their comparative estimation. Thus the approach was suggested called an Integrated Quality Concept (Boutoussov, 1988) to evaluate simultaneously (and in comparable terms) major positive and negative impacts of industrial activity at the level of industrial enterprise. These impacts correspond to economical, environmental and occupational components of the enterprise activity.

Since response times (specific reaction time of each of the corresponding processes- economical, environmental or occupational, to any change of conditions- say, amendments in technological process or in management) may vary quite strongly- from years to seconds, it is impossible to practically measure the degree of stability and sustainability of a given industrial enterprise vis-a-vis various deteriorations. It means, in turn, that an industrial manager should have adequate tools helping him to assess all positive and negative effects caused by industrial activity and its variations and to evaluate them in comparable terms.

Then, in order to optimize this industrial activity, all net negative outputs should be minimized by proper technical, managerial and other means. Essentially, this approach does not call for elimination of all negative manifestations related to industrial activities (which is not realistic) but, rather, represents an attempt to optimally balance all positive and negative factors related to industrial activity.

To do so, the UNIDO project in a form of a case study was launched on request of Bulgaria at the Aluminum Casting Plant in Pleven.

Project Formulation

As it became clear from the request of Bulgaria to UNIDO, the particular enterprise proposed as a test field for the UNIDO case study was meeting many requirements, i.e.:
-it was situated in the town which, by its size and features (including environmental situation), was a typical example not only for many regions on Eastern Europe, but as well, for many other developing countries;
-plant was of medium size and had a limited range of products thus making controllable the amount of factors to be assessed;
-UNIDO was assured that the staff and authorities of the Plant were willing to identify the main reasons of the productivity decline and environmental degradation at their enterprise.
The Integrated Quality Concept was used as a methodology for attaining following objectives:
-collecting and formatting all relevant information about the Plant activities and its situation;
-processing and verifying these data and, finally,
-converging different viewpoints into common conclusions and recommendations.
The UNIDO Concept, thus, was used as a method of formalization of diversified experience related to different facets of the Plant activity. Obviously, in order to elaborate objective criteria relevant for impartial consideration of all these components, not only the rigorous basic data were required but an extensive experience (formal and informal) was quite essential in assessment and optimization of various "interface problems" between industry and nature, industry and people, industry and market.

Project Implementation.

An expert team was recruited comprising specialists in following fields:
1.Environment impact assessment (team leader);
2.Aluminum casting technologies and their environmental aspects;
3.Environmental engineering;
4.Industrial toxicology;
5.Product quality and marketing;
6.Industrial management and personnel selection.
In order to provide versatility of opinions together with high competence, the experts were invited from different countries where environmental, managerial and economic problems of industrial development in a given industrial sector are better tackled- from Sweden, Hungary, Austria and USA.
The first expert's mission (experts 2 and 3) and then the mobile laboratory of contractor company, following this mission, performed assessment and rigorous measurements of major hazardous factors- phenol, formaldehyde, dust, noise, UV and IR radiation in relevant

areas in and outside the Plant. Contractor also took a responsibility to provide a study tour, for the local supporting team, to several companies and laboratories in Germany and Austria dealing with the same problems.
Further missions of experts were undertaken in order to assess, within the responsibility and experience of everyone and on the basis of provided data, the occupational impacts caused by indoor pollution, the environmental loads in- and outside of Pleven and the role of the whole industrial complex of Pleven in the degradation of nature. The following factors were studied: occupational health- immissions on working places (types, concentrations, influence onto the personnel, required protection measures and managerial decisions), environmental situation - air and water pollution (general situation in- and outside of Pleven), production facilities -level of technologies, energy and raw material supplies, logistics in location of the equipment, planning and marketing.They also suggested on the necessity of modernization of technologies involved, on the improvement of industrial management, etc. UNIDO TV film team also visited Pleven in order to record the situation at the Plant and to provide objective impressions to those members of the forthcoming Expert Group Meeting who did not visit the Plant.

Project Finalization.

Expert Group Meeting (EGM) was arranged in order to finalize the Project, comprising all experts, representative of contractor, relevant specialists from Austrian institutions and UNIDO and Bulgarian supporting team.
The first aim of the meeting was to elaborate, in a series of brainstorming sessions, the joint assessment of economical losses caused by environmental and occupational hazards, to compare them with the economical outputs related to the Plant production.
The second aim was to elaborate, on the basis of the assessed situation, one or several scenarios for the efficient improvement of the situation in view of merging, in future, with European market.
The economic assessment of the annual damages induced by the Plant gave following results : occupational impact onto personnel -$ 0.5 M, environmental damage - $ 2 M. The price of the Plant products at European rates was assessed as being about $ 12 M, out of which raw material should be acquired for $ 8 M. Taking into account salaries, energy and water costs, the mere existence of the Plant becomes problematic. This was the main, though quite striking conclusion resulted the application of the Integrated Quality Approach.
A plan of the Plant rehabilitation was elaborated which had to drastically improve the occupational and environmental status and quality of production and which required initial investments to be paid back in three years. Different investment schemes for the environmental and occupational upgrading were compared on the basis of acquired figures and estimations. Plant authorities received basic figures enabling proper evaluation, step by step, of the rehabilitation process to be performed and indicating the rate of

returns on these investments. This conclusion gave to Bulgarian authorities a solid foundation for the future rehabilitation programme.
A further request from Bulgarian Government to continue environmental quality management activities, but now- at the level of the whole industrial district was submitted to UNIDO.
Nowadays, the Project Document "Environment Quality Management Programme for Industrial District of Pleven" is prepared and submitted for financing.

References.

M.Boutoussov (1988) The Integrated Quality Concept. An Approach to Balanced Industrial Activity. Resources Policy , September: 218-220.

Environmental Tobacco Smoke in Public Places

Brynnel, M. and G. Löfroth
Environmental Health Unit, Nordic School of Public Health,
Box 12133, S-402 42 Gothenburg, Sweden

Introduction

Environmental tobacco smoke (ETS) is a major cause of indoor air pollution (Repace and Lowrey, 1980). Exposure to ETS is known to cause irritation (Cummings et al. , 1991), increased risks for respiratory infections (Samet, 1989), atopy (Kershaw, 1987), heart disease (Glantz and Parmley, 1991) and cancer (Pershagen, 1989). The exposure to ETS has often been evaluated by the presence or absence of smokers in the family. Exposure is, however, possible in all environments whenever smoking occurs. We have earlier analyzed ETS using the Ames Salmonella mutagenicity assay on extracts of particulate matter collected in homes (Löfroth and Lazaridis, 1986) offices (Löfroth et al., 1983) and public places (Löfroth et al., 1988). These latter studies have now been extended with comparisons between smoking and non-smoking areas and amended with analysis of airborne nicotine as a specific marker for ETS.

Material and Methods

Sampling. Indoor sampling of nicotine and of particles for mutagenicity was made with battery operated personal pumps with a rate of 2 or 4 L/min. Pumps were carried either in a brief-case (restaurants) or a shoulder bag (indoor plazas) with the air inlet at the height of the breathing zone. Outdoor sampling of particulate matter for mutagenicity was made simultanesously at a nearby location with larger pumps. Sampling times varied from about three to seven hours. Five experiments have been performed in a 19 cu.m room in which measurements were made for 150 min. Nicotine and particulate mutagenicity was determined for 15, 15, 30, 30 and 60 min and carbon mondoxide, isoprene and particle mass was determined every 7.5 min as described earlier (Löfroth et al., 1991).
Nicotine analysis. Nicotine was collected on 35 mm teflon-coated glass fiber filters (Pallflex Corp., USA) which had been pretreated with an aq. 4 % sodium bisulfate solution (Hammond et al., 1987). Filters were extracted with an aq. 10 M sodium hydroxide solution containing 5 % ethanol (Hammond et al., 1987). The basic aq. solution was then extracted into heptane containing 0.1 % triethylamine (Ogden et al. 1989). Isoquinoline was added as an internal standard. The extracts were analyzed for nicotine on a Varian Model 3400 gas chromatograph with a nitrogen detector.
Mutagenicity assay. Indoor particulate matter was collected on 35 mm glass fiber filters (MG160; Stora Filter Products, Sweden) whereas outdoor samples were collected on larger filters. The filters were extracted with acetone and the extract converted to samples in 5 µl dimethyl sulfoxide for the microsuspenson test as described by Löfroth et al. (1988) and based on the report by Kado et al. (1983). The samples were tested for mutagenicity with the Salmonella strain TA98 in the presence of S9 exogenous activation (LS-9; Molecular Toxicology Inc., USA). The response is expressed as revertants/cu.m air.

Results

Due to the failure to collect nicotine as a predominant gas phase behaving component (data not shown; cf. Discussion) nicotine concentrations reported here are either combined amounts collected on the particle filter and the following acid-treated filters (Hammond et al., 1987) or the amount collected on acid-treated filters directly facing the air as described in Material and Methods.

The results from two objects are presented in Tables 1 and 2. The two shopping malls (Table 1) are of about the same size and both have return air ventilation systems for energy conservation. They differ with respect to smoking policies. Smoking is freely allowed in the mall area of one of them whereas it is prohibited in the other although smoking is allowed in two cafeterias, the air of which can be mixed with the mall air. The presence of nicotine and a mutagenic response above that of outdoor air in the non-smoking mall can be explained by the smoke from the cafeterias and occasional violations of the smoking ban coupled with the use of a return air system. The concentrations detected in the smoking mall might be higher than expected from direct exposure from passing smokers due to the return air system.

Table 1. Airborne nicotine and mutagenic response in indoor shopping malls with different smoking policies. The results are given as average and range concentrations.

MALL TYPE (no. of visits)	MALL AIR Nicotine µg/cu.m	Mutagenicity Rev./cu.m	OUTDOOR AIR Mutagenicity Rev./cu.m
Non-smoking (4)	3.3 (2.4-3.9)	190 (120-230)	<50 (<20-130)
Smoking (6)	10.8 (7.6-14)	615 (260-1100)	<110 (<20-250)

The lunch restaurant (Table 2) has a larger area where counters are located and smoking is allowed and a smaller non-smoking room connected to the larger room with a wide opening. The relatively high concentrations in the non-smoking area can be explained by the fact that the ventilation system consists of an exhaust fan located in the ceiling of the non-smoking room. Air entering the building through doors, window slits and other openings is thus drawn into the non-smoking area.

Table 2. Airborne nicotine and mutagenic response in a restaurant.

LUNCH DATE	SMOKING AREA Nicotine µg/cu.m	Mutagenicity Rev./cu.m	NON-SMOKING AREA Nicotine µg/cu.m	Mutagenicity Rev./cu.m	OUTDOOR AIR Rev./cu.m
910206	13	830	6.6	880	40
910227	19	2300	11	1400	20
910306	14	930	8.6	660	90
Average as %	100	100	56±6	79±24	-

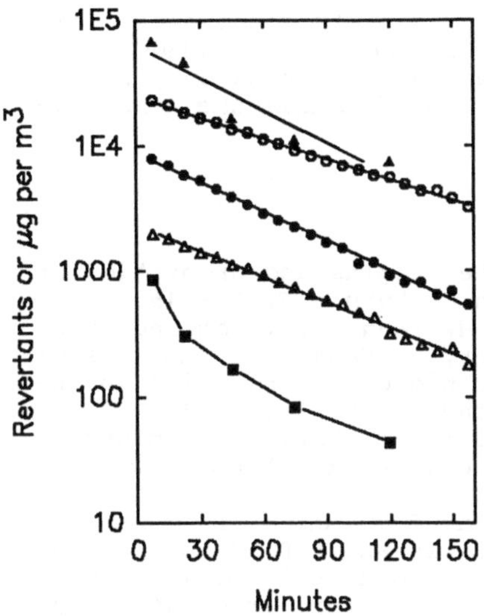

Fig. 1. The decrease of the concentration of ETS components in a 19 cu.m room with normal wall surfaces and furnitures after the emission of sidestream smoke from 10 cigarettes.

▲ particle mutagenicity

○ carbon monoxide

● particle mass

△ isoprene

■ nicotine

Experiments in a room (Fig. 1) were performed at a ventilation rate of 0.67 air changes per hour as given by the rate of the carbon monoxide disappearance. Other measured components decreased slightly or drastically faster than the ventilation rate in the order: ventilation rate < isoprene < particulate matter ~ particle mutagenicity << nicotine.

Discussion

At the initiation of this study there existed a promising method for the simultaneous determination of particle mutagenicity and nicotine with one sampler. As described by Hammond et al. (1987), particles could be collected on a first particle filter and nicotine, being predominantly in the gas phase, could be collected on one or two sodium bisulfate (e.g. acid) treated teflon-coated filters placed down-stream the particle filter. This method has been evaluated in an experimental chamber with clean air and inert surface materials (Hammond et al., 1987; Caka et al., 1990). No published data seems to exist from environmental studies in which the partioning between the filters are reported. In the course of the present study we discovered that variable and sometimes substantial amounts of nicotine is collected on the first untreated filter irrespective of it being an ordinary glass fiber filter or a teflon-coated filter. We believe that most of the nicotine collected on the particle filter is due to a more or less unspecific adsorption and not necessarily reflects amounts of particle-associated nicotine.

The environmental message from this study is that exposure to ETS can be encountered in public places in spite of the fact that they are thought to be non-smoking areas. Similar places with smoking give, of course, higher exposures. Measured as mutagenic response of particulate matter, the indoor pollution in locations where smoking occurs is always higher than that of ambient outdoor air in a busy urban center.

Nicotine seems to disappear faster than other ETS components. This phenomenon has also been mentioned by Eatough et al. (1987) and Baker and Proctor (1990). It indicates that nicotine only can be used as a qualitative tracer of ETS underestimating the true exposure to other ETS components.

References

Baker, R.R. and C.J. Proctor (1990) The origins and properties of environmental tobacco smoke. Environ. Intern. 16:231-245.

Caka, F.M., D.J. Eatough, E.A. Lewis, H. Tang, S.K. Hammond, B.P. Leaderer, P. Koutrakis, J.D. Spengler, A. Fasano et al. (1990) An intercomparison of sampling techniques for nicotine in indoor environments. Environ. Sci. Technol. 24:1196-1203.

Cummings, K.M., A. Zaki and S Markello (1991) Variation in sensitivity to environmental tobacco smoke among adult non-smokers. Intern. J. Epid. 20:121-125.

Eatough, D.J., C.L. Benner, J.M. Bayona, F.M. Caka, H. Tang, L. Lewis, J.D. Lamb, M.L. Lee, E.A. Lewis and L.D. Hansen (1987) Sampling for gas phase nicotine in environmental tobacco smoke with a diffusion denuder and passive sampler. Proc. 1987 EPA/APCA Symp, APCA Publ. VIP-8:132-139.

Glantz, S.A. and W.W. Parmley (1991) Passive smoking and heart disease. Circulation 83:1-12.

Hammond, S.K., B.P. Leaderer, A.C. Roche and M. Schenker (1987) Collection and analysis of nicotine as a marker for environmental tobacco smoke. Atmos. Environ. 21:457-462.

Kado, N.Y., D. Langley and E. Eisenstadt (1983) A simple modification of the Salmonella liquid-incubation assay. Increased sensitivity for detecting mutagens in human urine. Mutat. Res. 121:25-32.

Kershaw, C.R. (1987) Passive smoking, potential atopy and asthma in the first five years. J. Royal Soc. Med. 80:683-688.

Löfroth, G., L. Nilsson and I. Alfheim (1983) Passive smoking and urban air pollution: Salmonella/microsome mutagenicity assay of simultaneosuly collected indoor and outdoor particulate matter. IN: M.D. Waters et al. (eds.) Short-Term Bioassays in the Analysis of Complex Environmental Mixtures III. Plenum, New York. pp. 515-525.

Löfroth, G. and G. Lazaridis (1986) Environmental tobacco smoke: Comparative characterization by mutagenicity assays of sidestream and mainstream cigarette smoke. Environ. Mutag. 8:693-704.

Löfroth, G., P.I. Ling and E. Agurell (1988) Public exposure to environmental tobacco smoke. Mutat. Res. 202:103-110.

Löfroth, G., C. Stensman and M. Brandhorst-Satzkorn (1991) Indoor sources of mutagenic aerosol particulate matter: smoking, cooking and incense burning. Mutat. Res. 261:21-28.

Ogden, M.W., L.W. Eudy, D.L. Heavner, F.W. Conrad Jr. and C.R. Green (1989) Improved gas chromatographic determination of nicotine in environmental tobacco samples. Analyst 114:1005-1008.

Pershagen, G. (1989) Environmental tobacco smoke and cancer. IN: D.V. Bates et al. (eds) Assessment of inhalation hazards. Springer Verlag, Berlin. pp. 275-284.

Repace, J.L. and A.H. Lowrey (1980) Indoor air pollution, tobacco smoke, and public health. Science 208:464-472.

Samet, J.M. (1989) Environmental tobacco smoke: Adverse effects on respiratory infections, respiratory symptoms, and lung function. IN: D.V. Bates et al. (eds.) Assessment of inhaltion hazards. Springer Verlag, Berlin. pp. 265-273.

Carcinogenic Air Pollutants: Essential Legislative Regulations

Dobbertin, S.
Umweltbundesamt, Bismarckplatz 1, D-1000 Berlin, FRG

Introduction

In the federal countries of West Germany 160.000 persons annually die of cancer among them 25.000 of lung cancer. After subtraction of smoking effect and working place caused burden one has to assume that air pollution is an important factor for the etio-pathogenesis of lung cancer.
Although there is no evidence for a relationship between immissions and lung cancer, malignancy of this disease requires analysis and elimination of those facotors which are assumed to be mainly responsible.
Based on our current knowledge I want to emphasize factors which are essential in lung carcinogenesis.
Therefore we need data about emissions, immissions and the carcinogenic potential of diverse pollutants.

Emissions

Table 1 gives an overview of emissions of relevant carcinogenic air pollutants in tonnes per year.

Tab.1: Emissions of important carcinogenic air pollutants

Substances	Emission per year	
Arsenic	16 - 68	t
Asbestos	115	t
Benzene	47.000	t
Cadmium	25 - 48	t
Dieselparticles	70.000	t
BaP	4 - 16	t

Substances were mainly selected on their relevance for human health. According to this criterium nickel and chromium are not listed. Results of recently conducted animal experiments showed that only special compounds posess a cancerogenic potential.

Despite of its high toxicity, 2,3,7,8-TCCD was not listed, as quantity of emissions per annum amounts to 50 g. Nevertheless, table 1 gives only a very global overview, which should be commented in detail. The main sources of arsenic emissions are of industrial and occupational origin. The greatest amount of asbestos emission originates from weather-beaten unsheltered asbestos cement in big building industry amounting to about 100 t per year. Total emission of benzene comes up to about 47.000 t per year, of which emission of automobile traffic contributes to 41.000 t per year.

Cadmium compounds are mainly emitted in form of particulates produced by the non-iron metal industry at an amount of 12-12 t, followed by the iron and steel industry.

Emissions of diesel particulates are estimated to 70.000 t per year with at least 55.000 t produced by automobile traffic. According to the total emission of BaP the contribution of household and small quantity energy users is declining during the last years and is comparable to the BaP emission of automobiles.

Immissions: Table 2 demonstrates load of immission

Tab. 2: Immission Load

Substances	Mean value per year of Immission load				
	Country Side	City Area		Vicinity of Emission	Unit
Arsenic	1 - 5	5- 20		10- 30	ng/m^3
Asbestos	< 100	100-200		100-330	F/m^3
Benzene	< 1	5- 10	Coking plant	7- 15	$\mu g/m^3$
			Automobile	20- 30	$\mu g/m^3$
Cadmium	0.5-1.5	2- 5		5- 30	ng/m^3
Diesel-particles	0.5-1.5	5- 10	Automobile	15- 25	$\mu g/m^3$
BaP	0.5-1	1- 3	Automobile	3- 6	ng/m^3
			Coking plant	5- 10	ng/m^3

You may see that BaP concentration has clearly decreased during the last years. A slight reduction can be seen for arsenic and benzene.

Cancerogenic potential: Few years ago cancerogenic potential of a substance was calculated on the basis of experience of the researchers. Now mathematical models serve for quantitative risk estimation. However, this calculation requires sufficient animal, respective epidemiological data. Referring to the chosen cancerogens the following data base exsists:

1. there ist a sufficient epidemiological data base for arsenic, asbestos, benzene and BaP

2. data base from animal experiments exsist for cadmium and diesel particles.

With the aid of these data quantitative risk estimation can performed, whereby the Unit Risk Conception by the WHO and EPA has gained importance. It is defined as the additional probability to die in consequence of a special illnes when exposed lifelong to 1 $\mu g/m^3$ of a substance.
The Unit Risk conception has the advantage to estimate cancer risk for special environmental pollutants using relative simple calculation methods. However, this advantage is based on simplified assumption of a model.

1. Linearity: Important condition of the Unit Risk Conception is linearity of the dose-response relationship at low dose level. However, deviation to a sub-or supra-linear curve may be considered.
2. Time variation of the burden. The Unit Risk Conception takes only in consideration cancer risk in dependence of a total dosage, however not time-dependet exposure.

An always recurrent criticism of the Unit Risk Conception is due to variation of the calculated risk in dependence of the used mathematical model. However, this criticism had been refuted by Wahrendorf and Becher among other authors as published in the report of UBA 1/1990. Thus you can notice for benzene that Unit Risk estimation from individual cohorts lie close together.

Of course, all these derivations are estimations which should not lead to overrate mathematical accuracy. A mathematical model calculation can not at all compensate biological uncertainty. However, this concept gives us the possibility for a better risk estimation.

Valuation: Based on cancerogenic potential as well as emissions- and immission situation the following steps seem to have priority: Arsenic: For most kind of plants reduction is possible by improved cleaning of exhaust. As arsenic is mainly emitted as part of dust, cleaning of exhaust from dust ist most prominent. For arsenic and its componds an emission value of 0,1 mg/m^3 should be established on the calculation of arsenic.

Asbestos: For asbestos administrative orders are necessary to remove asbestos containing waste.

Benzene a) Lowering benzene content of petrol to 1 vol %,
b) introducing gas pendel conduit pipe,
c) built-in catalysator also for motory cycles

Cadmium a) Emission value for cadmium and its compounds should be 0,05 mg/m^3 calculated as cadmium,
b) use of cadmium in plastic matter should be prohibited.

Diesel: Improvement of motors are necessary for reduction of particles, f.i.
a) improvement of fuel injection systems,
b) reduction of not burnt lubricating oil,
c) retreatment of exhaust using filters for particles and catalyst for oxidation.

These steps indirectly lead also to reduction of PAH emissions.

Summary: With regard to cancer risk caused by air pollution automobiles belong to the primary problem. With regard to the steady growing traffic and its environmental pollution, improvment of the technique for automobiles is not sufficient for reduction of pollutants. Moreover, respective steps for traffic improvement are necessary.

Biological remediation processes - A challenge for the environmental hygiene

Dott, W. and P. Kämpfer
Fachgebiet Hygiene, Technische Universität Berlin, Amrumerstr. 32, D-1000 Berlin 65, FRG

Based on estimations of the German ministry of science and technology (BMFT), in the Federal Republic of Germany 70,000 sites exist which are in suspect of remediatory need. These old sites are typically found in industrial regions and urban areas and include undefined disposal of wastes, regions surrounding mining and industry sites, as well as old sites which used to harbor metal and coking works, chemical and power plants, refineries and gas work plants.
Furthermore, it is estimated, that the number of suspected sites in east germany in need of remediation lies in the same order. The urgent requirement of remediation of about 10% of the cases is a result of the direct and indirect risks on man through the different pollutants (Dott 1989).

A uniform evaluation of the pollutants is difficult. In addition to the different physical and chemical properties of the individual pollutants, knowledge on their sorption and desorption behavior, their solubility in connection with environmental changes, plant availability and their transport between the aqueous and gas phases is necessary. Furthermore, it is important to know the contamination route on which they may directly or indirectly reach man. In addition to the direct oral, pulmonary or cutaneous intake of pollutants, the indirect routes through drinking water and the food chain should be taken above all into account.

To limit the spreading of a contaminant, several techniques such as encasement, soil exchange, controlled disposal of wastes, fixation through supply of binding agents as well as immobilization brought about by reducing the solubility of the pollutants, are employed. However, all these techniques offer no final solution to the contamination problem. In the last few years, several remedial processes have been developed including chemical, physical and biological techniques such as thermal processes, microbiological clean up processes, and extraction techniques. Biological soil and groundwater decontamination techniques are receiving increasing attention in which a microbiological degradation of the pollutants is achieved. This can take place either in mounds following soil excavation (on-site) or in-situ (Dott, 1988).

The aim of in-situ techniques is to bring about the degradation of a soil pollutant by the available autochthonous microorganisms. In most cases, a water circulation is set up in which water withdrawal, treatment and reinfiltration into the underground is possible. The addition of electron acceptors such as oxygen, hydrogen peroxide or nitrate, as well as other nutrients like phosphate and ammonium, can optimize the conditions for the microbiological degradation of pollutants. Such in-situ techniques can be applied if:

(1) the pollutant is biodegradable,
(2) the microbial degradation is not inhibited by toxic effects of the pollutants, and
(3) if the pollutant in the underground is accessible.

The last point is the decisive factor for a successful application of an in-situ remediation because this is only possible with a soil permeability showing a KF-value of $>10^{-4}$ m/sec. In addition, the nonhomogeneous distribution of pollutants in soil, their adsorption at the fine structures of the soil particles and the fact that the whole process resembles a "black box", often prevents a successful remediation.

These uncertainties can be clearly seen considering some well-documented remediation cases, e.g. fuel oil and benzene contaminated sites. A comparison between the initial concentrations of these pollutants in soil and their concentrations in water following successful remediation, can be made. In most cases the extent of soil contamination has not been sufficiently determined. It must be stressed, that an overall balance of the biodegradation of a pollutant is indispensable.
Taking an example dealing with a diesel oil contamination, initial concentrations between 0 and 30 g/kg have been determined which reflects the heterogeneity of pollutant distribution in soil. A reduction down to 5 g/kg dry weight should be thus considered in view of the variable initial concentrations as insufficient.

For a successful in-situ remediation, an extensive chemical and microbiological analysis is necessary before the remediation process is started. The availability of the pollutants and their distribution in the subsurface have to be determined, in addition to some microbiological parameters. Among these, the presence of microorganisms showing degradation potential has to be determined. Furthermore, toxic effects of the pollutants on the microbial community have to be estimated. Such microbiological parameters may also include a general quantification of the different bacterial groups in soil (Kämpfer et. al, 1990).

Degradation experiments under laboratory conditions, but close to those of the location under study are indispensable. These studies can be achieved with a lysimeter set-up (Figure 1), in which an undisturbed soil sample containing the original pollutant and the autochthonous microflora is brought into a column.

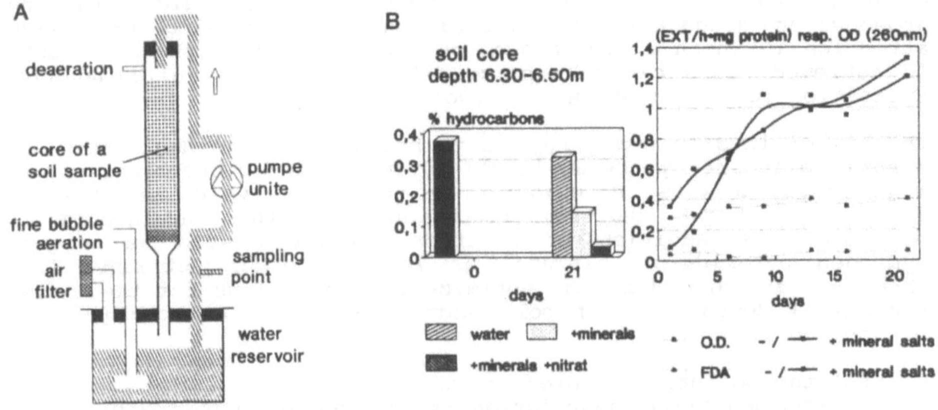

Figure 1: Degradation studies under laboratory conditions: (A) lysimeter installation, (B) hydrocarbons degradation, bacterial growth (O.D.= optical density) and actiactivity (FDA = hydrolysis of fluoresceindiacetate), (Dott, 1989)

Degradation experiments can be simulated, using a water circulation, whereby the water can be supplied with oxygen and some nutrients. The degradation can be documented by the decrease in the concentration of the pollutant, in addition to a simultaneous increase of biomass. Such investigations are useful to determine the optimal conditions under which a microbiological degradation of a contaminant is possible.
As can be seen in Figure 1, the addition of minerals such as ammonium and phosphates, or nitrate as an additional oxygen donor, enhanced the degradation of hydrocarbons and promoted the activities of microorganisms following 21-day incubation, compared to no nutrients added. Biomass measurements (protein) and enzymatic activities in soil samples can also give additional informations on the biological activities of microorganisms.

Although in laboratory experiments, an extensive degradation of hydrocarbons was observed after 21 days, in practice and under different conditions, an essentially slower and stagnant degradation of the hydrocarbons was observed after 30 weeks. Although about 70% of the hydrocarbons were degraded, several rest hydrocarbons were not easily degradable applying different remediation processes (Figure 2). Furthermore, the question of the production of metabolic products by the microorganism remains open.

Another biological remediation process is represented by the mound technique, which is based on the fact that the soil is excavated and treated above ground in a more controlled process. Investigations on different mounds also showed, that the time needed for degradation of the hydrocarbons might take weeks to months. In addition, the problem of rest hydrocarbons is also observed. However, a special feature of the mound technique should be also here mentioned. As can be seen in Figure 2, the BTX aromatics (benzene, toluene, xylene) seem to be quickly eliminated.

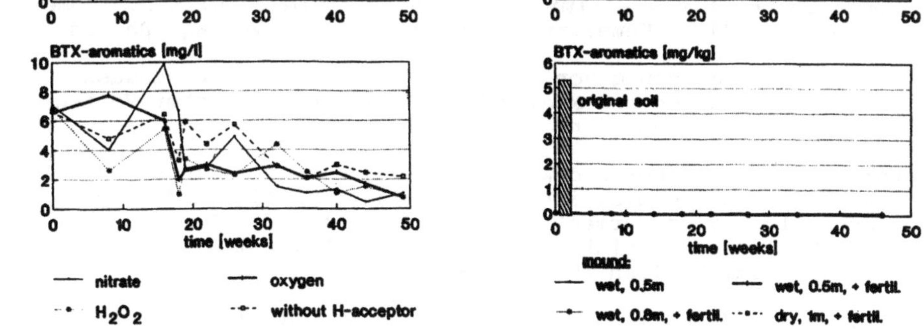

Figure 2: Degradation of hydrocarbons and BTX-aromatics under field conditions: (A) different in-situ conditions, (B) different mound techniques, (Kämpfer et al., 1991)

In spite of the initially high concentrations of more than 5 mg/kg, these pollutants were not detectable after few days of the mound set-up. It is evident that this cannot be due to microbial degradation. Volatile substances like the BTX aromatics have been quickly eliminated by stripping.

Summarizing, it can be concluded that many organic pollutants in soil and groundwater are biodegradable and generally, a discrepancy between the microbiological degradation potentials established in the laboratory and those in field studies can be found.
Careful chemical and microbiological analyses preceding and following a remediation process are necessary. In addition, the estimation of an overall balance of the degradation of a pollutant must be given. Furthermore, the formation of intermediary metabolites have to be assessed. However, such investigations are often neglected, in most cases because of the high costs.
Despite enormous expenditure in the last years for bioreclamation research, only a partial success has been realized. Several statements in literature are either misleading or incomplete. An objective and independent process control of remediation is indispensable. In addition, the determination of possible health risks on man resulting from rest contaminants or formed metabolites often fail due to lack of suitable methods.

The aims of the environmental hygiene are a combination of several methods and techniques, such as the analysis and documentation of the biological, chemical, physical and technical factors of pollution in the environment and their toxicological impacts. Based on these analyses, the direct (anthropogenic) and indirect ecological effects of pollutants has to be determined.
Furthermore, during development of techniques for remediation, the environment including water, soil and air, should be considered as a whole.
Strategies for preventive environmental protection should be also developed, and in view of the incorrect developments and estimations, a future improvement in remedial measures can be brought about through an increased cooperation in the different fields of medicine, science and engineering.

References:

Dott, W. (1988) The use of microorganisms for restoration of contaminated sites - limits and possibilities. in: K.J. Thomé-Kozmiensky (Ed.): Remediation processes 2. EF Verlag, Berlin

Dott, W. (1989) Sanierung von Altlasten im Boden- und Grundwasser -bereich - Grenzen und Möglichkeiten mikrobiologischer Verfahren. Forum Städte-Hygiene 40: 326-332

Kämpfer, P., P.M. Becker and W. Dott (1990) Investigations on changes in the composition of microbial communities during microbiological treatment processes. in: Arendt, F., Hinsenveld, M., van den Brink, W.J. (Eds.): Contaminated soil '90., Kluyver Acad. Publishers, 487-488

Kämpfer, P., L. Friedrich, P.M. Becker und W. Dott (1991) Beeinflussung der Bodenpopulation in Mieten durch vorgezüchtete Mikroorganismen. In 9. Dechema Fachgepräch Umweltschutz: Mikrobiologische Reinigung von Böden, Frankfurt

POLYCHLORINATED DIBENZODIOXINS AND DIBENZOFURANS
- MEASURES FOR THE PROTECTION OF MAN AND THE ENVIRONMENT -

Lebsanft, J. and Basler, A.
c/o Federal Ministry for the Environment, Nature Conservation and Nuclear Safety, 5300 Bonn, P.O.Box 120629, Federal Republic of Germany

INTRODUCTION

Polychlorinated dibenzodioxins (PCDD) and dibenzofurans (PCDF) - often refered to simply as "dioxins" - constitute a group of 210 closely related chemicals. They are mainly formed as byproducts in combustion or heating processes in the presence of substances containing chlorine as well as during the synthesis of chemical compounds structurally based on chlorinated aromatics.

This group of chemicals is of extraordinary interest equally to the scientific community, the competent authorities and the public. The enormous attention assigned to dioxins derives from the extremely high toxicity for laboratory animals, the ubiquitous occurence in the environment, the controversially disputed carcinogenic potency for man as well as from the relatively low safety factor when the average human intake is compared to the lowest observable effect levels determined in animal experiments.

The average daily human intake is estimated at 1-2 pg Toxic Equivalents* (TE)/kg of body weight, usually derived to more than 90 percent from food. WHO considers the maximum tolerable daily intake of dioxins to be 10 pg TE/kg. For reasons of prevention the Federal Health Office (BGA) and the Federal Environmental Agency (UBA) sets the value at 1 pg TE/kg/day. Since the actual intake of dioxins is already in the range of these levels, the German Federal Government has taken far-reaching legal measures in order to reduce considerably the input of dioxins into the environment.

BAN ON POLYCHLORINATED BIPHENYLS AND PENTACHLOROPHENOL

In the past, certain chemicals containing chlorine have been an important source of dioxins. In manufacturing processes of chlorinated aromatic hydrocarbons such as for example pentachlorophenol (PCP) and polychlorinated biphe-

*) In order to estimate the potential hazard of complex dioxin mixtures containing dioxin congeners of different toxicity, the concept of Toxic Equivalent Factors has been introduced expressing the amount of each dioxin congener in terms of an equivalent amount of 2,3,7,8-TCDD.

nyls (PCBs) comparably high levels of dioxins are formed. Since these dioxins contaminate the respective chemicals, they are cotransferred and may enter the environment together with the chemical.

PCP is used as a pesticide to protect a large variety of products including leather and textiles. After the treatment of leather and textiles, surplus PCP is removed by washing. The washing water is drained and reaches the sewage plant where, due to the fat solubility and the adsorption/desorption characteristics, most of the dioxins are transferred to sewage sludge. Since sewage sludge is predominantly used as fertilizer for agricultural purposes, the dioxins may enter the environment. Furthermore, PCP is widely used to protect wood against microbial decomposition. In the past, PCP-treated wood has been used among others inside of buildings. Thereby indoor pollution, caused by PCP itself as well as by contaminating dioxins, occured in certain circumstances. By banning the production and the use of PCP, a source that has accounted for an input of dioxins of 26 kg TE over the last 20 years has been eliminated. (Ordinance on the Ban on Pentachlorophenol of December 12, 1989). In consequence of the German initiative, the Commission of the European Communities proposed a directive restricting the use of PCP. This directive has recently been adopted by the Council of the EC. In contrast to the German ordinance, the directive allows the further production of PCP as well as its use in the treatment of wood and textiles provided this treatment is conducted in closed installations. Several member states including Germany voted against the directive demanding more stringent measures.

The other group of chemicals mentioned before, PCBs, have been widely used as dielectric in capacitors and transformers and as hydraulics oil. Heavy environmental contamination amounting to a total of 90 kg TE dioxins over the last 20 years was caused by the open application as hydraulics oil, improper waste management of capacitors/ transformers and by transformer fires. The production and the use of PCBs has been banned in Germany (Ordinance on the Ban on PCBs of July 18, 1989) and also in all other EC member states. However, problems are still being encountered regarding the safe and comprehensive management of existing facilities containing PCBs.

ORDINANCE ON DANGEROUS SUBSTANCES

The German Ordinance on Dangerous Substances sets stringent limit values for 8 PCDD/PCDF in substances, preparations and articles placed on the market. This regulation is currently reevaluated in order to extend the scope to presently unregulated dioxin congeners as well as to polybrominated dibenzodioxins/-furans.

COMBUSTION PROCESSES AS A SOURCE OF DIOXINS

Dioxins may be generated during any combustion of organic material in the presence of chlorine-containing compounds. The extent of the dioxin formation depends on the chlorine content, the composition of the material to be burnt as well as on the temperature, the pressure and the oxygen supply during combustion. According to a rough estimate, municipal incinerators contribute about 400 g TE/year in total in western Germany, the hospital incinerators and the special waste incinerators 1.5 g TE/year and 1.0 g TE/year, respectively. These dioxin emissions will be considerably reduced in the future according to the Ordinance on Incineration Plants for Wastes and Similar Combustible Substances of November 23, 1990. This regulation lays down a limit value for dioxins in exhaust gases amounting to 0.1 ng TE/m^3 and will decrease the dioxin emissions by a factor of 100.

PROPOSED BAN ON THE USE OF SCAVENGERS

A further significant dioxin source is the combustion of leaded petrol by motor vehicles. In order to achieve a transformation of lead oxide to volatile lead halogenides, leaded petrol contains so called scavengers as additives (1,2 dichloroethane, 1,2 dibromoethane). During combustion, brominated and mixed-halogenated dibenzodioxins/-furans are produced, amounting to about 50 g TE/year in total. The Federal Government has stipulated a ban on the use of scavengers in petrol. Before this ban can be put into force, however, it has to pass the notification procedure of the European Community.

PACKAGING MATERIALS

Dioxin contamination of milk caused by cardboard packaging may also contribute to the dioxin load of consumers. By reducing the dioxin content of packaging used for liquid foodstuffs (e.g. milk) to a maximum of 1 ppt TE according to a mutual agreement between the Federal Health Office and the industry, a further source has been eliminated.

GUIDELINES FOR DIOXIN VALUES IN SOIL

Although the knowledge about the transfer rate of dioxins from soil to cultivated feedstuffs and foodstuffs is incomplete, a substantial dioxin uptake by plants is rather unlikely. Nevertheless there is no doubt that cultivation on highly contaminated soil is not acceptable. Recommendations for the agricultural exploitation of soils are currently developed by a joint working group of the Federation and the Länder. For preventive reasons, this group considers that, as a long term target objective, a dioxin concentration should be reached below 5 ng TE/kg soil used for agricultural purposes. Limitations concerning the cultivation of certain feedstuffs and foodstuffs might be necessary if the dioxin contamination exceeds 40 ng TE/kg. Guideline values triggering redevelopment measures for childrens playgrounds and for urban areas are discussed to be 100 ng TE and 1000 ng TE/kg, respectively. Independent of the localization, soils containing more than 10 000 ng TE/kg are considered to be exchanged generally.

ORDIANCE ON SEWAGE SLUDGE

Sewage sludge may be a further source of dioxin-contamination as pointed out before. Accordingly, an ordinance on sewage sludge has been drafted prohibiting the fertilization with sewage sludge containing more than 100 ng TE/kg dried residue.

OUTLOOK

Altogether these measures represent a stringent programme to reduce the dioxin load for man and the environment. However, it is unrealistic to expect that the reduction of the total dioxin input will considerably reduce the dioxin load of the population in the short term due to the widespread distribution and the long half-lives of dioxins. In addition, we should be aware that a number of dioxin sources which have not yet been quantitatively assessed might also contribute significantly to the overall contamination. Among these sources, domestic fires of coal, oil, wood and natural gas, industrial power plants using brown coal and metallurgical processes are considered to be of possible relevance and will be subject to detailed studies funded by the Federal Minister for the Environment. Research projects concerning the elucidation of the behaviour of dioxins in the environment as for example the transfer rate of dioxins to feedstuffs and foodstuffs cultivated on contaminated soil are presently elaborated under the auspices of the Federal Minister for Research and Technology. Having the results to hand, the joint working group mentioned before will examine whether the implementation of guideline or limit values for dioxins in feeding and foodstuffs would be a suitable tool to reduce the human dioxin intake.

Measurements of the Environmental Pollution During Cleaning Up of Solid Waste Depositions of Chemical Weapons

Moriske, H.-J., M. Wiese, I. Block, R. Kneiseler, I. Trauer, H. Neumann and H. Rüden:
Institute for Hygiene of the Free University of Berlin, Hindenburgdamm 27, D-1000 Berlin 45; F.R.G.

Introduction

In Germany, we have a number of solid waste depositions out of World War I and II that are contaminated with high toxic chemical weapon agents (Spyra 1988). During the cleaning up of such depositons, contaminations of the total environment with the original agents ore with some decomposition products of it are possible and have to be measured.
According to this and according to legal reasons, it's also necessary to see what kind of environmental pollution we have before the the cleaning up begins.

Material and Methods

From 1988 up to 1990 we measured the contaminations of soil, water (ground- and surfacewater) and air before the cleaning up of a solid waste deposition of chemical weapons in Berlin began. The measurements were done again after the cleaning up process was finished. In soil and water the following parameters were measured:
Inorganics: nitrate, nitrite, ammonium, phosphate, cyanides, fluorides and chlorides, heavy metal elements: arsenic, cadmium and lead, organics: polycyclic aromatic hydrocarbons (PAH, 6 components), chlorinated organic components (COC, 4 components) and polychlorinated biphenyles (PCB, clophene A 30 and A 60), mutagenic activity of the total organic extracts on Salmonella typhimurium TA 98 and TA 100. In water, we also measured the concentrations of phenole, of chemical oxidation requirements and of biochemical oxidation requirements. In air, we measured the total suspended particulates, total organic extract (ether/cyclohexan-extractable matter), PAH, PCB, arsenic, cadmium, lead and volatile COC.
The investigations were done at an old fortress in Berlin where the chemical weapons were deposited after World War I and II.
The investigations were made at the centre of the deposition field and also in nearby surroundings, at a distance of 200-400 m around.

Results and Discussion

We demonstrate only those results where we got some differences -before and after cleaning up -in the centre of the deposition field and in nearby surroundings.
In ground water and in air, we got no significant differences.
In soil (figure 1), we found some inhances of arsenic in the centre of the deposition field (13-222 mg/kg) before the cleaning up began, in comparison to the other measurements (1-25 mg/kg). Also, we found some inhances for COC here (0.006 mg/kg (<0.0003 mg/kg others)).

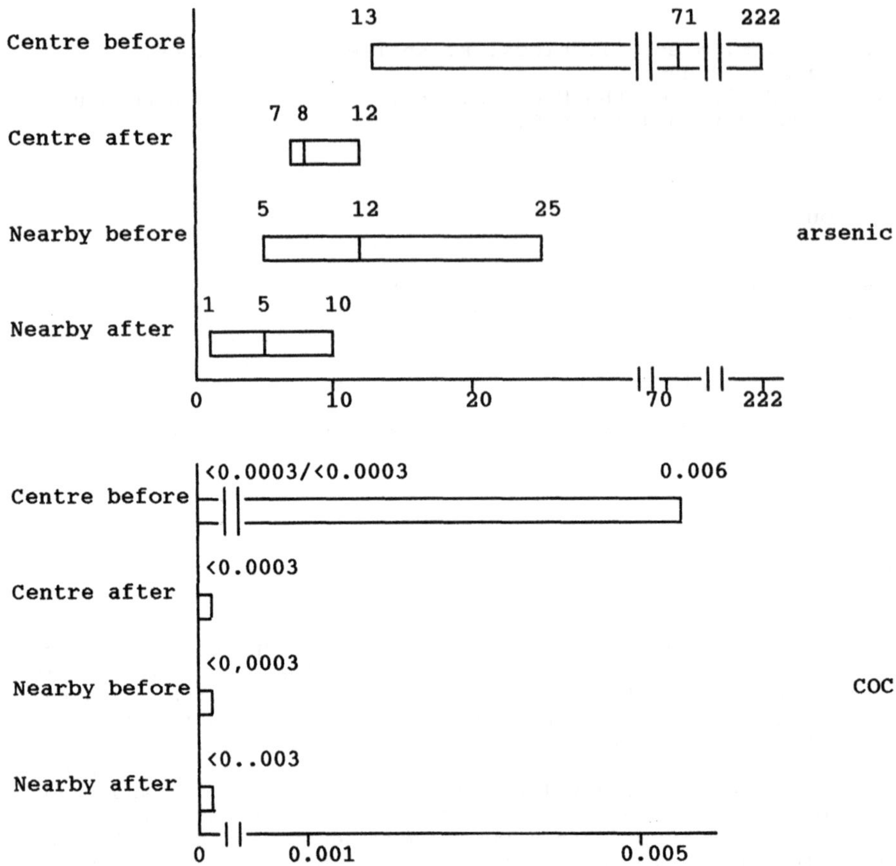

Figure 1: Contaminations of soil with arsenic and chlorinated organic components (COC) (mg/kg) in the centre of the deposition field and in nearby measurements, before and after cleaning up procedure; min. median point max.

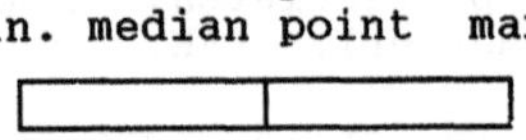

A positive mutagenic activity on S. typhimurium TA 100 was found with and without metabolic activation in the centre of the deposition field, before the cleaning up began, again.

In surface water (figure 2), we found inhances for arsenic in the centre of the deposition field before cleaning up (1,1-2,9 mg/l) and after cleaning up (0.4-0.5 mg/l), in comparison to the measurements in nearby surroundings (< 0.005 mg/l). Often, we also found inhances for phosphates (2.8-8.4 mg/l) in the centre of the deposition field in comparison to the other measurements (0.2-8.0 mg/l). A positive mutagenic activity was found, again, only in the centre of the deposition field, before the cleaning up process began.

The results showed that we got a contamination of the total environment, especially of soil and of surface water with arsenic before the cleaning up began. The cleaning up process took to a decrease of the total contamination.

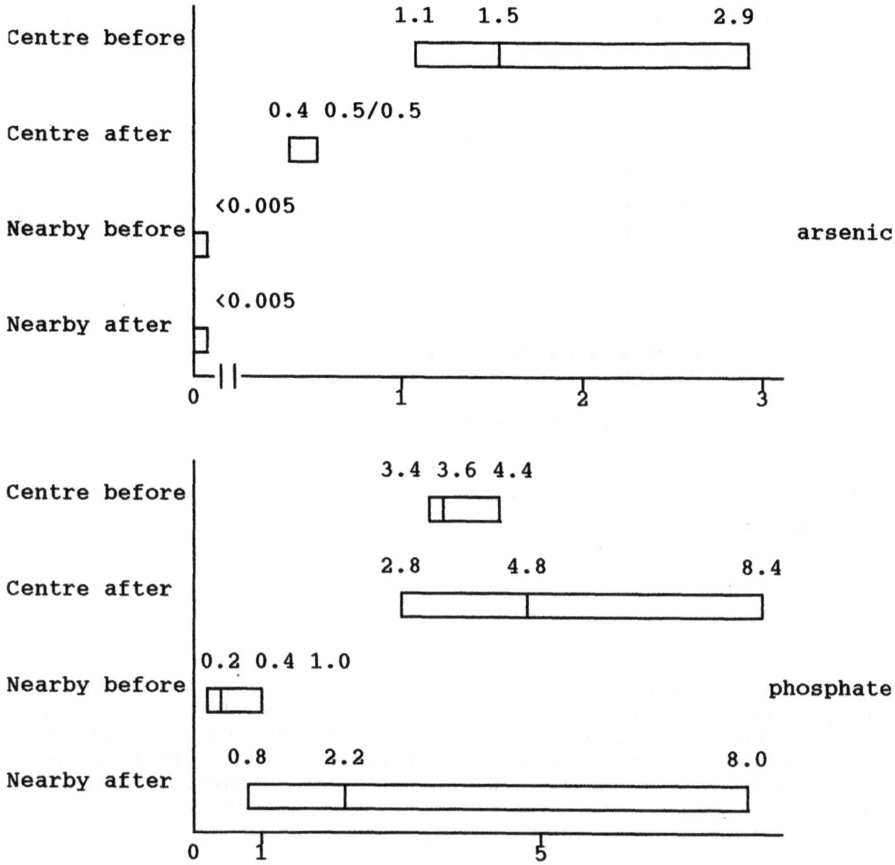

Figure 2: Contaminations of surface water with arsenic and phosphate (total) (mg/l) in the centre of the deposition field and in nearby measurements, before and after cleaning up procedure;
min. median point max.

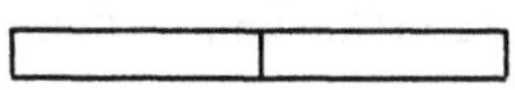

In table 1 results are summarized. The high arsenic concentrations in soil and in surface water (surface water means here an open historical well on the deposition field) in the centre of the depositions may be come from deposited skin penetrating chemical weapons, like lewisites which contain arsenic an which were produced mainly in World War I (Schuhmacher et al. (1984)).
The high concentrations of phosphate may be come from deposited organophosphates which are produced as neutro-toxic agents in World War II (tabune, sarine) (Schuhmacher et al. (1984)). Organophosphates are decomposed in soil and in water to phosphate and other single components. The higher concentration of COC in soil and in the centre of the depositions comes from decontamination agents which were used during the cleaning up, like trichlorethylene and tetrachlorethylene. At last, the positive mutagenic activity in soil and water in the centre of the depositons may be come from unknown organic components, especially oxidized components (because mutagenic activity was gained only for test strain TA 100) (Maron and Ames (1983)).

Sometimes, the high arsenic concentrations gave a toxic effect on the baterial strains.

components:	reasons for contaminations:
arsenic	skin-peentrating chemical weapons (lewisites)
phosphate	decomposition products of neuro-toxic chemical weapons
COC	decontamination workings during cleaning up
mutagenicity	unknown organic substances

Table 1: Contaminations of the total environment before and after the cleaning up procedure

Conclusions

The investigations showed that it's necessary to measure the environmental pollution during cleaning up procedures of deposited high toxic agents, like chemical weapons. For some components we got a high contamination of the total environment, already before the cleaning up began. This means that we have be vary careful during the cleaning up again, not to get furthermore contaminations from deposited components or from their decomposition products.

References

Brauch, H.G. and R.-D. Müller (1985) Chemische Kriegsführung - Chemische Abrüstung. Verlag Arno Spitz: 331-359

Maron, M.D. and B.N. Ames (1983) Revised methods for the Salmonella mutagenicity test. Mutation Research 113: 173-215

Schuhmacher, K., D. Tremmel, H. Liebmann, G. Sens and H. Vogler (eds.) (1984) Militärtoxikologie und Militärradiologie. Militärverlag der DDR, Berlin

Spyra, W. (1988) Altlasten besonderen Gefährdungspotentials - Möglichkeiten der Sanierung. In: Altlasten 2. ed.: K.-J. Thomé-Kozmiensky. EF-Verlag, Berlin: 357-369

Monitoring of an rDNA *B.subtilis* in environmental samples

Smalla, K.[1], M. Isemann[2], G. John[2], K.-H. Weege[2], K. Wendt[1] and F. Niepold[1]
Institut für Biochemie und Pflanzenvirologie, Biologische Bundesanstalt für Land- und Forstwirtschaft, 3300 Braunschweig;
Institut für Hygiene, 3040 Magdeburg

Introduction

For an industrial large-scale production of alpha amylase with an rDNA *B.subtilis* a case study was performed in the former G.D.R. from 1986-1990. Unintended releases of production strains are expected to occur in most cases of large-scale fermentations under safety level 1 conditions. Dominant release routes are via air (exhaust air from aerobic fermentation), biomass (after separation of the production strain from the extracellular enzymes) and waste water (Smalla et al. 1991). Our investigations were focussed on monitoring the extent and routes of release, characterization of microbial composition of bioprocessing waste, the fate of rDNA *B.subtilis* in sterile and nonsterile environmental media (potential recipient habitats such as soil, river water or waste water) and transfer of rDNA to indigenous microbes.
The production strain *Bacillus subtilis* ssp. *amyloliquefaciens* carried recombinant plasmids with a duplicated amylase gene insert. A resistance to macrolide, lincosamide and streptogramin B (MLS) antibiotics expressed by the rDNA production strain was used as selective trait.
Even though *Bacillus subtilis* strains are generally recognized as safe its potential selective advantage due to the presence of the streptococcal MLS resistance gene and the dissemination among indigenous microbes was an open question.
A prerequisite for our investigation was the development of reliable, sensitive and specific methods for the detection of the rDNA *B.subtilis* strain in different environmental samples.
This paper is focussed mainly on problems of detection of microorganisms in environmental samples. Problems of available detection techniques for microorganisms in environmental samples, their existing limitations and their suitability for routine work are discussed.

Materials and Methods

Organisms and plasmids:
The alpha amylase production strain *B.subtilis* ssp. *amyloliquefaciens* ATCC23844 pSB20 was received from G. Steinborn who cloned the amylase gene of *B.subtilis* ssp. *amyloliquefaciens* into pDB101 (4-6 copies, 18kb) a deletion derivative of the streptococcal MLS resistance plasmid pSM19035 (Steinborn and Hofemeister, 1984).
The plasmid pSB 128, a derivative of pDB101 used for the construction of the gene probe for rDNA *B.subtilis*, was also a gift from G. Steinborn.

Nutrient media:
Total counts were determined after plating onto nutrient agar I (SIFIN, Berlin) and 2-3 days incubation at 30°C. Gram-negative bacteria were enumerated on bile salt-lactose-agar or nutrient agar supplemented with vancomycin (10µg/ml) after one day incubation at 37°C. A starch (1%), phenylethylethanol (0.25%) and erythromycin (5µg/ml) containing nutrient agar was used for selective cultivation of rDNA *B.subtilis* at 37°C (Smalla et al., 1991). For a selective enrichment of rDNA *B.subtilis* from soil samples a nutrient broth (SIFIN, Berlin) supplemented with phenylethylethanol (0.25%) and erythromycin (5µg/ml) (Smalla et al., 1991) was applied.

Most probable number (MPN):
MPN was performed according to van Elsas et al. (1991). The wells of the microtiter plate were filled with 50µl TY broth supplemented with 5µg/ml erythromycin and 100µg/ml cycloheximide. After a two-day incubation at 37°C, the content of the wells was transferred to a nylon membrane using a dot blot apparatus, treated like colony blots and hybridized with the labeled gene probe.

Probe construction:
The small HindIII fragment (C-fragment) of pSB128 (2.1kb) was cloned into pUC18 and transformed by electroporation into *E.coli* JM 105.

DNA-Hybridization:
Digoxygenin labeling and hybridization were performed using standard protocols from Boehringer, Mannheim. Colony- and dot-blots were prepared according to standard protocols described in Maniatis (1982). The only modification was a lysozyme treatment step before SDS treatment for colony blots.

Immunological testing:
A polyclonal antibody against alpha-amylase from *B.subtilis* ATCC23844 pSB20 generated in rabbits was purified with the 65kD amylase protein according to Niepold and Nickel (in preparation). Colony blots were incubated with the monospecific antibody and detected after incubation with a secondary antibody (anti rabbit IgG coupled with alkaline phosphatase) and Fast Red staining.

Results and Discussion

Specific, sensitive and reliable methods of detection are required to monitor accurately the fate of rDNA microorganisms introduced into the environment. In principle rDNA microbes may be tracked in environmental samples at the level of the phenotype and the genotype. The most simple ways to determine whether a particular microorganism is present in an environmental sample are still traditional plating and enrichment techniques. Plating onto selective agar should permit an enumeration of rDNA microorganisms carrying a selective marker like antibiotic or heavy metal resistance. The selective cultivation can be impeded either when the selective marker is not unique in the habitat (background) or when the selective trait is not expressed under the specific environmental conditions. Furthermore it should be emphasized that plating techniques may lead to an underestimation of the total or a particular microbial population due to the fact that most microbes from natural sampling sites resist traditional cultivation techniques. It is expected that only 0.1-1% of bacteria from a soil or water habitat are culturable. Thus a plasmid transfer to indigenous microbes might not be detectable due to the nonculturability of the recipient. rDNA microbes introduced into natural habitats may exist in a viable but nonculturable state as a result of environmental stress (Roszak and Colwell, 1987; Colwell, 1985).

The detection of rDNA *B.subtilis* from environmental samples on nutrient media containing erythromycin was difficult or impossible due to the fact that most Gram-negative bacteria are intrinsically resistant to MLS-antibiotics. Detection of rDNA *B.subtilis* amongst considerable higher plating counts for Gram-negative bacteria became possible when we started to use a phenylethylethanol, starch and erythromycin containing nutrient agar. Phenylethylethanol inhibited the growth of Gram-negative bacteria. A further improvement of these selective media could be reached by addition of tellurite because we recently observed that Bacillus species tolerate high concentrations of tellurite. Amylase secretion seen as hydrolysis zones around the colonies plated onto a starch containing selective agar and colony morphology seemed to be influenced by environmental stress and could not be used as reliable properties for detection. The detection of the rDNA *B.subtilis* was difficult when its titer was at the level of the background (10^2-10^3 cfu/g soil).

Thus erythromycin resistant Gram-positive bacteria from environmental samplings were screened with the non-radioactively labeled gene probe and the monospecific antibody for the presence of DNA. The specificity of both tools was tested with 80 Bacillus species (22 different species) from a Bacillus culture collection.

None of the tested bacilli gave a positive signal after hybridization with the labeled C-fragment. Colony hybridization of 60 environmental bacilli with the C-fragment gene probe gave one positive signal in colony blots and three positive signals for dot blots. The three strains with DNA homology to the C-fragment belonged to *Bacillus licheniformis* and showed different phenotypic properties (antibiotic and heavy metal resistance patterns). Currently it is tested whether these signals correspond to hybridization with rDNA or with naturally occurring DNA sequences of the streptococcal plasmid present in *Bacillus licheniformis* as a result of genetic exchange.

The application of a monospecific antibody against alpha-amylase had different drawbacks because amylase is produced by many bacilli even though at a much lower level compared with the production strain and was not stably expressed by the rDNA B.subtilis exposed to environmental stress. Thus this antibody might be used at most in a prescreening method.

Investigations on survival of rDNA *B.subtilis* in soil and river water showed a rapid disappearance of the rDNA production strain which corresponded to our finding that the production strain was not detectable in soil from the surroundings of the enzyme factory (Smalla et al., 1990). Applying the MPN technique we were also unable to detect a positive hybridization signal resulting from the presence of rDNA *B.subtilis* in soil after 5 days. However, with the described techniques only culturable microbes were accessible. To obtain information also about rDNA which might be present either in free form adsorbed to soil particles or within nonculturable indigenous bacteria, we extracted total DNA from soil (results to be published). Detection of the rDNA on dot blot or after PCR amplification of target DNA is still under investigation.

Summarizing it should be emphasized that monitoring of introduced rDNA microbes in natural habitats, for instance in the responsibility of public surveillance laboratories/authorities, requires information about phenotypic and genotypic properties of the particular strain (availability of gene probe, specific antibody, primer for PCR amplification of target DNA). To ensure a reliable prediction of the fate and possible impact of introduced microbes a case-specific set of monitoring methods should be applied which in its simplest form may consist of tradional cultivation methods. The properties of the released rDNA strain and the resulting potential impacts on the natural microbial community at least determine how and to which extent the application of different monitoring methods is required. There still is research needed to improve or develop sampling and detection strategies which should be specific, sensitive and easy to handle. Excellent overviews of the problem of monitoring methods were given by Ford and Olson (1988), Trevors and Van Elsas (1989), and Sayler and Layton (1990).

References

Ford, S. and B.H. Olson (1988) Methods for detection of genetically engineered microorganisms in the environment. In: Advances in microbial ecology vol. 10 ed. K.C. Marshall. Plenum Press New York and London: 45-79

Maniatis, T., E.F. Fritsch and J. Sambrook, eds. (1982) Molecular cloning. Cold Spring Harbor. New York

Roszak, D.B. and R.R. Colwell (1987) Survival strategies of bacteria in natural environment. Microbiol. Rev. 51: 365-379

Sayler, G.S. and A.C. Layton (1990) Environmental application of nucleic acid hybridization. Annu. Rev. Microbiol. 44: 625-648

Smalla, K., M. Isemann, G. John, K.-H. Weege, H. Backhaus and K. Wendt (1991) A risk assessment of industrial production of alpha-amylase with an rDNA production strain. In: Biological monitoring of genetically engineered plants and microbes. Eds.: D.R. Mac Kenzie and S.C. Henry. Proceedings of the Kiawah Island conference, 205-220

Sommerville, C.C., I.T. Knight, W.L. Straube and R.R. Colwell (1989) Simple, rapid method for direct isolation of nucleic acids from aquatic environments. Appl. Environ. Microbiol. 55: 548-554

Steinborn, G. and J. Hofemeister (1984) Verfahren zur Herstellung von alpha-Amylase. WP DD 233 852 B1

Trevors, J.T. and J.D. van Elsas (1989) A review of selected methods in environmental microbial genetics. Can. J. Microbiol. 35: 895-902

Van Elsas, J.D., L.S.van Overbeek and R. Fouchier (1991) A specific marker, pat, for studying the fate of introduced bacteria and their DNA in soil using a combination of detection techniques. Plant and Soil (in press).

AIR POLLUTION IN THE FORMER GERMAN DEMOCRATIC REPUBLIC - CONSEQUENCES FOR POLITICAL ECONOMY AND HEALTH CONTROL

Thriene, B.
Landeshygieneinstitut Magdeburg, PSF 1748, Magdeburg O-3010, FRG

INTRODUCTION

The environment in the GDR was considerably damaged due to the structure of its source of energy which is unique in the world. The country covered 70 per cent of its primary energy consumption by the extraction of 320 million tonnes of raw lignite; oil and natural gas amounting to 12 and 10 per cent, respectively. As a result, since the oil crisis in the early seventies there has been a continuous increase in sulphur dioxide and dust emissions in the GDR, which included emissions from households and residential areas because of the strategy of replacing coke and gas by lignite.
In 1989, the mean life expectancy in the GDR was 2.6 years lower than in the Federal Republic of Germany. It was, however, lowest in the less polluted northern regions of the GDR. Thus the above discrepancy cannot be explained by different exposure to environmental noxes alone. Therefore, the necessity is stressed to perform further epidemiological studies in order to obtain more detailed and reliable data on the association and possibly causal relation of environmental pollution and certain human diseases and to provide a more substantial basis for control measures.

ATMOSPHERIC POLLUTANTS

With reference to the area of the country of 108 333 km² and a population of 16,709 million, pollutant emissions in 1989 were 5.2 million tonnes of SO_2, 2.0 million tonnes of dust, and 0.7 million tonnes of NO_x (1,2). The big territorial differences in pollution are shown in Table 1.

Hence the GDR was by far the leader among European countries with regard to SO_2 and dust emission per inhabitant and year at 316 kg and 125 kg, respectively.

The emission density for SO_2 was 49 tonnes/km² and for dust 19 tonnes/km². These emission data led to an unacceptably high immission load for 36.6 per cent of the population for SO_2 and 26.3 per cent for dust, which gave evidence that the legally prescribed Maximum Immission Concentrations (MIC values of 0,15 mg of SO_2/m^3 and of 15 g of dust/m² . 30 d) were exceeded in the land register areas over the calendar year (3). In the counties of Halle, Chemnitz and Leipzig as much as 73.3 per cent, 76.5 per cent and 87 per cent, respectively, of the population were subjected to an unacceptably high SO_2 load.

Pollution by CO, phenol and H_2S exceeded the limit values in the city centres, as a result of motor traffic, domestic heating and lignite processing plants for example in the Leipzig area. In the summer of 1990, the guide or warning value for ozone of 0.180 mg/m³ was frequently exceeded in the centres of the big cities for a period of several hours a day, the major cause being the massive increase in traffic after the opening of the borders and the purchase of cheap second-hand cars without catalyst technology.

THE NEW ENERGY POLICY

Due to the fact that the smog conditions in the cities of the former GDR were mainly caused by decentralised fireplaces, domestic heating and small industries, centralisation of heat supply, rational application of energy and optimum availability of low-emission energy sources are, therefore, also

Table 1: Pollutant emissions in the GDR by counties (1989)

County	Area km²	Inhabitants x 1000	SO_2 ktonnes	Dust ktonnes
Rostock	7 075	902	66	43
Schwerin	8 672	592	71	32
Neubrandenburg	10 948	620	58	40
Potsdam	12 568	1 121	114	56
Frankfurt/Oder	7 186	707	127	116
Cottbus	8 262	883	1 368	436
Magdeburg	11 526	1 252	180	133
Halle	8 711	1 791	1 054	446
Erfurt	7 349	1 236	221	119
Gera	4 004	741	167	53
Suhl	3 856	550	106	38
Dresden	6 738	1 776	396	216
Leipzig	4 966	1 378	954	225
Chemnitz	6 009	1 876	304	100
Berlin	403	1 284	70	27
GDR	108 333	16 709	5 256	2 018

demanded by the Public Health Service as an urgent requirement for prophylactic health care.

The last government of the GDR had adopted a new energy policy which aimed at reducing the energy demand by a change in the structure of energy sources and abolition of subsidies, i.e. a new price and tariff system.

Among the measures planned were a reduction of lignite production and an increase in coal and coke imports as well as increasing natural gas imports but also thermal insulation programmes for pipelines and buildings, energy saving household appliances and waste heat recovery which had to form part of a policy of saving energy.
Through the use of gas and district heating, the number of coalheated dwellings in the East German federal states is to be substantially reduced from the current 65 per cent. However, first of all simplest technical requirements for temperature control in dwellings and public buildings shall be provided in order to counteract the squandering of energy.
In 1990 the closure of plants and the general decline in production had the greatest share in the reduction in emissions.

EPIDEMIOLOGICAL ENVIRONMENTAL RESEARCH - A TASK FOR THE HEALTH SERVICE

In 1990, inter-German working groups agreed upon immediate environment-related epidemiological investigations in the five new federal states. Projects for research and development (health survey, environmental survey) have been presented to the respective ministries and in some cases have already begun with priority studies.

The development of an efficient environment-related epidemiological research brings with it a powerful apparatus for prophylactic medicine and the early detection of illness, which must be set up fas an attractive medical discipline and encouraged as such.
Thus epidemiological cross and longitudial section studies would be very useful to furnish proof of the exposure-effect relationship between SO_2 an other air pollutants and damage to health at various locations. The state of

health of the population and especially of risk groups can be examined and compared to values ascertained in the original federal states.
The life expectancy of a people presents the most important indicator of its state of health (Table 2).

Table 2: Development of life expectancy in the GDR from 1952-1989

Life expectancy from age	1952	1960	1970	1980	1985	1989	Increase in years
Male							
0	63,9	66,5	68,1	68,7	69,5	70,1	+6,2
1	67,3	68,5	68,5	68,7	69,3	69,8	+2,5
15	54,9	55,4	55,1	55,1	55,7	56,1	+1,2
45	27,6	27,7	27,4	27,3	27,8	28,1	+0,5
65	12,6	12,4	11,9	12,1	12,4	12,9	+0,3
Female							
0	68,0	71,4	73,3	74,6	75,4	76,4	+8,4
1	70,7	72,7	73,5	74,4	75,1	75,8	+5,1
15	58,0	59,6	60,0	60,7	61,4	62,1	+4,1
45	30,3	31,2	31,3	31,9	32,5	33,1	+2,8
65	13,8	14,3	14,4	14,9	15,3	15,9	+2,1

The GDR which could increase the mean life expectancy by 6.2 years for men and 8.4 years for women in 1989 as compared with 1952, still took place 17 for men and place 18 for women in 1985 among a listing of 24 European countries (4,5). The lead in life expectancy in the Federal Republic amounted to 2.3 years for men and 2.9 years for women in 1987 (6).
If we frankly discuss the causes today, then the overall social conditions play a decisive role .
Because of the better circumstances than in the original Federal Republic, as far as the number of road deaths is concerned, and because of a strictly organized programme of vaccination eliminating children's diseases to a large extent and the infant mortality rate being roughly comparable to that in the Federal Republic, the problems of overfeeding and false nutrition as well as failure to cope with stress, gain in importance.
At the same time the living conditions of elderly people are a crucial factor. Suicide rates twice as high as in the FRG draw attention to the social factors of quality of life and the living environment; they are an expression of social stagnation and social isolation, the latter being an aspect of mass housing construction and poor socio-economic conditions.

So-called "avoidable" deaths from cardio-vascular diseases, hypertension, respiratory tract infections, carcinomas, appendicitis among others were 4.6 times more frequent in the GDR than in the Federal Republic. This alludes to the quality of medical care including possibilities for the extensive use of diagnostic and therapeutical facilities, and in this connection especially the available equipment, the availability of pharmaceuticals, of various materials, benefit services, etc., that ist to say possibilities for therapeutical intervention for the prevention of mortality in these cases.
In any case, there existed noticable differences in life expectancy between the counties (Table 3), and the rural areas of the north have had the worse position for years compared with the conurbations of the central and southern areas.

Table 3 Life expectancy by counties 1988/89

Counties	Male (in years)	Female (in years)
Berlin (East)	71,19	76,24
Cottbus	69,33	75,81
Dresden	70,90	77,11
Erfurt	70,24	75,97
Frankfurt/Oder	69,08	75,83
Gera	71,11	76,54
Halle	69,66	76,11
Chemnitz	70,68	76,52
Leipzig	70,42	76,44
Magdeburg	69,40	75,50
Neubrandenburg	68,27	75,99
Potsdam	69,67	75,89
Rostock	69,13	76,23
Schwerin	68,8	76,41
Suhl	69,81	75,64
	70,03	76,23

In this respect, the contribution by environmental pollution can be speculated on, however, no comprehensive scientific statement can be made so far.
A cooperative relationship between the Medical Institute for Environmental Hygiene at Düsseldorf University and the State Hygiene Institute in Magdeburg has enabled cross and longitudinal section studies for the year 1990/91 aiming to quantify the correlation between air pollution and health problems among some 4 000 children and 100 people with previous health impairments living in Magdeburg, Wernigerode, Halle and Leipzig compared with a control area in the Altmark region. The results will be soon available. Participation of the people approached was over 75 per cent. And in my opinion, this is our big and, perhaps, last chance to use the territory of the former GDR with currently still high environmental pollution loads for an incontestable statement on their relevance to health.
The earlier results of investigations in Leipzig, Chemnitz, Rostock, Bitterfeld, Cottbus and Magdeburg, which have been carried out in spite of certain restrictions in order to assess the effect of air pollution on human health suggest their completion by further studies on a solid epidemiologic basis.

REFERENCES

1 Umweltbericht der DDR. Institut für Umweltschutz Berlin (Hrsg.): Verlag "visuell", Berlin, 1990
2 Berichte der Staatlichen Hygieneinspektion über die Situation auf den Gebieten der Lufthygiene, Wasserhygiene, Bodenhygiene, kommunalen Lärmbelastung und -bekämpfung 1989 in der DDR. Ministerium für Gesundheitswesen Berlin, 1990
3 Erste Durchführungsbestimmung zur Fünften Durchführungsverordnung zum Landeskulturgesetz - Reinhaltung der Luft - Begrenzung, Überwachung und Kontrolle der Immissionen vom 12.02.1987 (GBl. I, Nr. 7, S. 56)
4 Kant, H.: Zur Lebenserwartung in Europa - Fakten, Tendenzen, Ziele Z. gesamte Hyg. 34 (1988), S. 442
5 Statistisches Jahrbuch der DDR 1990 Rudolf Haufe Verlag, Berlin, 1990
6 Wiesner, G.E.: Zur Gesundheitslage der beiden Bevölkerungsteile DDR und BRD - ein Ausdruck sozialer Ungleichheit? in: W. Thiele (Hrsg.): Das Gesundheitswesen der DDR: Aufbruch oder Einbruch? Asgard-Verlag Dr. Werner Hippe KG, Sankt Augustin, 1990

Subject Index

Acid phosphatase 15
Adrenaline 53
Aerosol deposition 137, 141
Airborne particulates 27, 83, 231
Air pollutants 193, 249, 269
Air pollution 147, 185, 189
Alkylpurines 163
Allergy 185, 193
Aluminium 179, 241
Amalgam fillings 151
Ames test 83, 245, 261
Arsenic 249
Asbestos 171, 215, 223, 249

Benzene 249
Benzo(a)pyrene 63, 67, 71, 87, 91, 95, 249
Benzoquinone 71
Biodegradation 253
Biomonitoring 175, 265
Bladder carcinogens 95
Bronchial reactivity 185
Bronchoalveolar lavage 19, 105, 113

Ca^{2+}ATPase 7
Cadmium 7, 159, 167, 249
Cadmium sulfide 117
Carcinogenicity 11, 63, 71, 91, 95, 127, 203, 249
Cell differentiation 23, 219
Cell proliferation 227
Cell transformation 11, 71, 91, 99
Chemical weapons 261
Chick embryo yolk sac 53
Chrome 45, 127, 159
Chromosome aberrations 71, 155
Chrysene 87
Coal mine dust 227, 235
Cobalt 19, 147
Copper 37, 159
Croup syndrome 185
Cytoskeleton 15, 41, 49, 215
Cytotoxicity 27, 45, 49, 53, 67, 79, 99, 127, 215, 219, 235

Diesel particles 121, 249
Dioxin 211, 257
DNA adducts 59, 163
DNA fluorometry 37
DNA strand breaks 79

Electromagnetic radiation 23
Environmental medicine 3
Environmental pollutants 71, 197, 253, 261, 265
Enzyme polymorphism 75
Epidemiology 147, 159, 167, 185, 189, 193, 197, 207, 211, 269
Epigenetic effects 7, 71
Ethylene(oxide) 59

Fe(III)-NTA 79
Fibres 215, 223
Fibroblasts 23, 227

Hamster cells 63, 67, 91
Health effects 197
Heavy metals 37, 41, 159, 231
Helix pomatia 33
Herpes simplex virus 37
Hexachloroethane 105
High energy shock waves 49
HL-60 cells 219
Human exposure 137, 147, 151, 159, 163, 167, 179, 211
Hydroquinone 71
7(2-hydroxyethyl)guanine 59

Image analysis 15
Immunoglobulin E 185, 193
Indoor air pollution 245
Inflammation 113, 223
Inhalation toxicology 27, 63, 105, 109, 113, 117, 127, 133, 137, 141
Intercellular communication 83
Intracellular particle dissolution 19

Keratinocytes 11

Lead 33, 37, 159, 167
Legislation 249, 257
Leukocytes 113
Life expectancy 269
Lung cancer 121, 203, 207
Lung cells 45, 49, 63, 71, 79, 83, 87
Lung clearance 19
Lymphocytes 59, 155

Neurotoxicity 33
Nicotine 53, 245
Nitric oxide formation 231
Nitrosamines 63, 99, 133

Macrophages 19, 27, 171, 219, 223, 227, 231, 235
Magnetopneumography 141
Mercury 151, 159, 175
Methyl bromide 75
Micro-organisms 253, 265
Mutagenicity 67, 83, 99

Occupational exposure 155, 171, 175, 241
Ozone 113

Papillomavirus DNA 11
Pathogenicity 223
Phagocytosis 27
Phagolysosomal pH 19
Photocorrosion 117
Protein kinase C 7

Quartz 219, 227

Radon 203, 207
Respiratory disease 185, 197

Respiratory tract 109
Risk assessment 3, 121, 179, 207, 241, 249, 257, 269

Seveso 211
Sister chromatid exchanges 75, 79, 155
Sulfur(IV)aerosol 109
Surfactant 235

Tobacco smoke 245
TPA 219
Tracheal epithelial cells 99
Triethyl lead 71
Tumor promotion 83

Urine excretion 147, 159, 163, 175
UV-C 23

Vanadium 45

Waste deposition 261

Zinc oxide 105

MIX
Papier aus verantwortungsvollen Quellen
Paper from responsible sources
FSC® C105338

If you have any concerns about our products,
you can contact us on
ProductSafety@springernature.com

In case Publisher is established outside the EU,
the EU authorized representative is:
Springer Nature Customer Service Center GmbH
Europaplatz 3, 69115 Heidelberg, Germany

Printed by Libri Plureos GmbH
in Hamburg, Germany